Timeline of Einstein's Most Impor...

December 13, 1901: Einstein submits his first pu... ...tion, "Deductions from the Phenomena of Capillarity." This paper sought to explain the forces that operate between molecules.

March 17, 1905: The first of four great papers of 1905, "On a Heuristic Viewpoint of the Generation and Conversion of Light," is published. Einstein theorized that light was made up of packets of energy called quanta. These were later called photons.

April 30, 1905: "A New Measurement of Molecular Dimensions" is the first part of a two-part publication on Brownian motion.

May 11, 1905: "On the Motion of Small Particles Suspended in a Stationary Liquid" is the second part of what is considered to be a single publication analyzing Brownian motion. This motion is the constant random movement of small particles suspended in a liquid due to the colliding of their molecules.

June 30, 1905: The greatest of Einstein's four papers of 1905, "On the Electrodynamics of Moving Bodies," showed that space and time are relative, not absolute. This idea came to be known as Einstein's special theory of relativity.

September 27, 1905: "Does an Object's Inertia Depend on Its Energy Content?" was a supplement to the special theory of relativity that contained Einstein's most famous equation, $E = mc^2$. This formula shows that mass and energy are equivalent and can be converted into one another.

June 21, 1911: "On Gravity's Influence on the Propagation of Light" set the groundwork for Einstein's greatest work on general relativity. The paper reflects the happiest thought of his life, which was that gravity is equivalent to acceleration. This idea is known as the equivalence principle. It also predicted that gravity could bend light.

December, 1916: "The Foundations of the General Theory of Relativity" explained that gravity was no longer considered to be a force that attracted bodies, but was a field that bent space and thereby determined the motion of objects.

February, 1917: In "Cosmological Considerations of the General Theory of Relativity," Einstein tried to derive the size of the universe and introduced his cosmological constant to ensure the stability and finite boundaries of the universe. Later, when it seemed clear that the universe was expanding, Einstein called this paper "the greatest blunder of my life."

tear here

alpha
books

1925: Einstein publishes "The Quantum Theory of Single Atom Ideal Gases," which puts forth an argument for the wave character of matter.

1934: Einstein's *The World As I See It* is published in Amsterdam. It is a collection of his articles and essays on religion, pacifism, science, politics, and many other topics.

1935: The Einstein-Podolsky-Rosen, or EPR, paradox is published. It's entitled, "Can Quantum Mechanical Description of Physical Reality Be Considered Complete?" It was a brilliant attack on quantum mechanics and marks a certain conclusion to the debate about the foundations of quantum mechanics.

1938: In this book, *The Evolution of Physics* with Leopold Infeld, Einstein gave an insightful description of the history of physics. He wrote it in collaboration with a Jewish colleague to help ensure his friend's financial security.

1949: Einstein publishes his last paper, "A Generalized Theory of Gravitation." In this paper, he sought to develop his unified field theory, but the paper delved into unusual mathematical territory that was foreign to most everyone. When asked for experimental evidence to back up his theory, Einstein said, "Come back in 20 years."

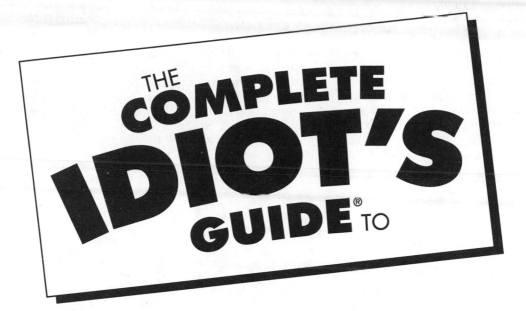

THE
COMPLETE
IDIOT'S
GUIDE® TO

Understanding Einstein

by Gary Moring, M.A.

alpha
books

Macmillan USA, Inc.
201 West 103rd Street
Indianapolis, IN 46290

A Pearson Education Company

International Standard Book Number: 0-02863180-3
Library of Congress Catalog Card Number: A catalogue record is available from the Library of Congress.

02 01 00   8 7 6 5 4 3 2

Interpretation of the printing code: the rightmost number of the first series of numbers is the year of the book's printing; the rightmost number of the second series of numbers is the number of the book's printing. For example, a printing code of 00-1 shows that the first printing occurred in 2000.

Printed in the United States of America

Alpha Development Team

Publisher
Marie Butler-Knight

Editorial Director
Gary M. Krebs

Associate Managing Editor
Cari Shaw Fischer

Acquisitions Editors
Randy Ladenheim-Gil
Amy Gordon

Development Editors
Phil Kitchel
Amy Zavatto

Assistant Editor
Georgette Blau

Production Team

Development Editor
Adam Weiss

Production Editor
Michael Thomas

Copy Editor
Heather Stith

Cover Designer
Mike Freeland

Photo Editor
Richard H. Fox

Illustrator
Jody P. Schaeffer

Book Designers
Scott Cook and Amy Adams of DesignLab

Indexer
Chris Wilcox

Layout/Proofreading
Angela Calvert
Donna Martin

Contents at a Glance

Contents

Appendixes

Foreword

In *The Complete Idiot's Guide to Understanding Einstein*, Gary Moring leads us on an exciting journey through time, from the world of the Greeks who first asked questions about the nature of things, to our present-day perspective of matter, time, and space. We experience the struggle between great opposing ideas in the birth of new paradigms, or worldviews.

To understand what science is about today, it's important to understand the interconnecting ideas that reach back into the beginnings of scientific inquiry. To put it simply, one needs to get the big picture. The reader of *The Complete Idiot's Guide to Understanding Einstein* has an invaluable opportunity to get the big picture—an opportunity that somehow evades our science majors in their university studies. Science courses are usually so narrowly focused that the relationship of the expounded material to the whole development of science is ignored. To fully appreciate the revolutionary ideas of Einstein, one must have some knowledge of the historical setting on which these developments unfold.

At the same time, Gary Moring does not skimp on the science he presents. Instead, he explains difficult scientific concepts with models and pictures that help the layperson grasp the important scientific ideas involved. Many of the present-day notions of space, time, and matter are far removed from our everyday experiences. In progressing toward the infinitesimally small, we encounter quantum weirdness that challenges our notions of classical Newtonian determinism. Our concepts of absolute space and time have been overturned by the Theory of Relativity. Much of this modern paradigm shift is due to the genius of one man, Albert Einstein. And that is why it is most appropriate to focus on the life and thinking of Einstein in writing about the development of science and our present-day worldview.

The early Greek philosophers asked questions about the physical reality they observed. This is what science is all about—asking questions. And obtaining an answer only leads to more questions. The Greeks depended on reason and observation to develop their worldview. It was much later that the scientific method, modeled on the work of Galileo, became fully developed. Today we believe that theories should have within them the means of their own testing. That is, a scientific hypothesis or theory, to be useful, should not just explain a given phenomenon but should also lead to other predictions that can be tested. Progress is only possible if the hypothesis or theory is open to being proven wrong. As an example, recent observations indicate that the expansion of the universe, instead of slowing down because of the gravitational attraction between its parts, is actually speeding up. This comes as a surprise but is welcome news for cosmologists, who now have an intriguing enigma to solve. Demonstrating shortcomings or errors in theories is what makes for progress in science. Science is never-ending; we will always have questions that remain to be answered.

Gary Moring leads us on an exciting adventure through time, matter, and space. His friendly prose and subtle humor will make the time pass quickly. Lucky reader, begin your journey.

—Thomas A. Weber, Professor of Physics, Iowa State University

Introduction

Most of us have heard of Albert Einstein. His face can be found on T-shirts and posters, and his name is associated with atomic energy, prestigious awards, and many young individuals who win science fairs. Some people have called him the greatest scientific genius of the century. He was welcomed in the United States as a hero and often treated as a Hollywood star. Who was this legendary figure who not only had great knowledge, but great wisdom as well?

Albert Einstein changed the face of physics. His revolutionary ideas would forever alter the way science viewed the universe. He presented ideas that would unify different areas of physics, making connections between concepts that had never been understood before. The insights that were expressed in his theories revealed a much deeper order to the universe than had previously been known.

Although Einstein is most often thought of in relation to his scientific contributions, he impacted many other areas of human endeavor as well. Before leaving Germany, he was very outspoken against the military regime that was coming to power under Adolph Hitler. His pacifist beliefs also led him to denounce the use of the atomic bomb at the end of World War II. He helped raise money and support for the victims of the Nazi concentration camps and was involved in helping Israel become a sovereign state.

To understand Einstein, it is also important to examine his philosophy of life. The fundamental philosophical questions that he asked went directly to the heart of cosmology. His beliefs about the nature of the universe strongly influenced his theories and prevented his full acceptance of quantum mechanics. Yet at the same time, these same beliefs gave him a deep compassion for those who suffered in the world, and he was recognized by many for his humanitarian efforts. In the end, it is impossible to separate his philosophy from his physics because in an individual such as Albert Einstein, the heart was equally as important as the mind.

Although the main focus of this book is Einstein's life and theories, it also explores different perspectives on how our own world views influence our perceptions. This book will try to provide you with a larger outlook, a bigger picture in which you'll see that to truly understand someone or something else, you must take a closer look at yourself. Here's a brief overview of some of the material we'll be exploring:

Part 1, "Ideas That Built Bridges in Physics," provides you with the basic building blocks of classical physics and the historical background through which it developed. The ideas and philosophy of the ancient Greeks, Copernicus, Galileo, and Newton provide a framework to understand how and why Einstein's theories changed the way the world was understood.

Part 2, "Energy Comes in Many Forms," describes the world of physics at the close of the nineteenth century. It explores some of the interpretations of what energy, light, electricity, magnetism, and motion are. You'll look at the birth of the scientific method and how it became the powerful tool that lies at the core of science today.

Part 3, "Into the Heart of Einstein's Mind," examines Einstein's life from his early childhood to the beginning of World War II. This time period is central to understanding Einstein, so it includes not only biographical information, but also his theories on relativity and some of his thought experiments that you can try for yourself.

Part 4, "Anybody Know a Good Quantum Mechanic?," explores some of the underlying principles of quantum mechanics. We'll explore the role of human consciousness in trying to understand the paradoxical nature of the quantum universe. You'll be introduced to some of the fundamental ideas found in psychology and consciousness studies and see why these ideas are an important part of any methodology that has observation as its main source of acquiring information.

Part 5, "Einstein, Man of the World," covers the rest of Einstein's life, including his trips and final move to America and his continued search for a unified theory, which he never found.

Part 6, "Worlds Beyond Einstein," takes a look at particle physics and astrophysics. You'll examine the mainstream theories, ideas, and beliefs of science and religion in relation to the creation of the universe. This section ends by looking at some Eastern perspectives on the structure of the universe.

At the end of the book, you will find three appendices. The glossary provides a listing of terms that will help you to deepen your understanding of physics terminology and will set you on the road to becoming another Einstein. The resources appendix lists some Web sites that contain information and photos on Einstein, as well as further reading both on Einstein and some of the other areas covered in this book. The timeline provides a snapshot of key events in Einstein's life.

Extras

In addition to providing you with information on Einstein's life and times and clear explanations of some pretty heady concepts, this book also contains some features that give you interesting, informative, unusual, deep, and sometimes baffling information.

Mind Expansions

Whoa. Some of this stuff gets really bizarre. This feature provides you with more detailed anecdotes about physics, philosophy, and unusual points of interest, and offers insights to other connecting ideas.

Albert Says

This feature gives you tips, trivia, and helpful hints, some of which are from Einstein, to help you better understand what's going on.

Relatively Speaking

You don't already know what a quark is? Are early Greek terms all Greek to you? Fear not. This feature will translate the specialized words of physics, cosmology, and philosophy in this book into terms you can understand. Remember, even Einstein didn't know what any of this meant when he first started.

Nuclear Meltdown

You probably have some misunderstandings about Einstein and physics in general stored away somewhere. Unless you're a nuclear technician, a little confusion probably won't hurt you. But this feature will help clear up some of the more common points of confusion.

Special Thanks to the Technical Reviewer

The Complete Idiot's Guide to Understanding Einstein was reviewed by an expert who double-checked the accuracy of what you'll learn here to help us ensure that this book gives you everything you need to know about Einstein and his theories. Special thanks are extended to Warren Holmes.

Acknowledgements

Any creative endeavor requires the support and insight of a number of people. This project would never have been started or completed without the help of the following individuals:

Sarah Shockley, who made the initial connection; Andre Abecassis, my agent, who believed I could do it from the beginning and coached me from start to finish; Gary Krebs, editorial director at Alpha Books, for taking a chance with a new author; Adam Weiss, developmental director at Alpha Books, for his constant support, witty comments, and creative deadlines; Warren Holmes, technical editor, for his insightful feedback on physics; Larry Abrahamson, for the primary pen and ink drawings; my family and friends, and especially Sam, who gave me words of wisdom and weekly encouragement; D.K., for inspiring me to take this test; and, most especially, my wife, Kathy, who supported me with acupressure, food, and guidance.

Part 1

Ideas That Built
Bridges in Physics

Welcome to the exciting world of ideas. Each idea that you'll read about will help lay the foundation for the next. Some ideas end up building walls that need to be torn down or scaled, and others build monuments to inspire insight and creativity.

Our journey to the twentieth century and the world of Albert Einstein begins in ancient Greece with the first individuals who asked questions about the world in which they lived. The roots of western civilization have their beginnings in many of the ideas that are explored in the following chapters, especially those of physics.

About 25 centuries ago, a Greek philosopher stated that the only thing anyone requires to be a good philosopher is the faculty of wonder. This is true for physicists as well, for without the insatiable curiosity to know what lies behind the veil of nature, we would still be stuck on the evolutionary scale somewhere behind pine trees.

The Early Greek Thinkers

In This Chapter

➤ From mythology to methodology

➤ Answering the big questions

➤ The Pre-Socratics and Socrates himself

➤ Plato's contributions

➤ Aristotle's impact

The word *physics* comes from the Greek word *physika*, meaning "natural things," or the study of nature. It has come to mean a little more than that in the last 300 years, as you will see as you read this book. But in the beginning physics was simply the study of nature.

Where do you start such a study? One way is by asking questions. (This is why most physicists are also good philosophers.) Physics all begins with a question, followed by a long process of hunting for solutions, some of which work, and others that don't. Let's look at some of the individuals and ideas that laid the foundation for this study.

A Star Is Born

All of the ancient civilizations tried to understand their worlds in terms of myth. Religion and mythology were two sides of the same coin. Natural forces such as wind, fire, and rain, or even rivers, mountains, and trees, were understood to embody sacred or divine life. Although these forces of nature had a tremendous impact on the lives of ancient people, no one understood how they worked. To help explain these mysterious forces, early people *anthropomorphized* them. For example, in Egypt, the Sun was the

[handwritten margin note: primitive beliefs]

[handwritten note at bottom: ① an interpretation of what is not human or personal in terms of human or personal characteristics. HUMANIZATION]

living embodiment of the god Ra. One of India's gods, Indra, was symbolized by fire, and in China, the gods of the four winds were powerful indeed. Stories about these gods were used to explain natural phenomena.

Relatively Speaking

Physics is the study of the properties, changes, and interactions of matter and energy. To **anthropomorphize** is to represent a god with human attributes or assign human qualities to nonhuman things. A **pantheon** is all of the gods and goddesses of a culture or people.

Nuclear Meltdown

It's a common misconception that mythology has nothing to do with reality. But mythical stories lie at the core of many of our religious traditions. The creation stories found in the Old Testament have their roots in myths that predate the writings by thousands of years. Even the story of the Flood can be found in stories from other cultures around the world. Mythical symbols, too, such as Trees, Mountains, and man-made symbols can also be found in the ancient creation stories. Across all cultures, these symbols are an important aspect of our psychological structure.

Often, the easiest way to understand something is to think of it within the context of concepts or ideas you are already familiar with. A family was something everyone participated in, and the interaction of natural phenomena was often explained in terms of family dynamics. For example, Mesopotamia's creation story was expressed as a family feud between the great mother goddess and her sons, who were making too much noise.

Mythological stories were important in explaining natural phenomena. Many of the most popular myths of ancient Greek society showed humans living intimately with their gods and other immortals. Distinctions between these groups often became muddled, as gods, immortals, and humans loved and bred together. Beings could change in this process, transformed from a mortal to a god, or to a star, mountain, flower, or even a song. It was a fluid, ever-changing world, and many stories are about the timeless moment of change. In the beginning, the Greeks believed, there had been no barriers between the different inhabitants of the world. When Cadmus abandoned his long search for his sister Europa and built Thebes, the first of all the cities in Greece, Apollo raised the walls for him and gave him the alphabet and the tones that make up the musical scale; all the Olympian gods came when he married Harmony.

Later the earth dwellers became arrogant and disrespectful; Zeus reprimanded them, but it did no good. Finally Prometheus, whose insolence knew no limits, stole fire itself from Olympus and taught humans how to use it, freeing them from dependence on divine generosity. Zeus was furious. He ordered Hephaestus to make a beautiful woman and send her to earth carrying a box which contained all the ills which now haunt us: hunger, strife, misery, exhaustion. He called the woman Pandora, which means All the Gifts, and instructed her not to open the box. She did, of course, and all the pains of mortality spread like a dark cloud over the world, settling and breeding everywhere. Pandora slammed down the lid of her casket just in time to keep hope safe.

After this, although humans flourished and the cities were founded, that simple intermingling of human and divine was over. Individual gods still loved and helped people, but they no longer all celebrated or ate and drank together. There was bitter competition on Olympus over which humans were deserving of favor and support. The gods fought among each other, too, and out of their partisanship most of the great Greek tragedies and great adventures were born.

The Greeks gave us striking visions of the gods in human form: Zeus with his thunderbolt, Poseidon with his trident, and Hermes with his winged feet and helmet. Greek culture developed vivid tales of family drama among these gods, and these myths have a strong presence even in today's world. Modern astronomers speak familiarly of constellations named after mythological Greek figures such as Orion, Casseopeia, and Andromeda, to name just a few. But this was all about to change!

Around 600 B.C.E. a group of philosophers known as the Pre-Socratics (because they came before Socrates) began thinking in a new way. Instead of seeing the world as a playground for Zeus, Athena, and the other divine beings of the Greek *pantheon*, they began to apply reason as a means to comprehend the ways of nature. Now this idea may not seem like any big deal to us today, but back then it was a significant shift in age-old traditions. This change was nothing less than the birth of a whole new paradigm, or world view. Religion, magic, and mythology now began to give way to human reason as the cornerstone for exploring and understanding the world.

What Is Reality?

The most fundamental question that was asked by these philosophers is really no different from the question Albert Einstein asked 2,500 years later: "What is the underlying order that is hidden in nature?" This question was to lead to others, such as, "What is the most basic substance of the universe?" and "Is the structure of nature based on mathematics, processes, or substances?"

The first great Pre-Socratic thinker was a Greek named Thales. He lived around 600 B.C.E. in Miletus, Ionia, which today is in Turkey. He thought that water was the primary and simplest element upon which nature is based. It's easy to see how he came up with this idea. After all, water is essential for growth in plants, and almost everything that is heated gives off moisture. According to evolutionary theory, life began in the oceans. Water also makes up a significant portion of our bodies. It is one of the only substances that is commonly observed in the everyday world as a solid, liquid, or gas, as water vapor or steam. So this idea wasn't bad for a first guess. Thales also constructed one of the first almanacs, predicted an eclipse of the sun, and taught sailors how to navigate a ship at night by using the Little Bear constellation.

Anaximander was a young friend of Thales. He viewed things a little differently. He thought that the world was composed of interacting, aggressive opposites, such as hot and cold, wet and dry, male and female, and light and dark. The sun dries up water, water puts out fire, a lantern lights up the dark, and darkness thwarts our ability to see.

Mind Expansions

Aristarchus of Samos, who lived in the first half of the third century B.C.E., postulated the heliocentric, or sun-centered, solar system a thousand years before Copernicus.

But how could a primary substance be something that was defined by its opposite? In other words, if a substance is primary, it should be something that stands alone and is the source or cause of other things, not something that equally interacts with and in some sense depends upon another material. So Anaximander came up with an idea for a substance he called the *apeiron*, which means "without boundaries." The apeiron was an indefinite, neutral mass of enormous, possibly infinite proportions that existed between opposites. This theory was pretty good, too, but it was tough to comprehend. You could probably think of it as something like a vacuum, but then again thinking of a vacuum as a substance presents its own problems.

A few other philosophers belong to the Pre-Socratic club:

➤ Anaximenes, like Thales, thought that the universe was based on a primary substance, but he believed it was air, not water.

➤ Empedocles thought that four kinds of matter existed: earth, air, fire, and water.

➤ Parmenides wasn't concerned as much with substances as he was with processes. He believed that whatever the world is made of can neither come to be nor pass away. In other words, matter doesn't pop in and out of existence. It always exists and can't be destroyed. Change is an illusion. The natural world is complete and enduring—it's a oneness, a wholeness.

➤ Pythagoras is remembered by most of us for his *Pythagorean theorem*, the one formula that stuck in our heads from high school geometry. (You remember, good old A squared plus B squared equals C squared in a right triangle.) He defined the world in terms of mathematics, asserting that numbers are the essence of things and the source of what is real.

Relatively Speaking

Apeiron is a Greek word describing something that is boundless and limitless, literally almost infinite in size. The **Pythagorean Theorem** states that in a right triangle, the square of the hypotenuse (the side opposite the right angle) is equal to the sum of the squares of the two other sides.

➤ Leucippus and his successor Democritus both felt that everything was made up of elementary particles. These *atoms*—a term still used in twentieth-century physics—were indestructible, indivisible, without parts, and infinite in number. They came in three geometric shapes. Their various combinations and continuous movement accounted for all things in existence, and also controlled how these things changed from one thing into another.

There were many more philosophers who won't be mentioned here, because the purpose of this book is not to provide you with a comprehensive history of ancient schools of thought and every theory that was ever developed. But it's good to know how old many ideas in physics are and the names of some of the significant individuals who contributed to the fundaments of modern physics.

Look Ma, No Math

Are the original ideas developed by early Greek thinkers considered science by today's standards? Not really. The *scientific method*, which forms the cornerstone of all the physics done today, didn't come into existence until the seventeenth century. The next chapter will introduce you to some of the great minds of the Renaissance that gave birth to this methodology.

Physics begins with ideas. These ideas are based on reason and logic and, in almost all cases, follow observations of the natural world. This is called *natural philosophy*, which is what science was called 2,500 years ago. Instead of relying solely on observation and reason to come to a conclusion, modern science utilizes experimentation and mathematics to establish proofs of the ideas. You could say that mathematics is the language of physics.

For many individuals, the mention of math sends a cold shiver down the spine. But you can study physics without the use of complex formulas. This book will familiarize you with some basic equations. They are important, but the key is to understand what the formulas mean, not to memorize them.

Albert Says

As you read, try to see the big picture of how ideas are connected to one another. Often these connections are the most important thing to understand. It's the bridges that take you across the empty spaces.

Relatively Speaking

Natural philosophy is the branch of ancient philosophy devoted to the study of the natural world; today it's known as science and more specifically, physics. The **scientific method** is the process in science in which someone develops a theory or hypothesis based on data that has been collected, and then tests that theory through experimentation and/or the use of mathematics.

The Greeks would have used present-day math if they could, but it just hadn't been invented yet. Simple trigonometry and geometry were available, but algebra (originated by Islamic scholars in the eighth and ninth century) and calculus (independently invented by Isaac Newton and Gottfried Leibniz) wouldn't be around for a long while.

There are many ways to understand physics without the use of mathematics. One way is through pictures or graphs. Another way is through the use of analogies, or correspondences between ideas. For example, you can think of the structure of an atom as being similar to the structure of our solar system, with electrons playing the role of planets revolving around the nucleus, which is playing the role of the Sun. Analogies make it possible to draw upon common knowledge to shed light on difficult concepts, and they have played a very important role in the development of physics. Analogies and models were two of Einstein's favorite tools.

It is possible, although difficult, for a talented person with limited mathematical abilities to become a great physicist. Michael Faraday, a famous nineteenth-century British scientist I will tell you more about in Chapter 6, was such a person. A self-educated orphan, Faraday made many important contributions to both chemistry and physics. He read a great deal, and with his limited mathematical abilities, he found analogies extremely useful. He originated the concept of lines of force to represent electric and magnetic fields, using an analogy with rubber bands. He is generally credited with using this analogy to develop the idea that light is an electromagnetic wave phenomenon. James Clerk Maxwell, whom you will learn about later in the book, developed the formal mathematical theory of the electromagnetic nature of light and acknowledged that without Faraday's intuition he could not have developed his more rigorous theory.

You Can't Judge a Book By Its Cover!

Our journey back to ancient Greece comes to a close with a look at three philosophers who were more influential in the development of western civilization than any other individuals in history. Their ideas affected science, physics, ethics, art, theology, politics, and metaphysics.

The Man and His Method

Socrates (circa 470–399 B.C.E), whose trial is the most well-known in history outside of those of Jesus and O. J., is the first member of our triumvirate. He laid the foundation for a method of inquiry that would come to be known as the Socratic method. This method was a question-and-answer process that would help both individuals involved in the discussion to arrive at the underlying truth of a topic.

Unfortunately for Socrates, many of the politicians he questioned in public didn't like the fact that he showed them to be wrong about most of the things they talked about. Not a good idea! He was put on trial for his life, facing trumped-up charges about corrupting the youth and not believing in the gods officially recognized by the government.

Socrates did come up with some of the best one-liners in history. Here are a few of them:

"Know thyself."

"The unexamined life is not worth living."

"The wise man is the man who knows that he doesn't know."

Mind Expansions

To learn more about the trial of Socrates, read the *Apology* and the *Phaedo*.

Although these and other sayings are attributed to him, Socrates never wrote anything down. What we know of him comes from his most gifted pupil, Plato (circa 427–347 B.C.E.), the second member of our hearty trio. In Plato's collection of dialogues, Socrates is the main character. We learn about Socrates' philosophy in these works of Plato.

Plato and the Cave People

Plato examined how we acquire knowledge about the world in which we live. He represents a transition point between purely philosophical thinkers such as Socrates and natural philosophers such as Aristotle, who is considered by many to be one of the first scientists.

Plato felt that the first task of a philosopher is to determine the reality or the truth behind the way things appear. His famous Allegory of the Cave illustrates this idea clearly. In this allegory, humans are envisioned as existing like slaves chained together in a deep, dark cave that is dimly illuminated by a fire burning some distance behind and above them. There is a wall both behind and in front of them. They are chained and shackled in such a way that they can't turn around to see what is going on behind them.

People moving back and forth on the other side of the wall behind the slaves hold various objects over their heads and make sounds and noises unintelligible to the slaves. The slaves can see only the shadows of these objects, cast by the firelight onto the wall in front of them, and can hear only muffled echoes of the noises being made. The slaves have seen these shadows all their lives and know nothing else. In this analogy, the shadows represent the appearance of things as we understand them. The reality of things, represented by the objects that are making the shadows, is inaccessible to us.

One day, one of the slaves (who will later become a scientist) is released from the chains and brought out of the cave into the real world. This world is bright and beautiful, with blue skies, green grass, and trees. At first, he can't see what is around him because his eyes are unaccustomed to the bright light. Once he can see, he finds it very painful to contemplate what he sees in the real world. However, he eventually becomes accustomed to his newfound freedom. He has no desire to return to his former slavery, but a sense of duty compels him to return to try to enlighten his fellow beings. (Sounds pretty tough!) He must readjust to the darkness and try to explain the shadows in terms of the objects that the other slaves have never seen. They reject his efforts and those of anyone like him, threatening such people with death if they persist (a common theme throughout most of our history).

To understand how Plato applied this notion of appearances to the real world, consider the appearance of the heavens, which is the subject of astronomy. The Sun rises and sets daily, as does the Moon. The Moon also goes through phases regularly. The Sun is higher in the summer than in the winter. Morning stars and evening stars appear and disappear. These events are the appearances of the heavens, but the philosopher or scientist must discover the reality underlying these appearances.

Most objects in the sky appear to move in circular paths around Earth as the center. If you didn't know that Earth revolved on its axis, it would appear that Earth was stationary, and everything circled around it. For much of history, this idea was what most people believed. But according to Plato, it doesn't matter how things appear. Using the cave analogy, humans can perceive only the shadows of reality. The task of the philosopher or scientist is to show how the perfect nature of heavenly motion is distorted by human perception. The goal of discovering reality is still one of the major goals of physics, even though the modern definition of reality is different than Plato's.

But Seriously, Folks

The final member of our golden group, Aristotle (circa 384–322 B.C.E.), had the most serious historical impact. He was a student of Plato and the teacher of Alexander the Great. Of the three philosophers, he was the closest to being a modern scientist. Aristotle's writings and achievements take up volumes of books. His ideas formed the foundation of most of what followed in the fields of biology, logic, ethics, and physics. Let's take a brief look at this brilliant and innovative thinker.

While living on the island of Lesbos, Aristotle became fascinated with the life of the inshore waters. Aristotle had a much greater interest in biological questions than most of the earlier thinkers. He thought that human biology could be understood only through the investigation of simpler creatures. No one else before the nineteenth century developed such a broad knowledge of marine biology. He also categorized plants and animals of the land. He asked Alexander to continually send him plants and animals from the far corners of the world.

Aristotle began work on an encyclopedia of all existing knowledge including information about the universe, the natural environment, and human life, both private and public. Every entity would be defined in such a way as to show its particular characteristics. Central to Aristotle's thinking was the concept that facts should always prevail over what might be supposed to follow from general principles.

Albert Says

Ideas can be understood in many ways. The use of analogies is one way to comprehend difficult concepts. Two other ways are through the use of reason and intuition. Aristotle embodied the use of reason and logic.

Aristotle was one of the first people to make a serious attempt at developing a unified physical theory. In other words, he tried to set down a complete system to rationally explain all types of known physical phenomena. Specifically, he turned his attention to understanding three important areas:

➤ Earth and its place in the universe

➤ Physical phenomena

➤ Motion

The Movement of Those Great Heavenly Bodies

Aristotle adopted a *geocentric theory of the universe*, meaning that he considered Earth to be the center around which the heavenly bodies revolved. He also divided the universe into two domains, the astronomical (heavenly) and the sublunar (Earth). The astronomical realm was composed of ether, which is not found on Earth. The sublunar realm was made up of four substances: earth, air, fire, and water. (You'll remember that Empedecles, one of the Pre-Socratic thinkers noted earlier, originated this concept.) The four sublunar substances interacted with each other to form all of the material objects around us.

Aristotle's astronomical system was linked together by a series of interconnecting spheres, 56 in all. The outermost sphere was called the *prime mover*, because all the other spheres were linked to it and derived their motion from it. He never explained what moved the prime mover to get the whole thing started, but later church leaders considered this theory reasonable proof for the existence of God, who moved the prime mover. Aristotle's astronomical system was a bit confusing, and many philosophers disagreed with it.

Relatively Speaking

The **geocentric theory of the universe** is the idea that Earth is at the center of the universe and that everything revolves around Earth in perfect circles.

Material Bodies

The second important area of Aristotle's thought is related to physical phenomena. As was mentioned previously, he believed that the earthly realm was composed of four substances. Qualities such as weight, hardness, and motion could be described in terms of these simple substances. Heavy objects were composed more of earth and water; lighter objects contained more fire and air. The motion of objects could also be explained by the characteristics of these substances. Air bubbled up through water, fire heated water into steam, and earth mixed with water to make mud, which is about how clear this all is. The point of mentioning these ideas is to demonstrate how Aristotle used simple assumptions to try to create a comprehensive system.

Mind Expansions

Hippocrates (circa 460–377 B.C.E.) set forth the first principles of medicine, based on the four elements of Empedocles. This notion of four elements pervaded many areas of thought.

Motion

The third and final important area of Aristotle's thought is related to the four basic types of motion he observed in the world around him:

➤ Alteration

➤ Natural local motion

➤ Horizontal or violent motion

➤ Celestial motion

Alteration isn't considered a type of motion anymore; it's now a process of chemistry. The rusting of an iron object, leaves changing color, or colors fading are some examples of Aristotle's idea of alteration.

The second type of motion was natural local motion. If an object is released, it naturally falls to the ground; other substances, such as smoke, naturally rise. These movements up and down were observed to occur without the objects being pushed or pulled, so they were considered natural movements.

Aristotle also observed how objects of different weight fell in different mediums. Heavier materials sink quicker in water than light ones. He also saw this difference to be true in air, but the difference was not as great. By reason, he concluded that all objects would fall at the same rate inside a vacuum, which was an accurate theory. However, he didn't believe a vacuum could exist, because he reasoned that the speed of an object falling in a vacuum would be infinitely fast. This speed would mean that the object could be in two places at the same time, because it would take literally no time for the object to move from one place to the other. This idea of the nonexistence of a vacuum would cause major problems down the road.

The third type of motion was horizontal motion, or violent motion. He broke this type of motion down into two categories: objects that are pushed or pulled, and thrown objects or projectiles. Aristotle considered both of these unnatural, because they do not occur by themselves if the object is released. The aspect of his theory that he had a hard time explaining was: What kept a projectile moving after it was released? He thought that possibly a column of air was created as the object moved and that this column carried the object along. This ingenious explanation was way off-base! The fourth type of motion, celestial motion, involved the 56 interconnecting spheres discussed previously.

Aristotle's Lingering Impact

Aristotle's views and arguments about the structure of the universe would go unchallenged for almost 1,800 years. His theory that Earth was at the center of the solar system and that everything revolved around it was put into a mathematical system of circular orbits by Ptolemy of Alexandria in the second century. This system held back the advancement of astronomy until Copernicus rediscovered Aristarchus's system in the sixteenth century. It was also ultimately responsible for all the trouble that Galileo would get into with the Catholic Church in the seventeenth century. Some things shouldn't be taken so seriously!

The Greeks, especially Aristotle, were some of the earliest contributors to the science of *mechanics*. The science of mechanics is concerned with the description and the causes of motion of material objects. It explains the motions not only of celestial bodies, but also of earthly objects, including falling bodies and projectiles. Mechanics is involved to some degree in all physical studies and is sometimes called the backbone of physics. Mechanics is one of the areas in physics that Albert Einstein had a dramatic impact on, but at this point in our story he's 2,000 years away. For now, prepare yourself for the greatest rebirth mankind has ever known: the Renaissance!

Mind Expansions

Eratosthenes (third century B.C.E.) used trigonometry to show that Earth was round and to measure its circumference. He came very close, too—within 400 miles!

The Least You Need to Know

➤ The ancient Greeks were responsible for changing the way nature was understood.

➤ Ideas are the beginning of all physics.

➤ Philosophy and physics can support one another.

➤ It's always important to question any theory—you never know where that questioning might lead.

➤ Plato was the first to form a bridge between ideas in philosophy and natural philosophy.

➤ Aristotle developed some of the first theories in the study of mechanics.

Great Minds of the Renaissance

During the 200 years between 1350 and 1550, a number of significant shifts in the way people understood the world altered the fabric of society, which had remained unchanged for almost 1,000 years. The tremendous flowering of creativity in the arts, philosophy, and science during this time, which is called the Renaissance, forever changed how people understood themselves and their place in the universe. Such a change is called a *paradigm shift*. As you will see in this chapter, before a paradigm shift occurs collectively in society, it happens to a few special individuals first.

This time of rebirth, which is what the word *Renaissance* means, is unparalleled in the history of western civilization. But as you will see, change never seems to come easily. With the challenges to and changes in traditional religious beliefs came centuries of intolerance, persecution, and war.

Seeing the big picture helps one understand how ideas evolve from one into another. However, sometimes the collective force of a number of ideas is so great, as it was during the Renaissance, that a huge change occurs in a very short period of time, rather than gradually. You can think of the force of ideas as a couple of snowballs that begin to roll down a hill and suddenly start an avalanche. This chapter is going to focus on a few of those snowballs. You will learn about individuals such as Copernicus, Brahe, and Kepler, whose ideas were so powerful that no obstacles placed in their path could stop them. Put on your snowshoes.

Paradise Lost and Found Again

In the 1,800 years between the time of Aristotle and the beginning of the Renaissance, not much changed in physics, but other areas went through major changes. Shortly after Constantine the Great (at least that's how he liked to think of himself) made Christianity the state religion of the Roman Empire in 310 C.E., problems that had been brewing for some time came to a head. No specific date is associated with the collapse of the Roman Empire, but in the third and fourth centuries, a number of Germanic tribes began pressing against the borders of the Roman Empire in northern Europe.

Around the beginning of the fifth century, entire populations of Germanic and Mongolian peoples migrated into Roman territory, no longer held back by the Empire's legions. The army was in a state of turmoil because the soldiers were not being paid and their leadership was poor. Therefore, it was easy for the invading cultures to set up permanent settlements on the borders of the Roman Empire. In 410 C.E., the city of Rome itself was sacked by the Goths.

The period between 450 and 750 C.E. is referred to as the Dark Ages. With the fall of the Roman Empire came a breakdown in what used to be an organized society. All of the knowledge of the ancient Greeks that had been passed down to the Romans and all of the cultural heritage that Rome had added to it disappeared from its birthplace. The light of classical civilization was kept lit by countries in the eastern Mediterranean, especially by Islamic cultures. If no one had kept the flame of classical civilization burning, the Renaissance could never had occurred.

They were called the Dark Ages for a reason. The sun was working just fine. The problem was that the amount of information that a person acquired in an entire lifetime was about equivalent to what you can read in a daily newspaper. No need to worry about using too many brain cells. The majority of the population became illiterate, with only wealthy families and church clergy able to read Greek and/or Latin.

Relatively Speaking

Some of these migrating Germanic and Mongolian peoples were the Gauls, Visigoths, Franks, Huns, and Vandals. We get our word *vandal* from this last group, who were presumably not the best of neighbors.

Nuclear Meltdown

It's commonly thought that the Crusades were conducted to reclaim the Holy Land from the Muslim Turks. This was the cover story at the time. However, behind the scenes, the Crusades became a political power play for Pope Urban II. He had been looking for a way to unify Europe under the banner of the church, so he sent a proclamation throughout the lands announcing that anyone who fought in a Crusade would have all of his sins washed away and would go directly to heaven upon death. Smart guy!

Slowly, during the Middle Ages (750 to 1350 C.E.), ideas, inventions, philosophy, and art began to develop in greater abundance throughout Western Europe. The Roman Catholic Church became the unifying force that gave stability, a belief system, and meaning to people's lives.

A number of factors were coming together that would act as catalysts to begin the Renaissance. Foremost among them were the Crusades. There were 11 Crusades altogether, but the first four were the ones that set up the religious, economic, and political factors that ushered in the Renaissance.

When the Crusaders arrived in the Islamic (or Muslim) countries, they were astonished by what they saw. Never mind that they were at war. The food (spices became one of the main reasons that trade routes were set up in the first place), the art, the architecture, and the science of this culture were beyond anything the Western Europeans had ever experienced. Islam had inherited all that the Greeks and Romans had to offer and had taken this knowledge to new heights.

Here's some of what made its way back to the West from Islamic countries in the Middle Ages:

➤ The concept of zero in mathematics

➤ The mathematical system of algebra, which was necessary for the development of physics

➤ Anatomical drawings of the skeletal, circulatory, and nervous systems of the human body

➤ Advanced charts of the stars, planets, and constellations

➤ Pre-Copernican theories of the sun as the center of the solar system

Mind Expansions

Did you know that the children's nursery rhyme "Ring Around the Rosie" refers to the Plague? When people first became sick, they became covered in red marks ringed with circles. "Pocket full of posies" refers to the practice of carrying posies to smell, because it was thought that the Plague was airborne and that the horrible smell caused the disease. In the context of the Plague, "Ashes, ashes, we all fall down" becomes a lot more sinister than it seemed when we were kids. Leave it to children to maintain their ability to play under any circumstances.

The arrival in Western Europe of all this knowledge, which had been lost for centuries, was one of the biggest factors responsible for the beginning of the Renaissance. But why did the Renaissance begin in Italy, and not somewhere else?

In the summer of 1347, the most devastating event that has ever hit mankind ravaged Europe. The Plague, otherwise known as the Black Death, took the lives of almost a third of the population of Europe. Approximately eight million people died in four years. The bubonic and pneumonic plague was caused by the spread of bacteria in fleas carried by rats. It was called the Black Death because the victim's decaying flesh blackened in the final hours before death. (I hope you haven't just eaten.) The Plague spread rapidly throughout the Mediterranean ports in Italy, Sicily, North Africa, Spain, and France and finally reached the European mainland.

In the early 1400s, Europe began showing signs of recovery. People were in a hurry to put the ravages of the Plague behind them. One of the first countries to fully recover was Italy. In the early 1300s, Italy was already heavily engaged in trade with the Middle East and the Orient. They imported spice, silk, perfume, and other commodities and sold them to northern European countries. Because of their head start in trade, Italy was among the first countries to develop banking practices, managerial systems, and sound financial principles after the Plague.

Florence became the world leader in international banking and trade. Economic growth requires inventions and new technologies to keep it expanding. This demand made Italy preeminent in scientific inquiry and advancement. Mathematics was used in navigation, surveying, cartography, and construction. The refinement of geometry, trigonometry, and algebra (thanks to the Islamic countries) led to uses for mathematics that no one had ever dreamed of. Thus, the Renaissance was (re)born.

Winds of Change

During the Middle Ages, the most dominant force in peoples' lives was the Catholic Church. The church governed what people believed and how they lived their lives and gave sanction to kings to wage war in its name. *Scholastic philosophy* dominated Western thought. This system of thought was based on religious ideas rather than observations, because God's word was considered to be a more authoritative guide than man's experience or reason. But this way of thinking was about to change.

Albert Says

A paradigm is a world view. Each of us has one. It's made up of all the individual beliefs we have about the world in which we live. At different times in history, changes or shifts in world views have happened on a wide scale, throughout an entire culture. One such paradigm shift occurred in ancient Greece around the time of Aristotle and his contemporaries; the Renaissance brought about another paradigm shift in Europe. We will see a few more by the time we get to Albert.

The paradigm shift that occurred during the Renaissance was both simple and profound. Humanity began to realize that not only were they a creation of God, but they also could be creators themselves. This idea seems self-evident to us today, but during the Renaissance, it was a radical new concept. With this realization, people were free to ask questions and think in ways that had lain dormant since the golden age of Greek philosophy. Another way of understanding this paradigm shift would be to say that God had created a world of order, and now it was humanity's responsibility to ask questions and to use reason to discover the rules that governed the divine order. This is pretty similar to what happened back in ancient Greece.

This new realization of freedom and responsibility produced a flourishing of genius that was unprecedented in European history. Da Vinci, Michelangelo, Erasmus, Luther, and Copernicus were all born between 1450 and 1475.

Erasmus and Luther each fought the intellectual domination of the Catholic Church in his own characteristic way. Erasmus, a man of reason, fought against the limits of doctrine, pleading for man's ability to pursue truth to its logical conclusions regardless of church dogma. Luther's focus was different. He disagreed with the need for a hierarchical clergy and felt that each man could find his own salvation through faith in God and the Bible. He was the catalyst for the Protestant Reformation. Their common ground was that each attempted to give man a central place in the scheme of things and each was deeply religious.

Michelangelo and Da Vinci made man the central object of art. Medieval art dealt with man only in generalities. Most art dealt with religious themes. Michelangelo created art that pictured not only a particular man, but more than that, a heroic man. His art cried that man could be like the gods. Da Vinci, the greatest representation of the *Renaissance ideal*, created art that captured ordinary reality so extraordinarily that it caused viewers to realize that there was much more to seeing the world than they thought. Both, according to their own styles, were bringing God down from the heavens and placing divinity in the world and within each person.

Origins of Astronomy

Before we discuss Copernicus and his role in the evolution of ideas that gave birth to the scientific method, let's take a look at the astronomical system that the Renaissance inherited and another important individual who aided the transition from the ancient world view to the new one.

Ptolemy's System

Chapter 1, "The Early Greek Thinkers," briefly examined Aristotle's ideas about how the universe was set up. His theories were put into a mathematical system by an Egyptian astronomer named Claudius Ptolemy (100–170 C.E.). He reduced Aristotle's 56 rotating spheres down to eight perfect crystalline spheres. One sphere carried the

Relatively Speaking

Scholastic philosophy was a branch of philosophy practiced during the Middle Ages. Much of it dealt with using reason to enhance faith. Some of the writings that had the most influence on theology were produced by scholastic philosophers like St. Anselm, St. Thomas Aquinas, and Boethius. The **Renaissance ideal** was the concept that knowledge of the world is based on human perception and that perceptions need to be questioned. What you observe isn't necessarily accurate and must be tested.

Mind Expansions

In Greek, the word *planet* also means wanderer. This term was assigned to the spheres that were thought to orbit around Earth. Why? Because although the stars in Aristotle's system were fixed and unchanging, a number of spheres appeared to wander around in the heavens, without a clear destination. These spheres were named planets for their wandering.

sun, one carried the moon, and five carried each of the planets known at the time: Mars, Mercury, Jupiter, Venus, and Saturn. The eighth sphere carried the stars.

The seven "Sacred Planets," which included the sun and the moon, followed this order because they reflected the seven days of the week. Each was also associated with a Roman god, and therefore was considered to be sacred. Here's how it breaks down:

➤ The Sun corresponds with Sunday, which is pretty obvious.

➤ The Moon corresponds with Monday. In Spanish, the word is *Lunes*, which is a cognate with the word *lunatic*, because madness was thought to be influenced by the phases of the moon.

➤ Mars corresponds with Tuesday. There's not much similarity there, but in Spanish the word is *Martes* and in French, *Mardi*. The English word comes from Norse mythology: Tiwes Day.

➤ Mercury corresponds with Wednesday, although in English the words aren't even close. In Spanish, the word is *Miercoles*, and in French, *Mercredi*. Again, the English word comes from Norse mythology: Wodin's day, or Odin's day.

➤ Jupiter corresponds with Thursday. Can you guess? Norse mythology again, this time it's Thor's day. In Latin, it's *Jovis*, or Jove's day, which is equivalent to the Roman god Jupiter. In Spanish, the word is *Jueves*; in French, it's *Jeudi*.

➤ Venus corresponds with Friday. This is a tough one. Again, Norse mythology comes into play: thank God it's Frig's day. In Latin, it's *Veneris*, in Spanish, *Viernes*, and in French, *Vendredi*.

➤ Saturn corresponds with Saturday; this is an easy one.

The order of the days and planets doesn't reflect the planets' positions in the sky. Instead, planets are ranked according to their influence based on ancient astrological systems. Astrology was the forerunner of astronomy in the same way that alchemy was the precursor to chemistry. The system developed by Aristotle and Ptolemy was not only an attempt to describe the natural world, but it also developed into a powerful set of cultural beliefs.

Mind Expansions

Astrology dates back to ancient Egypt, with possible origins in Babylon. It was a part of almost all ancient cultural belief systems. Ideas and theories were developed by observing the heavens. Astrologers tried to place humanity within the universal order that they observed in the skies. The primary purpose of study was to understand how the heavenly bodies influenced the course of human actions.

Alchemy was a physical as well as philosophical practice that sought to transform base elements into purer forms. On the physical level, alchemists sought to transform lead into gold, which led to chemical discoveries that in turn led to chemistry. On the philosophical level, alchemists tried to transform their less noble qualities, both physical and mental, into a spiritually perfected form.

Modern science has calculated that Earth takes 365 days, 5 hours, 48 minutes and 46 seconds to make its annual trip around the Sun. That's pretty accurate. On the other hand, the ancient Egyptians calculated the year to be 365 days and 6 hours. That's about 11 minutes too long. The first Emperor of Rome, Julius Caesar, adopted the Egyptian calendar (which is why it's called the Julian calendar), and it stuck. But that 11-minute error accumulated over time, so by the fifteenth century, the calendar was about 10 or 11 days off. This caused a big problem for the church, because Easter had been fixed at March 21, the vernal equinox (the first day of Spring). By the fifteenth century, the vernal equinox was occurring around March 11 on the calendar, moving Easter further away from Easter because of the fixed date on the 21st.

This same problem with medieval astronomy was also discovered by farmers and sailors. Astronomical observation is critical both for agriculture and for finding your way on the sea, and this is one of the reasons why early civilizations spent so much time observing the heavens. Farmers needed to know the positions and movements of heavenly bodies to ensure correct planting and harvesting times. Ship navigators relied on accurate astronomical information to get from place to place on open waters, out of sight of the shore. Astronomy was indispensable to both professions.

By the fifteenth century, people knew a great deal about astronomical observations, and mathematics was being used in everyday life. For centuries, observation had to fit theory, instead of theory fitting observation. Unfortunately, calculations based on Ptolemy's perfect spheres didn't fit observations. More complex explanations had to be made to preserve Earth's position at the center of the universe.

Johann Müller

One of the first concepts about the nature of the world to change was the idea that Earth was flat. Instead of a flat disk comprised mostly of Europe and Asia, Earth was found out to be a sphere covered mostly by water. (Pop quiz: Who originally figured out that Earth was round? If you said the ancient Greeks, you get half a point. If you said Eratosthenes, pass Go and collect $200.) This change was due to the generation of ocean explorers who ventured out and failed to fall off the edge of the world. (Believe me, this was a very real fear and the cause of more than one mutiny.)

In 1475, Pope Sixtus IV asked a German-born astronomer, Johann Müller (1436–1476), to investigate the problems people were having with the calendar and the changing dates of Easter. Müller had written a book called the *Epitome*, which examined weaknesses in Ptolemy's Earth-centered model of the universe.

Ptolemy's model explained the movement of the planets as perfect circles, but within these perfect orbits were *epicycles*, which were smaller perfect circles within each planet's orbit. To understand epicycles, consider that the Moon orbits Earth, and

Nuclear Meltdown

Today, everybody knows the world is round, right? Not the people in the Flat Earth Society—they really exist!

Earth orbits the Sun, so you have an orbit within another orbit. The epicycles were just like that. Not only did each planet revolve around Earth, but they each also revolved in an orbit. The epicycle concept was an attempt to account for the unusual heavenly movements that were observed.

Müller tried to show that all of these circles within circles were the cause of the problems people were having with the calendar and astronomy, but ultimately he failed. With the tools at his disposal, it was nearly impossible to collect enough accurate data to refute this ancient and unquestioned system. Nonetheless, Müller was one of the first astronomers to question the belief that Earth was the unmovable center of the universe.

Copernicus's Snowball

Nicolaus Copernicus (1473–1543) was born in Torun, Poland. He studied in the University of Cracow in Poland and the universities of Bologna and Padua in Italy. He acquired a substantial amount of knowledge in mathematics, astronomy, theology, medicine, canonical law, and Greek philosophy. His favorite area of study was astronomy. When his duties as a minor church official were through, he spent his time studying and observing the heavens.

Copernicus knew that Ptolemy's writings were consistent with the teachings in the Bible, but he basically agreed with Müller's opinion that Ptolemy's theories didn't agree with observations. Copernicus also had a bone to pick with a few of the other beliefs that were popular at the time. Some of these ideas, which could all be traced back to Aristotle, were as follows:

➤ Objects fall to the earth because they are naturally attracted to the center of the universe, which is Earth. If the Sun were the center of the universe, objects would fall toward the Sun. (No one knew anything about gravity yet!)

➤ Earth does not move. When a person is moving, for example on a horse, he or she can feel a breeze. If Earth were moving, we would feel a continuous wind. Also, if Earth were moving, an apple falling from a tree wouldn't fall straight down. It would land somewhere off to the side, because the tree and the ground beneath the falling apple would have moved in the time it took for the apple to hit the ground.

Relatively Speaking

Epicycles are the smaller circular orbits that planets traveled around as they orbited Earth in Ptolemy's convoluted system of planetary orbits.

The **Gregorian calendar,** named after Pope Gregory XIII, is the calendar we use today. Each year has 365 days, and every fourth year is a leap year, with 366 days. Why do we have that extra day every four years? Because Earth travels around the Sun every 365 $\frac{1}{4}$ days. The previous calendar, called the Julian Calendar, didn't account for this extra quarter of a day, so astrological events, such as the vernal equinox, drifted further and further away from their assigned dates.

Heliocentric means sun-centered. The development of the heliocentric model of the universe was a major advance in astronomy.

If you think about it, these ideas are pretty reasonable. It would take another century to refute these arguments. Later on, you'll meet some of the individuals responsible.

Despite these arguments, Copernicus was convinced of the accuracy of his observations and calculations. He went public in 1540 by publishing his work, commonly called *The Revolutions*. He gave a detailed explanation of the *heliocentric*, or sun-centered, theory of the universe. This theory helped solve the calendar problem and showed why Ptolemy's theories were wrong. In 1582, his calculations were used to determine the new *Gregorian calendar*, which is the one we still use today.

The church didn't officially denounce Copernicus's publication, but the publication did produce a strong negative reaction. His view seemed sacrilegious to sixteenth-century church-men, who were convinced that God had placed us at the center of the universe. They weren't about to allow Earth and themselves to be displaced from that center without a decent fight. They believed that the world was static and unchanging since its creation, with known exceptions recorded in the Bible, such as the Flood. They and most educated Europeans took Ptolemy's model of an Earth-centered universe to be a scientific explanation of what they already knew to be true according to Bible. Vanity, fear, and theological dogma would not allow humanity to be placed on an insignificant planet orbiting the Sun.

Copernicus's ideas were revolutionary in themselves, but they led to even more astounding conclusions. He had the brilliant realization that perhaps movement was, in part, the perception of the viewer. His theory suggested that the nature of how the world and its movement are perceived might depend on the position of the observer. This idea was a key concept in Einstein's theory of relativity. This insight was counterintuitive to Copernicus's contemporaries, but it laid the foundation for much of what was to come in science.

Albert Says

In his 1953 speech commemorating the 410th anniversary of Copernicus's death, Albert Einstein said, "Copernicus not only paved the way to modern astronomy, he also helped to bring about a decisive change in man's attitude toward the cosmos. Once it was recognized that the Earth was not the center of the world, but only one of the smaller planets, the illusions of the central significance of man himself became untenable. Hence, Copernicus, through his work and the greatness of his personality, taught man to be modest."

Albert Says

The dramatic shift in the understanding of astronomy during the Renaissance occurs on a smaller scale within our own consciousness on a regular basis. That is to say, our beliefs act to filter our perceptions of reality. If you change your belief, you change your perception of reality!

All with No Telescope!

On August 21, 1560, there was a partial eclipse of the Sun. This event is not earth-shattering in itself, but it altered the life of at least one 13-year-old boy who witnessed it. That boy was Tycho Brahe (1546–1601). Although he had been studying the traditional courses required of a young member of an aristocratic family, he became intrigued by man's ability to predict such astronomical events. This partial eclipse had been predicted by Copernicus's astronomical tables. However, Tycho's family and the church authorities had different plans for him, so it wasn't until his completion of university studies that he found the time to sneak away and pursue his passion.

In November 1572, while watching the night sky, Tycho observed a very bright object that had never been seen before. It was brighter than any other object in the sky, so bright that you could even see it during the day. History had recorded the existence of comets, so most people thought that's what this light was. Besides, Aristotle's teachings, which were still predominant, said that the heavens are fixed, and no new stars are ever created. Tycho decided to watch this new object with his shiny new sextant (a navigational instrument used to measure the altitude of celestial bodies) to see if it moved. If it did, it was just another comet. But if it didn't, there was another kink in Aristotle's chain. Guess what? The light didn't move. It was a brand new star. Today this astronomical phenomenon is called a *nova*.

Over the next 20 years, Tycho spent all his time at an incredible observatory he had built. The King of Denmark had granted Brahe title to a 2,000-acre island in the sound off of Copenhagen. On this island and with state funding, Tycho built a large home, a laboratory, an observatory, clocks, sundials, globes, a corn mill, 60 fish ponds, a paper mill and printing shop to produce his manuscripts, and a host of incredible devices that were to become wonders of the world. He employed a staff of artisans to build his inventions and professional astronomers to help with the recording. His community was called Uraniborg (heavenly castle).

Relatively Speaking

The term **conjunction** comes from astrology and refers to the passing of any two planets within 30 degrees of each other as laid out in a horoscope. In scientific terms, it refers to when planets line up in such a way that they appear very close to one another when observed from Earth.

One of the instruments that he built was a huge sextant that you could sit in. This apparatus allowed him to make the most accurate measurements to date about the positions of the Sun, Moon, and planets. Most other observers were content with measuring planets at specific points, but Tycho tracked them through their entire orbit. The most astounding thing about his work was that his observations were all made without the aid of a telescope (which wouldn't be invented until 1608). Tycho made his observations with various sighting tubes and pointer instruments that he invented.

With the very accurate measurements he accumulated, he found that neither the Copernican nor the Ptolemaic theory agreed with his new data. In one example, he observed a *conjunction* of Jupiter and Saturn that he had predicted quite accurately. The Copernican tables predicted that it would occur several days later, and the Ptolemaic tables predicted that it would occur a month later. Tycho must have been feeling pretty good about himself.

Tycho was also impressed with the various arguments against the motion of Earth. (Can you remember them? If you can, you're on your way to becoming a physicist. If not, read "Copernicus's Snowball" again.) His observations and beliefs led him to create a new theory about how the universe was laid out. This one was a combination of the previous two.

Like Copernicus, Brahe proposed that the Sun was at the center of some orbits, and like Ptolemy, Brahe proposed that they orbited in perfect circles with a few epicycles mixed in. In Tychonic theory, Earth was the center of the universe, with the Sun and Moon orbiting Earth and the other five planets orbiting the Sun. Most astronomers liked this theory. Ptolemy's model was slowly being thrown out, and the Tychonic theory agreed to some extent with Copernicus's theory. It also left Earth where people liked it: as the center. Tycho never worked out his theory completely. He left his assistant and successor the job of performing the necessary calculations to work it all out. That person was Johannes Kepler (1571–1630).

Did You Say Elliptical?

The volumes of data that Tycho had accumulated needed to be unlocked by a finer mind than his to discover their hidden secrets. That mind belonged to Johannes Kepler. He is an important link in our pursuit of connecting ideas. He brought astronomy closer to physics, more than anyone else before.

Kepler was born about 28 years after Copernicus's work was published. He had an unhappy childhood and pretty poor health, but these obstacles didn't hold him back. Kepler was considered a mathematical genius in his teens. He obtained a teaching position very quickly, but he turned out to be a poor instructor who didn't have many students. That was fine with him, for it gave him the time he needed to devote himself to his astronomical studies. (I think we've all had our share of Keplers for math teachers!)

Nuclear Meltdown

Each astronomer covered so far has made a step in the right direction, and a few took a step or two in the wrong direction as well. Their mistakes are as important to consider as their contributions. Very few ideas are right the first time, and the old adage that we learn from our mistakes was as true then as it is now.

Belief vs. Reason

Before getting into what Kepler did, we should discuss a broader point. All of the astronomers covered in this chapter were concerned not only with mathematical accuracy, but also with philosophical implications. Why do you think that was?

Remember that the basis for the study of the universe begins with asking questions. When you develop a system of thought that answers these questions, you're on your way to a greater degree of understanding. Whether this system survives depends on how well it works. This is as true of science today as it was 2,500 years ago. We use what works and discard it when we find something that seems to work better.

Belief systems were inextricably intertwined with philosophical ideas and scientific models, so even if a thinker developed mathematics that helped to explain the physical mechanics of how something operated (such as the information collected by Tycho), he would run into trouble if that explanation was in conflict with the cherished beliefs that went along with the old system. That is why the philosophical implications were such a concern for astronomers.

Look at how threatening the idea of the heliocentric theory was. The calculations based on a heliocentric model were simpler and more accurate, but this model went against widely held beliefs that people were reluctant to discard. (As you will see with Kepler, and in the next chapter with Galileo, one's personal beliefs can interfere with the emergence of new insights and information as much as the beliefs of others.) This new conflict between beliefs and the reasoning used to question them had devastating repercussions. Faith and reason, which once acted symbiotically, became enemies.

Relatively Speaking

The **Pythagorean** or **Platonic solids** were the five perfect solid shapes besides the sphere, according to Greek mathematicians. They were the tetrahedron, which has four faces, all equilateral triangles; the cube, which has six faces, all squares; the octahedron, which has eight faces, all equilateral triangles; the dodecahedron, which has twelve faces, all pentagons; and the icosahedron, which has twenty faces, all equilateral triangles. Each of these shapes can contain a sphere that touches all of its sides, and each can be contained within a sphere that touches each of its points.

The main source of conflict arose within the teachings of the Catholic Church. The Protestant Reformation led by Martin Luther brought with it a Counter-Reformation put forth by the Catholic Church. As a result, there was little tolerance for deviation from established doctrine, and this intolerance led to centuries of religious wars that are still being played out in today's world. It is no wonder that thinkers who challenged existing doctrine were concerned about the philosophical implications of their work. I would be, too. What's more important: your life or the truth?

Kepler's Convictions

Kepler accepted the heliocentric theory and believed that Earth was one of six planets that orbited the Sun. He was very curious as to why there were only six planets, instead of some other number. His fascination with numbers led him to wonder about the spacing and speed of their orbits.

One of the areas in mathematics that interested him most was geometry. (A little math humor: What did the acorn say when he grew up? "Gee, I'm a tree.") Greek mathematicians had proved that there were only five perfect shapes, known as the *Pythagorean* or *Platonic solids*, so Kepler tried to connect these with the five planets other than Earth that orbited the Sun. In an ingenious method, he built wooden models of each of the solids that showed their spacing and the order of the planets when nestled inside each other.

Kepler was a believer in the more mystical aspects of Pythagorean philosophy. The Pythagoreans thought that the structure of the universe was mathematical in nature and could be explained using numerical relationships. These relationships could best be described using music. In one sense, music is purely mathematical in nature, so you could say that when you hear music, you're listening to the mathematics of the universe. Well, the Pythagoreans thought so anyway.

When you listen to musical sounds, they are much more pleasing to the ear when the component notes are harmonic, which means that they are multiples of each other. The Pythagoreans also discovered that the tone emitted by a taut string is related to its length. If you change the length, you change the tone. They applied this idea to the heavens, proposing that each planet had its own distinct vibration or sound and that the spaces between them also reflected a specific musical interval. This was known as "the music of the spheres."

Kepler firmly believed in and adopted this system of thought. Actually, he became pretty obsessed about the whole idea, convinced that he had achieved the ultimate insight into God's plan of the universe. Too bad it turned out to be wrong, especially considering that there are nine planets and not six. But Kepler didn't know this at the time.

The point is that his belief that he had found the ultimate answer stopped him from looking else-where for a long time. It wasn't until he began working for Tycho Brahe that he discovered the true mathematical workings of the planets. Even then, he never gave up his search for the relationship between the geometric figures and perfect orbits.

Mind Expansions

This comment is just entertaining: Mars has the largest ellipticity of the planets studied by Tycho and then by Kepler. In fact, Tycho's measurements were not precise enough to determine that any other planet, except Mars, had an elliptical orbit. He was too far north to ever see Mercury, and Neptune, Uranus, and Pluto weren't known. If Kepler had focused his analysis on another planet, he may never have concluded that planetary orbits were elliptical.

The Laws of Planetary Motion

For several years, Kepler had been studying the tremendous amount of data that Tycho had amassed. He could find no solution to fit a perfect circular orbit. Because of his familiarity with geometric shapes, he knew he had other options besides a circle for an orbital shape. This brings us to Kepler's laws of planetary motion (published between the years 1609 and 1618):

➤ **First law:** All of the planets revolve around the Sun in elliptical orbits.

➤ **Second law:** A line joining the Sun and the planet sweeps over equal areas in equal time as the planet moves through its orbit.

➤ **Third law:** The square of the time a planet takes for one complete revolution around the Sun is proportional to the cube of the planet's distance from the Sun.

The discovery of all three of these laws came from Kepler's study of the planet Mars. He discovered the second law, which deals with the speed of the planet in its orbit, before the first. Shortly after, he discovered the third law, which relates the size of the planet's orbit to its speed.

These laws are the first mention of purely mathematical concepts in this book so far. In Chapter 1, it was noted that a good way to understand math is through pictures or graphs. So let's take a look at a picture and table so you can understand what Kepler's three laws mean.

Kepler's first and second laws.

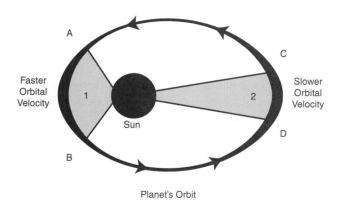

• Time of orbit from A to B = C to D
• Area 1 = Area 2

Kepler's Third Law: Planetary Motion for the Solar System

Planet	T	D	T²	D³
Mercury	0.24	0.39	0.058	0.059
Venus	0.62	0.72	0.38	0.37
Earth	1.00	1.00	1.00	1.00
Mars	1.88	1.53	3.53	3.58
Jupiter	11.9	5.21	142	141
Saturn	29.5	9.55	870	871

The picture describes Kepler's first law. You can see that the orbits of the planets around the Sun are elliptical, or oval in shape. That's a pretty easy concept. The same picture describes his second law as well. Kepler noticed that as a planet approached the Sun in its orbit, its speed increased, and as it moved away again, it slowed down. The cool thing about this is that the planet covers the same area in the same amount of time, even though it travels at different speeds. The areas of the shaded portions are equal.

The table illustrates Kepler's third law. It lists the six planets that Kepler knew about. Look at the first planet listed, Mercury. The number in the first column is the time it takes to revolve around the Sun. The second column is the distance from the Sun. (The 1.000 for Earth distance represents 93,000,000 miles, the distance from Earth to the Sun. The distances in the chart are shown in proportion to the distance from Earth to the Sun.) The third column is the time it takes for Mercury to revolve around the Sun, squared. The fourth column is the distance of Mercury from the Sun, cubed. If you compare columns three and four, they are very close to being equal. In some cases, the difference is in the thousandths, the third decimal place.

In his efforts to try to understand God's plan through geometric shapes, Kepler ended up discovering mathematical principles that described the relationship between each planet's distance from the Sun and the length of its year. That gives us the last two snowballs of the chapter:

➤ Kepler was the first person in history to explain the planets' motion and distance from the Sun in terms of physical principles rather than divine control.

➤ Kepler's laws joined physics and astronomy together for the first time.

The replacement of the geocentric theory with the heliocentric theory is known as the Copernican Revolution and is one of the greatest advances in the history of western thought. The Copernican Revolution has been viewed as a prime example of the conflict between science and religion, even though its proponents were deeply religious men.

But revolutions don't happen overnight. Over 150 years passed between the publication of Copernicus's theories and the publication of Newton's gravitational theories. In between, another significant individual would help usher in the scientific method. In the next chapter, you will be introduced to the genius of Galileo.

The Least You Need to Know

➤ A paradigm shift occurred during the Renaissance that led to questioning about humanity's place in the universe.

➤ The ancient Greek astronomical system put forth by Ptolemy was first questioned by Johannes Müller.

➤ The Copernican Revolution replaced the geocentric, or Earth-centered, theory with the heliocentric, or sun-centered, theory of the solar system.

➤ Tycho Brahe provided the most accurate astronomical data available, and this data led Kepler to develop the laws of planetary motion.

➤ Kepler's laws show the stability of the solar system and the mathematical relationships between the time it takes for a planet to revolve around the Sun and its distance from the Sun.

Chapter 3

Galileo Gets in Hot Water with the Church

In This Chapter

➤ The religious wars of Europe

➤ The martyrdom of Giordano Bruno

➤ Galileo's telescope

➤ The trial of Galileo

➤ The dynamics of motion

Have you ever looked through a telescope at the Moon? It looks quite a bit different from the way it looks to the naked eye. For one thing, the "man in the moon" disappears. You can see craters, mountains, plains, and valleys. Galileo probably thought that seeing all these surprising features on the Moon's surface would make him famous. It did, but probably not in the way he had hoped.

The rise of scientific inquiry ultimately brought a number of individuals into direct conflict with established cultural and religious doctrines. Galileo was one such individual. The stage had been set for the separation of certain belief systems that would not find common ground again until the second half of the twentieth century. This chapter examines an individual whose discoveries, innovative ideas, and experimental observations once and for all established the *Scientific Revolution*. Ladies and gentlemen, I would like you to meet Galileo Galilei, a sought-after heretic of the Inquisition.

I'm OK, But You're Not

To truly appreciate and understand Galileo's accomplishments and conflict with the Catholic Church, you need to understand the era and environment in which he lived. The end of the Renaissance marked a transition in the development of ideas in many areas of human endeavor. These ideas established patterns of thought that permanently altered long-held religious, philosophical, and artistic views of humanity's place in the universe. This book's focus so far has been mostly on astronomy, because these studies eventually led to the birth of physics.

Relatively Speaking

The **Scientific Revolution** was the period in human history when reason, rational thought, and experimentation replaced the ancient Greek understanding of how nature operates. It began in the mid-sixteenth century and culminated in the early eighteenth century.

Albert Says

The beliefs we hold come from our life experiences. We consider these beliefs to be true, but are they the Truth? Philosophers have argued that many beliefs are relative in nature and change according to our experiences. The opposite is also true. If you change your belief, you also change how you experience the world.

Such paradigm shifts in one area of thought are always accompanied by changes in other areas. Picture a wheel, with a hub at the center where the axle goes through and spokes that extend out from it to the rim. Your world view or paradigm is the hub, and the spokes represent different beliefs you have that are attached to this hub. Each spoke helps to support the wheel. If a spoke is removed or damaged or somehow compromised, the stability of the entire wheel is brought into question. If you change your paradigm, the hub, you have to get a new set of spokes.

Replacing old beliefs with new ones or having an existing belief conflict with a new belief is not an easy situation. More often than not, this situation results in a struggle that has a number of possible outcomes. Although other possibilities exist, the following outcomes are the main ones we are concerned with:

➤ Your desire to change or acquire new knowledge outweighs the possibility of a new belief causing you inner turmoil.

➤ You feel safe with the beliefs you already have, so you have no desire to question or change your existing paradigm with anything new. Change is not a priority.

➤ Experience has shown that what you believe is true—otherwise, you wouldn't believe it. When conflicting experiences occur in your life, the outcome is eventually resolved. Resolution occurs by totally rejecting the new experience, destroying its source because you perceive it as a threat, or embracing it because it offers a better understanding than the one you had.

The dynamics of change we go through as individuals as we live and accumulate new experiences is also reflective of how groups in society react to new ideas and therefore explains something about the dynamics of history. Galileo and many others during the late Renaissance were caught up in this volatile process of change, and it cost some of them their lives.

Saints, Heretics, and Martyrs

The degree to which opposing forces manifest themselves in human history is truly amazing. At the end of the High Renaissance, a pinnacle of cultural achievement, Europe also underwent one of the bloodiest eras in its history when religious wars broke out between Protestants and Catholics. The conflict began back in the early 1500s with the Peasant's War in Germany, reoccurred in the Thirty Years War in central Europe, and appeared once again in the seventeenth century with the English Puritan Revolution.

The violence peaked between 1560 and 1600. During this time, a bitter war broke out in the Netherlands between the Calvinist Protestants and Spanish Catholics, an equally horrific conflict occurred in France between the Huguenots and the Catholic League, and the Anglo-Spanish naval war was going on in the seas. During these passionate and violent outbursts, Protestant mobs vandalized Catholic cathedrals, and the Inquisition tortured and burned at the stake anyone deemed a heretic. Into this cauldron of intolerance was added anyone who questioned the beliefs of either side. This atmosphere of intolerance forms the backdrop for the story of the martyr Giordano Bruno.

Giordano Bruno (1548–1600) was an ordained priest and a well-known writer in the fields of theology, astronomy, and philosophy. He wrote quite a bit about the Copernican theory of the solar system and thought that the universe was infinite. (He was one of the first people to expound this belief.) He also had a few other ideas that were considered heretical:

➤ If Earth is like the planets, then the planets must be like Earth. So the planets must also be populated by people who have experienced similar historical and biblical events.

➤ The Sun is a minor star, and other planetary systems exist. Therefore, we are not a unique creation of God, and the religious practices incorporated in our traditions are also not unique.

Mind Expansions

On Friday, April 13, 1307, Philip IV of France captured and brought to trial a group known as the Knights Templar. He charged them as heretics. The ones who survived seven years of imprisonment and torture were burned at the stake. From this event came the superstition about Friday the 13th being unlucky.

Mind Expansions

One of the first groups to use the telescope was insurance companies. They would send an agent to the top of a cliff overlooking the ocean. With the telescope, he could see what ships were approaching the harbor before they arrived. Then he would run back and tell his company which ships were coming. The company could adjust its rates before its competitors had this valuable information and make a bigger profit.

Although Bruno had no proof that any of these theories were true, some of them turned out to be valid. (Not the one about people on other planets in our solar system, unless of course you've seen the movie *Mars Needs Women*.)

If these theories weren't enough to get Bruno into trouble, he also felt that the Bible shouldn't be followed for its astronomical teachings, but only for its moral teachings. He openly criticized Aristotle's physics and rejected the Protestant principle of salvation by faith alone. It wasn't a particularly good idea to publicly disagree with the only two choices around. Bruno had no place to hide.

In January 1593, Bruno was arrested by the Roman Inquisition and put on trial. The trial would end up lasting seven years, and at the end, the only thing that could save him was a complete retraction of his theories. But Bruno refused to retract anything, so in February of 1600 he was bound to a stake and burned alive. He became a martyr for self-expression and freedom of thought. His writings would influence Galileo and the entire future of scientific thought.

The Starry Messenger

Galileo Galilei (1564–1642) is considered by many to be the founder of the Scientific Revolution. He led the way out of natural philosophy by combining three areas of study in which he was gifted: mathematics, astronomy, and the new physics. Kepler applied geometry to astronomy and also combined physics with astronomy; Galileo applied mathematics to astronomy and mathematics to physics. This application had to happen in order for a new form of physics to develop.

The ability to create something new by putting together familiar ideas in new ways is often the mark of genius. Galileo had this ability. Galileo gave science two important new ways of looking at things. The first was his utilization of the newly invented telescope, with which he studied the features of celestial bodies. The second was his application of mathematics to falling bodies. His explanations would become the foundation of classical mechanics.

Because we've already been following the development of astronomy in physics, let's cover Galileo's contribution in that area first, although chronologically it came after his theories about falling bodies. Around 1609, Galileo was about to publish his book on uniformly accelerated motion (more about that later), but he was sidetracked by a startling new invention, the telescope. Some people think that Galileo invented the telescope, but that's not true. It had been invented in 1600 by a Dutch spectacle maker named Hans Lippershey.

By 1608, the telescope was being used for navigation and in the military. Galileo had already developed a reputation as skilled instrument maker and was asked by the Venetian Senate to build a telescope for its military. He ground his own lenses and through experimentation devised a telescope three times more powerful than any that had been produced before. By the end of the year, he had built a 30-power instrument, which tripled even his own previous efforts. That's why Galileo is often associated with the invention of the telescope.

The first object that he turned his powerful new telescope toward was the Moon. To his delight, he found that the Moon was not the shiny, polished, silver pearl that most people thought it was. It looked very much like Earth, with mountains, valleys, seas, and oceans, even though he later realized that there was no water on the Moon. Galileo even measured the height of the mountains, some of which are four miles high. Incredible as it may seem, his measurements turned out to be extremely accurate.

Galileo also discovered that there was a secondary illumination of many dark areas of the Moon. Using geometry, he showed that the only object capable of reflecting light to the surface of the Moon was Earth. He had discovered *Earthshine*. Until this discovery, it was widely thought that planets shine from internal light, like the Sun and the stars. Showing that the planets shine by reflected light as they circle the Sun, and not by their own light from within, was another of Galileo's great discoveries.

The following sections look at some of Galileo's discoveries that proved Copernicus's heliocentric theories. The publication of these discoveries marked him as a target of the Roman Inquisition.

Relatively Speaking

Earthshine (no relation to moonshine) is the faint illumination of the dark areas of the Moon by sunlight reflected from Earth. The Apollo crews who landed on the Moon saw it in person.

So Many Stars

When Galileo examined the night sky with his telescope, he found that even within a very small area, he was able to see over 500 more stars than with the naked eye. He also discovered that viewing stars through the telescope didn't bring them closer in proportion the way that viewing planets did. Stars looked the same, but planets appeared perfectly round and were flooded with light. In other words, the telescope makes the planets look like discs, but it can't do the same for stars. This observation confirmed Copernicus's theory that the stars are much, much further away from Earth than the planets are.

The Moons of Jupiter

While Galileo was observing Jupiter, he discovered four new planets, at least that's what he thought they were at first. But then he found that as Jupiter moved in its

Mind Expansions

The moons of Jupiter that Galileo discovered were not originally called moons. He named them the Medicean Stars, after the Medici family in Italy. They were one of the most powerful and wealthy families in Europe. Galileo also discovered sun spots, and by studying them demonstrated that the Sun rotated on an axis. He also found that Saturn appeared to have ears (that's just what they looked like); his telescope wasn't quite strong enough to see the rings.

Nuclear Meltdown

Although Galileo made great astronomical discoveries, he missed an important point. He either didn't read or simply didn't agree with Kepler's laws of planetary motion. He stuck to the belief that the orbits of the planets around the Sun were perfect circles, not ellipses as explained by Kepler. Even a genius makes mistakes sometimes.

orbit, these four bodies moved along with it. Sometimes they were on one side, and then they would appear on the other. He concluded that they were moons that circled Jupiter, just like the Moon circles Earth. Jupiter and its moons was a small-scale model of the Copernican system.

If Jupiter with its four moons could circle the Sun, this implied that Earth could revolve around the Sun with the Moon. The fact that Earth also was not the only planet in the solar system with a moon made it less unique. Earth seemed to be becoming less and less important.

The Phases of Venus

Of all of the arguments that Galileo presented in support of the Copernican system, the phases of Venus was the clincher. Galileo found that as Venus orbits the Sun, it goes through phases just like the Moon does. This discovery proved that Venus shines by reflected light, and as it moves closer to or further away from Earth, the phases change in size. None of these observations would have been possible according to the old Ptolemiac geocentric system. None of these discoveries would have been possible without the aid of a good telescope either.

Galileo published his astronomical findings in a book entitled *The Starry Messenger*. He never asserted that they proved the heliocentric theory; he merely presented them as new discoveries about the heavens. But the table was now set for him to go public with an endorsement of Copernicus's theory and, in doing so, incur the wrath of the Inquisition.

Oh No, Galileo!

Upon the publication of *The Starry Messenger*, Galileo was praised by many fellow philosophers and astronomers. He was even invited to Rome to meet the pope and to give a demonstration of his telescope. He became quite popular and was free for many years to report his observations without too much worry about whether they conflicted with widely held beliefs of his time.

The first indictment against Galileo came in 1614. A priest had denounced the Copernican system and mathematics, all mathematicians, and Galileo's reports as being religious heresy. Galileo's response was similar to that of Bruno, in that he asserted the Bible should be followed for its moral teachings, but should not be used as a source of astronomical teachings.

One of Galileo's most important arguments was against the literal interpretation of the Bible. He believed that if people weren't allowed to question and interpret the meaning of scripture for themselves, why would God have given humanity a mind with which to think? This position went right to the core of the religious conflict at the time. What was ultimately at stake was whether the Catholic Church would remain the sole authority for interpretation of the Bible. Galileo's views became more of an issue of individual freedom, and the political position this put him in made it look like he was on the wrong side.

Galileo decided that if he could present enough evidence to prove that the Copernican system was indeed correct, enough important people would become convinced of the truth of what he was saying, and he would not be condemned for heresy. He first invited a group of Jesuit priests to look through his telescope at the night sky so that they could see his discoveries for themselves. Most of them refused, but one brave soul looked. Although this priest saw what Galileo told him he would see, he refused to accept it, rationalizing that the telescope altered reality. Talk about being blinded by your beliefs!

Having failed to convince church representatives, Galileo decided to publish a book that would lay out his arguments. This work, *Dialogue on the Two Chief World Systems*, is his most famous book. It was praised by scholars throughout Europe, but ultimately the publication of this book got him in the most trouble.

In 1632, Galileo was arrested and brought to trial by the Holy Office in Rome for defending the Copernican system. He was not allowed to have a defense attorney, nor was he allowed to see the evidence against him or hear the charges. Faced with the same choice as Giordano Bruno, to publicly retract his theories or die, he chose life and never taught Copernican theory again. He lived the rest of his days under house arrest and died on January 9, 1642.

In 1757, the church removed Galileo's works from its list of banned publications. In 1992, 359 years after Galileo was forced to recant his theories, Pope John Paul II formally acknowledged the church's error to the Pontifical Academy of Sciences. Better late than never!

Mind Expansions

In 1620, the church put out a publication that listed its corrections to Copernicus's theories. In 1622, the Institution for the Propagation of Faith was created. That's where our word *propaganda* comes from.

The New Physics

The other area in which Galileo had a significant impact was mechanics; he worked in this area before he made his discoveries in astronomy. As you'll recall, mechanics is the science that describes the causes of motion of material objects. Galileo's use of mathematics to explain his theories of the motion of material bodies was the birth of what is sometimes called the "new physics." It was new because the theories were totally different from any previous explanation of falling bodies. (If you remember who developed the first theories of moving bodies, I promise I won't ask you to remember anything else for a few pages. Here's a hint: his name starts with an *A* and rhymes with *bottle*.)

At the age of 17, Galileo was studying medicine at the University of Padua. The story goes that one Sunday, while attending mass, he noticed a large chandelier swinging overhead. As it swung back and forth, each swing getting shorter and shorter, he wondered if the time of each swing also shortened. He didn't have a stopwatch (it hadn't been invented yet), so he measured the time by counting his pulse. To his surprise, he found that even though the swings got shorter, it took the same amount of time for the chandelier to complete each swing. This discovery intrigued him so much that he switched majors and began to study mathematics instead.

Mind Expansions

Before Galileo left his medical calling, he developed a type of pendulum that could be used to measure the pulse of patients. It was called a **pulsometer** and was the forerunner of the stopwatch used by nurses to check pulses today.

Essentially, Galileo had discovered how a pendulum operates. He tried out different lengths of string with various weights attached to the ends. As long as the length of string remained constant, it didn't matter whether he used a heavy weight or a light weight. A big swing took the same amount of time as a short swing. This discovery led to one of the first conflicts with Aristotle's teachings on falling bodies.

If you release a heavy object attached to a string and then release a lighter object tied to a same length of string, it will take the same amount of time for each to reach the lowest position of the swing. Each of these is a pendulum. Taking this a step further, this means that these two objects would take the same amount of time to hit the ground if they were released from the same height. This idea is in direct conflict with Aristotle's theory that if two objects are dropped from the same height, the heavier one will hit the ground before the lighter one.

To test his new idea, Galileo went to the Leaning Tower of Pisa and dropped two spheres, a wooden one and an iron one. He observed that they both hit the ground at the same time. In the bottom of the ninth, the score is Galileo 1, Aristotle 0.

Galileo is considered to be responsible for establishing the Scientific Revolution because he introduced a fundamental methodology that is the cornerstone of science. Based on the Leaning Tower of Pisa incident, what do you think he did? (Alex, I'll take scientific experimentation for $800.) He established the idea that an observation or theory needs to be tested by experimentation. The validity of a theory can be determined only by collecting information from an experiment.

I Fall, You Fall, Freefall

Armed with his new idea of experimentation, Galileo began to develop the elementary laws of dynamics. Aristotle asked the question, why does a body move? Galileo asked a similar question, but with a slight difference. He wanted to know *how* a body moves more than why it moves. He did many studies to answer this question.

Acceleration

When you drop a stone, it falls faster and faster until it hits the ground. Galileo wanted to understand the mathematical principle that governs that accelerated motion. Objects fall too fast to be able to study them in detail, so he devised a method to slow the fall down. He built an inclined ramp, which is called an inclined plane in physics. The flatter he made the ramp, the slower the ball rolled. He needed a method to accurately time the ball as it rolled the distance of the ramp. At first he counted the beats while a musician played, but he later used a water clock that measured time by filling a small container with water.

His results showed that the velocity, or rate, at which the ball rolled down the plane was proportional to the amount of time that went by. More precisely, the distance traveled increases with the square of the time that has passed. In the first unit of time, the ball rolls a certain distance, one unit. After two units of time, the ball has rolled four units of distance. In three units of time, it has rolled nine units in distance, after four units of time, it has rolled sixteen units of distance, and so on. The speed increases at a uniform rate. Galileo called this *uniform accelerated motion*.

What is important to understand about this principle is that for the first time, a relationship had

Nuclear Meltdown

Some of the popular stories about Galileo's discoveries may not have happened. He may never have sat in church watching the chandelier swing and might not have dropped the balls from the Leaning Tower of Pisa. Stories like that crop up around many famous figures of history. Think of George Washington cutting down the cherry tree or Isaac Newton getting hit on the head with an apple. (Or how about the story that Clinton didn't inhale?) These stories are cute, but they're not necessarily reality.

Relatively Speaking

Abstraction refers to a way we think, called abstract thinking. This type of thinking deals primarily with pattern recognition or seeing similarities between different kinds of patterns. An example of this type of thinking is seeing how the movements of the pieces form distinct patterns in the game of chess. It can also come into play when learning a foreign language. Abstraction is especially useful for doing algebra. If you learn to think more abstractly, you'll do better at math.

been established among three different types of quantities: time, distance, and speed. Galileo went from thinking about the nature of motion to showing how one can derive principles from observation and experimentation. This process is called *abstraction*, and as you will see, it forms the cornerstone of the development of ideas in physics.

Steel yourself: you're going to learn a couple of equations. You need to understand these equations before you meet Newton in the next chapter. The first equation describes uniform accelerated motion. Keep in mind that letters are used to represent abstract quantities. In algebra, such representative letters are called *variables*.

Galileo's inclined plane experiments demonstrated that the distance the ball traveled was proportional to the square of the time it took. Mathematically, this idea is expressed this way:

$$D \propto T^2$$

In this equation, the letter D stands for distance, and the letter T stands for time. The sign in the middle means "is proportional to." The following chart shows exactly how the equation works when you plug in values for D and T.

Total Distance Traveled after Different Times for the Inclined Plane Experiment

Time Elapsed	Distance Traveled
0	0
1	1
2	4
3	9
4	16
5	25
.	.
.	.

The second equation describes the relationship among distance, velocity, and time:

$$D = VT$$

In English, this equation means that distance equals velocity multiplied by time. Put another way, this equation also tells us that velocity equals distance divided by time:

$$V = D/T$$

Velocity is the same as rate and is usually expressed as miles per hour, or feet per second, or in metric terms as kilometers per hour or meters per second. This means that there is a change in distance, as well as a change in time. There are a few more derivations you can get from this equation that deal with acceleration, but we'll save that for another time. These equations were just to warm up for upcoming ideas about time, distance, and velocity, which are central to some of Einstein's biggest ideas.

Friction

Remember the story of Galileo dropping the two balls from the tower? Although he basically observed that two different weights both hit the ground at the same time, if one of the objects was very much lighter than the other, the heavier object did in fact hit the ground first. Why do you think that was? It was because the object was meeting air resistance; such resistance is known as *friction*.

That realization led to the discovery of something called terminal velocity. When an object falls, it accelerates only until it reaches a certain speed. Because an object falls through a medium such as air or water, it eventually ceases to fall faster, and plateaus at a constant speed. For example, when a skydiver jumps from a plane, he accelerates to about 130 miles per hour, and then continues to fall at that speed.

Galileo decided to use an inclined plane for his experiments because he realized that the objects would roll slow enough that the friction of the air wouldn't be a factor. This experiment led to his understanding of uniform motion. On a flat surface, a ball will continue to roll downward forever. The motion of the ball is constant or uniform. Today, we call this property that the ball exhibits *inertia*.

Imagine you're in a spaceship orbiting Earth. You burn your engines to accelerate so you can break the orbit and head for Mars. Suppose you reach a speed of 25,000 mph, and then shut off your engines. The ship will continue on indefinitely at that speed, unless something comes along to slow it down or speed it up. That's inertia, and it is best understood in a vacuum, because there is no friction-causing medium in a vacuum. (Remember that Aristotle didn't believe in vacuums? It must have been difficult for him to get the lint out of his carpet.)

Albert Says

Some terms used by physicists are very similar and are often used interchangeably. First, there are dynamics and mechanics. **Dynamics** is *how* things move or evolve in time. **Mechanics** describes *why* things move or evolve in time. Second, there are rate, speed, and velocity. Rate is the change in anything over time. Speed and velocity are rates. **Speed** is the distance traveled over a given time interval. **Velocity** is almost the same as speed, but it includes direction. The moon travels in nearly a perfect circle around the earth. It has a constant speed. However, since the moon is traveling in a circle, it's constantly changing direction. Therefore, its velocity is not constant.

Projectile Motion

The last of Galileo's important discoveries that we'll discuss has to do with objects that are thrown—projectiles. They combine both horizontal uniform motion and vertical uniform accelerated motion. This combination has been called *complex motion*, or the *superposition principle*. Each motion can be analyzed separately and then combined to show the final result.

For example, if you shoot a cannon with the muzzle parallel to the ground, the cannonball will travel horizontally until gravity pulls it to the ground. If your friend drops a cannonball from the same height as the cannon muzzle, the cannonball will hit the ground at exactly the same time as the one shot from the cannon. The one shot from the cannon just covers a larger vertical distance before it lands. The ball shot from the cannon has a large horizontal velocity that allows it to travel a greater distance, but that speed doesn't keep it in the air any longer than if it were just dropped. Pretty amazing!

Galileo used this principle to refute one of the arguments against the motion of Earth covered in the last chapter. Remember Copernicus's beef with the argument that an apple falling off a tree should land some distance away from the tree because Earth is moving under it? The reason this doesn't happen is because the apple maintains its forward motion imparted by the movement of the tree on the moving Earth, and therefore it doesn't get left behind.

Relativity

Galileo's principle of relativity is a very important idea, and we'll be studying Einstein's theories of relativity before you know it. Galileo's theory comes from his book *Dialogue on the Two Chief World Systems*. In it, Galileo discusses the role of the observer in performing an experiment. It's exactly like Copernicus's discovery that how we perceive the world depends on our positions while observing it.

Galileo's principle goes something like this: Suppose you're in a moving car with super shock absorbers, and the car is moving at a constant speed. You can't tell whether you're moving or standing still. The windows are all covered so you can't see outside. No matter what kind of motion experiments you do to try to figure out whether you're moving, the results are the same as if you were standing still.

What this problem means is that all steady motion is relative and can't be detected without reference to an outside point. This concept may not sound like much, but it's very important. The key to understanding it and to understanding Einstein's theory of relativity later is that the observation of anything (such as speed, time, or movement) depends upon (or is relative to) the position of the person doing the observing.

That's about it for Galileo's contributions to the study of falling bodies. He explained how objects fall and proposed the idea of inertia for moving ones. He showed where Aristotle had gone wrong and why projectiles keep moving. The only thing missing was the reason why something falls in the first place. Figuring this question out required a higher form of mathematics and someone to ask the right question. This task would be left to Isaac Newton, whom we're going to look at in the next chapter.

The Least You Need to Know

➤ Giordano Bruno was the first scientist to be executed for challenging the authority of the church. He stood for free thinking and rational thought.

➤ Using an improved telescope of his own design, Galileo made significant discoveries that proved the validity of the Copernican, heliocentric theory of the solar system.

➤ The first principles that govern the dynamics of motion were developed by Galileo.

➤ The relativity theory advanced by Galileo laid the foundation for Einstein's theories on relativity.

➤ The distance an object falls is proportional to the square of the time it takes to fall.

Chapter 4

Newton's Clockwork Universe

In This Chapter

➤ Newton combines inductive and deductive logic

➤ The development of calculus

➤ Newton's laws of motion and gravitation

➤ Descriptions of the nature of light

➤ Living in a Newtonian world

The year is 1687, and the greatest work in physics has just been published. You run down to your local bookseller and pick up a copy of "The Mathematical Principles of Natural Philosophy," or *Principia* for short. Unfortunately, it's written in Latin, as were all scientific works of the time, so unless you know the classic tongues, you'll have to wait until 1727 for it to be translated into English. Most physicists agree that this work by Isaac Newton is the single most significant book ever written about physics.

In *Principia*, Newton brought together the knowledge about physics that had been discovered so far and expanded it. He combined and synthesized ideas that would remain unchanged for almost 300 years. Even then, the alterations made to his theories would be minimal.

Newton's principles of celestial mechanics gave society the knowledge to place satellites in orbit, put men on the moon, and plan excursions into planetary space. Through the brilliance of his gravitational theory, he explained why Kepler's laws work. He also invented the necessary mathematics that would allow physics and other areas of science to develop into the great systems of knowledge we have today. No other individual in the history of science has accomplished what Newton has, which is why this entire chapter is devoted to his accomplishments.

The Mental Tools of Science

Logical thinking in and of itself is nothing new. Philosophers have used it to establish and refute arguments for centuries. Most of us like to think of ourselves as logical people. (But if you're ever in the middle of an emotional situation, you know how difficult it is to be logical all the time.) Science, especially the ideas of Isaac Newton, used logic in new ways.

Basically, logic involves using two types of reason. *Inductive* reason is based upon probability and inference. If you observe or experience a certain event over and over, you will infer that given the same conditions, it will happen again in the same way. Another example of inductive logic is drawing general principles from observation. Many discoveries have been made in this way. Aristotle used it, as did many other natural philosophers. The problem is that inductive logic is not conclusive. Room for error exists when you don't have a way of testing your discoveries.

Deductive reason allows you to use basic principles or theories to reach new conclusions. Suppose you know that your spouse likes to read science fiction, so you deduce that he or she would love to go see the latest Star Wars movie. You have used deductive logic to move from a basic premise to a conclusion.

One of the things that made Newton so brilliant was the way he combined inductive and deductive reason. He realized that using both was necessary to understand and describe the laws of nature. By starting with specific observations, Newton generalized the principles of a new theory. Then he would deduce results from this new theory and predict phenomena that he could test against the original observation. If what he predicted didn't work, he would continue to adjust the theory until it could account for all possible phenomena. This constant testing is the heart of the scientific method!

Relatively Speaking

There are two forms of logic that you can study. Newton studied and incorporated **natural logic.** Natural logic involves applying two types of reason: **inductive** and **deductive.** The other form of logic is called **symbolic logic,** or formal logic. This form of logic involves translating a natural language such as English into symbols, and then using the symbols in formulas to describe the operations that occur in a sentence.

In the last chapter, Galileo made the conceptual leap of combining his experiments with mathematics. The physical principles that he discovered were put into formulas. In the period of time we are looking at now, roughly 1600 to 1750, significant progress in mathematics went hand-in-hand with the strides in physics. Newton could not have developed his new theories without the help of new areas of mathematics.

In addition to the advances in mathematics, another important process was taking place at this time. Scientific knowledge was beginning to accumulate and build on itself. Science is a cumulative process, much like building a house. To know where you're going, you need to know where you've come from. (How Zen!) Newton said, "If I have seen further, it is by standing on the shoulders of giants," and he was not a modest man. This statement revealed a belief that defined the new approach in science, and this same spirit of scientific approach is used today.

Care for an Apple or Fig, Newton?

Isaac Newton was born on Christmas 1642, the same year that Galileo died. He was born several weeks prematurely and wasn't expected to survive. Overcoming the poor health of his early years developed his character in ways that would greatly benefit him in years to come.

In his teens, he wasn't particularly gifted in math or science, but like Leonardo da Vinci, he had a keen, intuitive mechanical ability. He built working models of windmills, sundials, and a water clock and had an innate knowledge of geometry.

He enrolled in Trinity College at Cambridge when he was 18. There he met a gifted mathematician, Isaac Barrow, who encouraged him to study math and optics. Upon graduation, he left Cambridge to seek refuge in the countryside because the Plague had returned. This time it didn't devastate Europe as badly as it had in the fourteenth century, but it still reduced the population of London by 16 percent (about 75,000 people).

Newton spent the next 18 months in almost total seclusion. During this time, he laid the foundation for the work he developed over the rest of his life. No other scientist in history accomplished so much in such a short period of time. During this time, he made these breakthroughs:

➤ Developed the binomial theorem of mathematics

➤ Developed both forms of calculus, integral and differential

➤ Began his study of mechanics

➤ Developed the universal law of gravitation

➤ Analyzed the decomposition of light into its spectrum

And he did all this by the age of 25!

In 1667, Newton returned to Cambridge to teach. His friend Isaac Barrow was now head of the math department. Barrow was so impressed with his

Mind Expansions

The first thing that pops into mind when many of us think of Newton is the story of him getting hit on the head with an apple and in a flash discovering the law of gravity. In reality, an apple did fall, but not on his head. On May 16, 1666, while contemplating how the Moon revolved around Earth, he saw an apple fall off of a nearby tree. At that moment he realized that the same pull that made the apple fall to the ground also applied to the Moon.

former student's works and abilities that he turned over the very prestigious position of Lucasian Professor of Mathematics to Newton. (Stephen Hawking holds this position today.)

Newton decided to publish his first work in the area of optics. Unfortunately, it wasn't very well received by his fellow scientists. He became very depressed and was reluctant to publish anything else. If his friend Edmond Halley hadn't intervened, the *Principia* would never have been written. Even so, it took Newton 20 years to put all his notes together and get the book published.

Mind Expansions

Edmond Halley was secretary of the Royal Society and played a major role in convincing Newton to publish his book *Principia*. He applied Newton's theories to the study of comets and discovered that one returned at regular intervals. This comet was named after him: Halley's comet.

A brilliant mind is usually not the sign of an equally sparkling personality. Newton was easily jealous and very egotistical and dealt with fits of depression. More than once he sought revenge against his critics. He was a confirmed bachelor, taken care of by his niece. In his day, Newton was both loved and despised. Voltaire and Alexander Pope admired him, but Jonathan Swift and William Blake disdained him.

After the publication of the *Principia*, Newton withdrew from science for almost 15 years. The work of putting together his monumental work left him tired and indifferent to science. Most scientists can't leave their interests for very long, but Newton often went for years showing no interest in physics whatsoever. He would retreat into his study of alchemy and chemistry, theology and biblical prophecy. He left behind quite an extensive library and many rare alchemical treatises from the Middle Ages.

Can You CalQL8?

Before delving into the *Principia*, you should understand the evolution of ideas that led to the development of calculus. First, what exactly is mathematics? Pythagorean philosophy described it as the study of order and relationships. As mathematical concepts are applied to the physical universe, the world becomes more and more ordered, and relationships between its different parts become clear. For now, we'll look at mathematics from this perspective.

As was noted previously, advances in physics often reflect developments in mathematics. Kepler's knowledge of geometry allowed him to correctly explain the orbits of the planets. Galileo used algebraic formulas and graphs to explain the nature of falling bodies. So far, so good. The next advancement required the combination of algebra and geometry.

A French philosopher and mathematician named René Descartes (1596–1650) provided the next step forward in mathematics. He may be best known for his famous one-liner, "I think, therefore I am." If you took algebra in high school, you used the Cartesian coordinate system to graph lines on the X and Y axes. When you use mathematics to convey physical motion and shapes, you have to combine algebra and geometry to do it. In other words, to describe geometrical figures with algebraic formulas or to do the reverse, you need to use *analytical geometry*, which was the name given to this new form of mathematics.

Developments in mathematics are very similar to those we have seen in science in that they are cumulative. Each new system of thought builds upon and often synthesizes ideas that came before it and in turn lays the foundation for the next major breakthrough. After analytical geometry, the next major breakthrough in mathematics was calculus.

Newton developed calculus to explain how his theories and principles operated. It allows physicists to apply the laws of Newtonian mechanics, understand general relativity theory, and utilize quantum mechanics. This practical application is the one that most engineers and physicists are familiar with. But what does calculus do that other forms of mathematics can't? The answer to that question is at the same time easy and difficult to explain.

Voltaire once said, "Calculus is the art of numbering and measuring exactly a thing whose existence cannot be conceived." That's pretty out there, but it has an element of truth. Can you comprehend infinity? Probably not. No one can. (Although I know some people who claimed they did after their last acid trip.) Calculus uses infinite numbers to put limits on a quantity being measured. It can define relationships between things such as instantaneous speed and area. You can measure the rate of change of a quantity through time. If you're not getting any of this, that's all right. Just know that calculus allows you to do some great things with your mind without the use of drugs.

Nuclear Meltdown

Although Newton has received most of the credit for inventing calculus, it was also invented independently at the same time by a German philosopher named Gottfried Wilhelm Leibnitz. Newton accused him of plagiarizing his work, and Leibnitz remained unrecognized until after his death. As a sort of justice, the symbols we use in calculus today are actually those developed by Leibnitz.

Relatively Speaking

Newton didn't call his mathematical system calculus. He called it **fluxions.** The name comes from the idea of flow. He called the quantities that are generated by motion **fluents** and the rate at which they change **fluxions.** I'm sure posterity is glad the name was changed!

The Laws of Motion

Now that you're familiar with the mathematical ideas that led to Newton's breakthroughs, it's time to examine what his great book was all about. *Principia* is divided into three parts:

➤ In Part 1, Newton develops the general principles of the dynamics of moving bodies.

➤ Part 2 deals with the elements of fluid mechanics, the theory of waves, and some other aspects of physics.

➤ In Part 3, he applies the principles developed in Part 1 to the operation of the universe.

To understand Newton's ideas, you need to know a few basic definitions of some quantities that Newton introduced in this book. These quantities form the fundamental ideas that he uses to explain his laws. As a matter of fact, his laws are directly derived from these quantities.

➤ *Mass*, in Newtonian terms, is a quantity of matter that is a product of the volume and density of an object. Think of it as how much stuff there is in an object.

➤ *Velocity* is the rate at which an object travels. This rate is defined by the distance the object covers over a period of time, so for example, it can be expressed as miles per hour.

➤ *Momentum* is an object's mass multiplied by its velocity. A 100-ton train traveling at 50 mph has a greater momentum than a Volkswagen Bug traveling at 20 mph.

The fundamental quantities—time, mass, and length—are the building blocks of science. Newton wanted to describe science using the smallest number of basic ideas as possible. Fundamental quantities are measurable and independent of the observer. In scientific units, they are expressed in seconds, kilograms, and meters, respectively. Part 2 of this book adds two more measurements to this list: *coulomb*, used to measure electric charge, and *Kelvin*, used to measure temperature.

In *Principia*, Newton describes the six basic laws he discovered. The following sections look at four of them and cover the other two only briefly.

First Law of Motion: the Law of Inertia

A body in motion will tend to stay in motion, and a body at rest will tend to stay at rest. As covered in the last chapter, Galileo already knew about the properties of inertia. Whatever an object is doing in terms of motion, it will continue to do that unless acted upon by an outside force, such as friction. Chapter 3 used the example of an object moving in outer space to describe this concept.

Second Law of Motion: the Law of Acceleration

The force acting on an object is equal to the object's mass multiplied by its acceleration. What is acceleration? In terms of physics, it's change in velocity over time. When an object is falling or is acted upon by a force, it continues to speed up as time goes by, up to a certain point. And what is a force? A force is a push or a pull. For instance, friction is a push, and gravity is a pull.

Here is the most popular formula in all of physics:

F = ma

A force (F) is equal to an object's mass (m) times its acceleration (a). This law enables you to calculate what the acceleration of an object will be if you know the force applied to it and the object's mass. Scientists, engineers, and physicists all use this formula. It applies to rockets, artillery, cars, spaceships, baseballs, and subatomic particles—quite a few things you deal with every day. It is the most frequently used formula in physics.

Third Law of Motion: the Law of Action and Reaction

The first two laws dealt with the motion of just one body. The third law brings into play the interaction between bodies: For every action, there is an equal and opposite reaction. One of the best examples to illustrate this law is that of a jet or rocket engine. Very high-speed gases are exhausted from the engine, usually measured in pounds of thrust. These gases are forced out of the back of the engine by the weight of the ship. In return, they exert a force against the weight of the ship and push it in the opposite direction. Liftoff!

The Universal Law of Gravitation

The force of attraction between two objects is proportional to the product of their masses and inversely proportional to the square of the distance between them. No, this isn't love! Close, but not quite. Basically, this law means that objects with higher masses have a greater force of attraction, as do objects that are closer together.

Mind Expansions

It's a good thing that our numbers are different from our alphabet, or things would get pretty confusing. However, our numerical system, which we adopted from Arabic, has been in use only for about 900 years. Before that, we used Roman numerals. Imagine trying to do math with all those letters. Ancient cultures didn't use numbers. Both the Greek and Hebrew alphabets were used to represent numbers, just like Roman letters. As a result, the spelling of each name also had a numerical value. In Judaism, the esoteric study of the relationship between names and their numerical values led to the teachings found in the mystical Cabala.

Relatively Speaking

Gravity is a word Newton adopted from *Gravitas*, a Latin word meaning heaviness or weight.

You're already familiar with the idea of proportion. Galileo used this concept to define the rate at which an object falls. That phenomenon was called directly proportional because as one unit increased (distance), so did the square of the other (time). Inversely proportional means that as one unit increases, the other decreases, in this case by its square. So if the first value goes up by two, the other goes down by four. Get it? This law is the inverse square law.

If you apply this relationship to the idea of *gravity*, it goes something like this: As the distance between two objects increases, the force of gravity between them goes down proportionally as an inverse square. So if two planets are two units of distance apart, the force of gravity between them is one-fourth. If they're three units apart, the force of gravity is now one-ninth. See?

Newton used these four laws to mathematically prove Kepler's laws of planetary motion. He explained the workings of the solar system. He calculated the masses of the Sun and the planets based on the mass of Earth. He explained the exact motions of the Moon and that the tides of the oceans are caused by the gravitational pull of the Sun and the Moon. The law of gravitation can be applied everywhere in the solar system, and astronomical evidence shows that it even works beyond its bounds. That's why the law of gravitation is universal.

Besides these four laws, Newton derived two others: the law of Conservation of Mass, and the law of Conservation of Momentum. The first one simply states that mass can't be created or destroyed. The matter contained within mass *can* be transformed from one kind into another, as from a solid to a gas when burning, but nothing new is created. Einstein showed this idea to be incorrect with his theories on mass and energy.

The second law shows that whenever two bodies interact with each other, their total momentum is always the same, even though their individual momenta will change as a result of the forces they exert on each other. For example, when two objects collide, their paths, speeds, and direction immediately after the collision will be different than they were before the collision. But when you add the momentum of the first object before the collision to the momentum of the second object before the collision, you'll find that the sum of the momentum of the two objects after the collision will be the same as before the collision.

The Bearer of Light

In 1704, Newton published his second important book, *Opticks*. (No, it's not misspelled; that's the Old English spelling of the word.) This work made him even more popular, because it was written in English so that nonscientists could read it too. It discussed how rainbows were formed, which inspired writers and artists of the time. He explained that light isn't just white; it's made up of all the colors of the visible spectrum. He showed this by passing sunlight through a prism. His studies led to a significant improvement in how lenses were made for spectacles, microscopes, and telescopes.

Albert Says

To remember the colors of the spectrum easily, just think of Roy. Roy G. Biv, that is. R for red, O for orange, Y for yellow, G for green, B for blue, I for indigo, and V for violet.

Newton also put forth his theory of the nature of light. What is light made of? He proposed that light consists of small particles called *corpuscles*, which are shot out from the source of light. You can think of these particles as being something like pellets ejected from a shotgun. This idea was in direct contrast to an existing theory that stated that light consists of waves. Because of Newton's status, his theory won out, but by the nineteenth century, it no longer held water. Experiments done by scientists in the late 1800s determined that the nature of light could best be explained as a wave.

The publication of *Opticks* marked the first time that anyone made any serious attempt to study the nature of light. The purpose of introducing it here is to mark the beginning of another group of ideas that will be covered in the next few chapters. The study of the nature of light will ultimately lead us to Einstein's theories on light and will reveal the significant role that the understanding of light plays in our understanding of the universe.

Newton's achievements in the study of light also had a metaphorical impact. The word *light* came to be used differently, to mean truth and revelation. Newton has been called the "bearer of light." The time period surrounding his life has been called the "Age of Enlightenment." Newton is considered by many to be the torchbearer who ushered in this new era. As we've done in previous chapters, let's look at this period from a larger perspective, so that you can see the connections of ideas.

Faith in science and progress, along with developments in human rights based on reason, became the cornerstones of the eighteenth century. Politics and business began to supersede religion and were wrestling away the leadership of the churches of all denominations. Toleration increased, and persecution for religious and political offenses decreased.

The rapid increase in scientific discovery that followed Newton's work resulted in the development of new areas of study. The new sciences of botany, zoology, and mineralogy were born. A technological boom was also occurring. The barometer and thermometer were invented, along with the air pump and the steam engine. The invention of the steam engine gave rise to other machines and paved the way for the emergence of the Industrial Revolution by the end of the century.

Mind Expansions

Another of Newton's significant contributions was the invention of the reflecting telescope. Before Newton, astronomers looked through telescopes that used lenses to magnify the images. These telescopes were called refracting telescopes, and the images they showed were always inverted. Newton used a curved mirror to reflect the light and magnify it. All modern, large, astronomical telescopes are this kind.

Mind Expansions

Because of the development of a regular postal service in Europe (toward the end of the eighteenth century), many scientists were now able to write to each other to share their findings. Small groups would meet to discuss information that they had received in a letter from a fellow scientist abroad. These groups formed into the many prestigious scientific societies that continue to exist today.

53

Living in a Newtonian World

This time period, often called the Age of Reason (or roughly the 18th century) witnessed another major paradigm shift. As you saw in the Renaissance, a large change occurred both individually and collectively. It began in England, for a number of reasons:

➤ This is where Newton's works first made an impact.

➤ The spirit of invention and free thinking was enthusiastically encouraged there.

➤ The wealth that England had acquired because of its colonial empire was supporting a new economic class.

➤ Industry was developing hand-in-hand with a new economic system, the free market system.

These are just a few of the most important changes that affected the shift in consciousness. This list also highlights some of the main differences between England and continental Europe, where the shift was a little slower in getting started.

On a strictly philosophical level, Newton's theories raised a number of important questions. It was much easier before Newton to imagine the universe as finite and closed (with boundaries). All you needed to do was look up, and what you saw was what there was. Newton had hypothesized that the universe was infinite. How could the popular mind grasp that concept and really understand it? This concept also posed a major theological question: If the universe is infinite, where was Heaven situated? Before Newton, Heaven was understood to exist on the other side of the boundary of the universe.

Nuclear Meltdown

By definition, a law in physics is a rule or principle that's expected to work under most if not all of the conditions to which it is applied. But laws of physics should not be taken as absolutes. If laws are believed to be infallible, they become taken for granted and are never questioned. And as we all know, laws were made to be broken.

The publication of Newton's works also highlighted a paradox. Newton's laws, like all natural laws in physics, *can't be proven*! Does this mean they don't exist? No. Newton's first and third laws, governing inertia and action/reaction, are still considered to be hypotheses, subject to experimental testing. The second law, governing acceleration, is simply a definition. All of Newton's laws fall under what's called the *axiomatic property*. What that means is that the laws in physics are axioms. They can't be proven; they can only be presumed and tested. An *axiom* is a universally accepted principle that's taken to be true.

Chapter 1 stated that the science we have and use today is a methodology that works. It just plain works, and that's why we use it. But the very fact that it works has given us the idea that science somehow defines objective reality better than anything else. The truth is that science defines reality better than anything else only because we have become so used to it working for us that we never question it. We take it for granted that science provides us with the truth. It defines reality the way that we want it to and expect it to. After all, science is a product of our minds. But this doesn't mean that there aren't other systems that can interpret the world in which we live in a valid way. This idea will be explored in Part 6.

Newton's laws explained motion, force, and gravity in straightforward ways that lent themselves to practical application. His world was a world of absolutes: absolute space and time and perfect indivisible particles moving in that absolute space and time. But a world made up of such absolute things is a construction of the mind. Neither Newton nor any one of us has ever experienced absolute space or absolute time. All of us experience space and time as a particular moment in space and time. (This is a warm-up for Albert's theories.) Absolute space and time are concepts that Newton used in order to develop general theories of nature.

Great new theories are born of what's called the *zeitgeist*. That's German for the "spirit of the times." A great new theory develops because there is a need for such a theory. If the need is not great enough, the energy to produce such a reinterpretation of reality is just not present. The ideas that Newton presented were not his ideas alone, but were the visions and ideas produced in him by his time.

Medieval humanity had seen itself in a world where everything was meant to serve God's purpose, in an unchanging static universe. Renaissance people wanted change and separation. They wanted to be free to pursue their own ideas, unencumbered by religious limits. Now that's not to say that the people of the Renaissance were not religious; they were quite devout. But they wanted to separate their thoughts from spiritual needs, to pursue their thoughts to wherever they might lead. That was the zeitgeist that Newton fulfilled.

Mind Expansions

Newton was the most recognized scientist of his time. He was held in such high esteem that in 1705, he was knighted, an honor never before bestowed on anyone for achievements in science.

Have you figured out why this chapter was called "Newton's Clockwork Universe"? Newton's laws explained that all motions and interactions of bodies follow a few simple laws. Accordingly, if you know the location, mass, and speed of any object, you can predict its future position in time and space. So that means that you can predict the future position of any object in the universe if you know its initial location. Whatever the position an object is in right now determines where it will be in the future. Another way of putting this concept is that everything is already predetermined, evolving through time to a fixed destination, like a giant clock. Do you agree? Were you predetermined to read this book? (The soundtrack from the *Twilight Zone* fades in the background.)

The Least You Need to Know

➤ Newton's book, the *Principia,* is the most important single volume ever written in physics.

➤ The development of calculus was a significant advancement in mathematics that gave Newton the ability to apply his formulas.

➤ The three laws of motion and the universal law of gravity are Newton's most significant contributions to physics.

➤ White light can be broken down into the spectrum of color that you see in a rainbow.

➤ Natural laws can't be proven; they can only be tested and applied.

Part 2
Energy Comes in Many Forms

If you open any freshman text on physics and look at the table of contents, you might say to yourself, "Wow! There's a lot of stuff to cover." You'll see chapters on heat, electricity, magnetism, sound, light, mechanics, laws, forces, and much more. What do all of these things have in common? They all deal with energy.

In the first part of this book, I defined physics as the study of nature. That study has brought us to the realization that everything in nature is a manifestation of some form of energy. The discoveries in physics over the last 200 years that have given us that knowledge are the focus of this part of the book.

Chapter 5

The Heat Is On

In This Chapter

➤ The first thermometers

➤ Caloric theories of heat

➤ The various states of matter

➤ Entropy and heat

➤ The laws of thermodynamics

Although Newton's laws successfully explained lots of different kinds of natural phenomena, quite a few processes in nature could not be understood using his laws. Scientists were eager to find a mechanical explanation for these invisible processes. The first one they tackled was heat. Many attempts had been made since the time of Aristotle to understand heat, but without much success until the nineteenth century. The ideas and theories that were developed to understand heat and how it's produced eventually led to the current concept of energy.

It's a Question of Temperature

Heat and cold are two names given to how we perceive temperature. They are imprinted on our consciousness at a very early age, and we can usually tell the difference between the two very easily. In some cases, however, the difference isn't clear. If you were blindfolded and touched with a hot iron and with a piece of dry ice, you'd have a very hard time distinguishing them. (Note: Don't try this at home.) This example demonstrates why physiological response isn't a dependable method for measuring temperature.

Galileo invented the first scientific instrument for measuring temperature in 1592. He used a glass flask with a very narrow neck that was half filled with colored water. He placed the flask upside down into a bowl of colored water. When the temperature changed, the air in the flask would expand or contract. The column of water in the neck would then move up or down. This invention was called a thermoscope. It didn't have a temperature scale to give a measurement of how warm or cold it was, so it wasn't called a thermometer. That invention would come almost a half a century later, around 1640.

A group of scientists in Italy built the first thermometer. This group's prototype of the modern thermometer used mercury in a partially sealed tube. This thermometer involved the expansion and contraction of gases. Thus, it was natural that some of the first studies of heat dealt with the nature of gases.

Mind Expansions

The first thermometer was calibrated by an eighteenth-century Danish astronomer named Olaus Roemer. He chose two convenient, useful, and reliable fixed points: the melting point of snow and the boiling point of water. He assigned the melting point of snow as $7^1/_2$ degrees, and he set the boiling point of water at 60 degrees. These numbers sound arbitrary, but they weren't. Zero degrees was the temperature of an ice and salt mixture that was widely believed to be the lowest temperature possible. Water melted exactly $7^1/_2$ degrees above that temperature. But Roemer's scale of temperatures didn't catch on very well!

It's a Gas, Gas, Gas

Around the time that Newton was at Cambridge, another Englishman, Robert Boyle, was at Oxford investigating the properties of compressed air and other gases. He made a U-shaped glass tube with one end open and the other sealed. He found that as he poured mercury into the open end, the volume of air contained in the tube decreased. When he doubled the amount of mercury, the volume of air was cut in half. When he doubled the amount of mercury again, the volume of air became one quarter of what it was. Is this starting to sound familiar?

Scientists love looking for laws. Laws help to establish rules of conduct that will always pertain to the material, object, or force under study. What Boyle discovered is now known as Boyle's law (go figure). This law states that the volume of a given amount of any gas at a set temperature is inversely proportional to the pressure applied. Chapter 4 explained the mathematical principle of inverse proportions. (Time for another pop quiz: Which one of Newton's laws is described in terms of an inverse proportion? Gravitation.)

About a century later, Jacques Charles discovered a second law about gases, which was called, surprisingly enough, Charles' law. Charles found a similar relationship between temperature and pressure. But instead of keeping a set temperature and increasing the pressure, Charles fixed the pressure and increased the temperature. Stated in proper science lingo, the law is: The pressure of any gas contained in a set volume increases by $1/_{273}$ of its original value for every degree the temperature is raised.

Boyle's law and Charles' law were combined to form what's called the ideal gas law. This law is handy for chemistry and physics. Collectively, these three laws answer one question and raise an even tougher one. The question they answer is, "What is the best method to standardize temperature measurements?" The answer is to make a gas thermometer, because it doesn't matter which gas you use; all gases expand and contract in the same way. Different liquids expand and contract differently, so they wouldn't make a good instrument for calibrating other thermometers.

The question these three laws raised was, "Why do gases expand and contract so much as the temperature and pressure varies?" Liquids and solids don't change their volume anywhere near the extent that gases do. The reason for this difference is the molecular structure of gases. A gas consists of a huge number of tiny molecules, with great empty spaces between them; liquids and solids have a lot less space between their molecules. The large spaces between gas molecules provide a lot of room for the molecules to expand and contract. But no one knew about atoms or molecules when these gas laws were developed. Such knowledge wasn't to come until chemistry began making the same advances as physics. Chemistry didn't develop as a science until the mid-eighteenth century and nineteenth century.

Watch Those Calories!

Heat had never been thought of as a measurable quantity until a Scottish physician named James Black (1728–1799) came along. He thought that heat was a colorless, invisible fluid that was able to permeate or penetrate whatever it came in contact with. He called this liquid *calor*, which is Latin for heat.

If you mix a gallon of boiling water with a gallon of ice water, the temperature of the mixture will be halfway between the two temperatures you started with. Black thought that when the two waters were mixed, the calor in the hot water was equally distributed between the two portions. After a number of experiments in which Black mixed different liquids at different temperatures, he defined a unit of heat as the amount of heat it takes to raise the temperature of one pound of water by 1 degree *Fahrenheit*. Today, that unit is called a *calorie* and is equal to the amount of heat it takes to raise 1 gram of water 1 degree *Celsius*. (The kind of calorie that you would eat would raise 1000 grams of water 1 degree Celsius!)

Relatively Speaking

The **Fahrenheit** scale, created by Daniel Fahrenheit, is based on the melting point of ice and the temperature of the human body and is the most popular temperature scale in America.

The **Celsius** or Centigrade scale was created in 1742 by Anders Celsius. It's based on a division of 100 equal degrees, with 0 being the freezing point of water and 100 being the boiling point. Most scientists and most of the world use this scale.

A **calorie** is the amount of heat it takes to raise 1 gram of water by 1 degree Celsius.

To illustrate how heat acts like a fluid, a Frenchman by the name of Sadi Carnot compared a water wheel to a steam engine. For a water wheel, as water falls over a wheel, the wheel turns. In a steam engine, the hot steam flows through the engine, turns a shaft, and comes out cooler. So the water fell from a high point to a low point and, in the process, turned the water wheel. The calor, or heat fluid, fell from a high temperature to a low temperature to turn the steam engine shaft.

Nuclear Meltdown

An analogy is a great tool to understand the relationship between two dissimilar things. It can also be a good teaching tool. But analogies don't often work well for developing scientific theories, especially if your whole theory is based on just one analogy. An analogy can be so easy to understand that it misleads people into thinking that a theory is correct. Carnot's caloric fluid theory was one of these. It clouded the true nature of heat for decades.

Albert Says

Many of the terms you'll be coming across, such as **work, energy,** and **conservation,** have distinct meanings in physics. Although these words are commonly used, it's important to understand their specific meanings in physics so you can understand what the heck's going on.

Although Carnot was getting close to understanding what heat was, it just plain wasn't a fluid. A steam engine transforms some of the heat that flows through it into mechanical energy, and the amount of heat that comes into the condenser is less by the amount of heat that is thus transformed. The amount of heat that is no longer found in the condenser is the amount that was used to do the work of the steam engine. Heat can do work.

Working in the Heat

Benjamin Thompson grew up in Massachusetts during the Revolutionary War. He was the first to realize that heat is a form of internal motion, not a fluid as James Black and Sadi Carnot thought it was. Thompson became the Minister of War of Bavaria and was given the title of Count Rumford for training and reorganizing the German army. As the director of the arsenal, he supervised the boring of cannons for the army. He noted that the water that was used to cool the cannon during the boring process continually needed to be replaced, because the boring tool boiled it away. What was causing the cannon to get so hot? Heat was being introduced into the boring process by the friction caused by the movement of the boring tool against the metal cannon; heat wasn't being transferred as a fluid from some other source.

The ideas of Count Rumford were developed in the 1840s by James Prescott Joule. Joule performed a number of experiments with an ingenious setup. He attached some weights to a string that ran over a pulley to a paddle wheel inside an insulated container of water. As the weights fell, the paddle wheel turned, and the friction caused by this turning heated up the water. The temperature change corresponded to the amount of heat put into the water. Joule discovered that the same amount of heat was put into the water whenever the same weights fell the same distance.

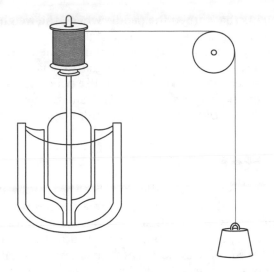

Joule's experiment on transformation of mechanical energy into heat.

Understanding everything that was happening in this experiment will provide you with the background necessary to understand related concepts of energy and will prepare you for the laws of *thermodynamics*, which are the foundation for later developments in physics. But before you can comprehend the relationships in energy systems, you need to know how physicists use these three terms:

➤ *Energy*, as it is understood in science, refers to the ability to do work.

➤ *Work* is not something to groan about on Monday morning. It pertains to a force applied over a given distance.

➤ *Conservation* is the ability to preserve or protect from loss or waste. In physics, this term means basically the same thing that we mean when we use this term every day, but it relates to systems of energy or matter.

Joule's experiment involved three types of energy:

➤ *Potential energy* is also called the energy of position. It is a form of energy that has potential to do work. It's energy waiting to happen.

➤ *Kinetic energy* is also called the energy of motion. This type of energy actually does the work.

➤ *Total energy* is the sum of the potential and kinetic energy.

Relatively Speaking

Thermos is the Greek word for heat, which is why you use a thermos to keep your coffee hot. **Thermodynamics** involves the study of the reversible transformation of heat into other forms of energy, such as mechanical energy, and also covers the laws governing those transformations.

A **joule** is a unit of energy in physics, named after James Prescott Joule for his discovery of the conversion of work into heat.

Before Joule released the weights that turned the paddle wheel, the weights had potential energy. They were ready to fall. As the weights fell and turned the paddle wheel, they lost potential energy as the work was being transformed into motion, or kinetic energy, and from there into heat. During that transformation process, no energy was lost. Energy was conserved! This discovery of the conservation of energy is known as the first law of thermodynamics. This law states that energy is neither created nor destroyed; it can change only in form. You can't get something for nothing!

Three States of Matter

Do you remember the Greek philosopher who thought that the world was made up of basic building blocks called atoms? (Here's a hint: His name rhymes with Democritus. Oops! That *is* his name.) Democritus thought that there was an atom for each kind of material in world: fire, rock, wood, air, and so on. We know now that hundreds of thousands of different chemical substances exist. These substances are called *compounds*. The smallest amount of a substance is called a *molecule*. *Atoms* make up molecules, and each atom is associated with an *element*. Chemists made these discoveries, which became very useful knowledge for physicists.

Let's make sure that we really understand the nature of what heat is. We know that it's a form of energy, that it can do work, that it's not a fluid, and that it's different from temperature. So...what produces heat?

In Joule's experiment, the agitation of the paddle wheel in the water put heat into the water. But what was being agitated, and how did this create heat? The key to answering both of these questions is to consider the molecules that make up the water. The molecules of water, good old H_2O, are being agitated. The more they are agitated, the warmer the water becomes. So the temperature of any substance is related to the movement of the molecules that it's made of. This molecular movement changes when a substance is in different forms: gas, liquid, and solid.

Gases

When a substance is a gas, the molecules in it are free to move in any direction and are not bound to each other because there's plenty of room between them all. I'm sure that you have sprayed an air freshener in a room. I have, especially after ordering Mexican take-out. The molecules in the air freshener don't have to go far in any direction before they collide with air molecules. In a

Relatively Speaking

An **element** is the simplest form that matter can exist in by itself. It was once thought that there were only four elements. Today, there are 113 known elements. If you want to see what they are, check out the periodic table of elements. I'm sure you have one on the wall somewhere.

A **compound** is a combination of two or more elements, in which the individual properties of the elements are lost but the combination has new properties. For example, table salt is a compound of sodium and potassium. Individually, sodium and potassium are extremely poisonous, but together they form a compound we can't live without.

short period of time the nice clean scent permeates the air, as the air freshener mixes completely with the rest of the air molecules.

When you pressurize a gas, its temperature goes up because pressure forces more movement and collisions among the molecules. More molecules of gas now occupy the same space as before, so there are more collisions. When a gas cools, the molecules move more slowly and don't collide as frequently as they do when the gas is warmer. When the gas cools enough, it condenses into a liquid.

Liquids

In a liquid state, at room temperature, the molecules are still free to move around, colliding with each other as they do so, but they're not as free as those in a gaseous state. A force of attraction exists between the molecules in a liquid. This small force adds up, basically preventing any single molecule from escaping the others and leaving the liquid. Escape does occur on a small scale, in the process known as *evaporation*. A molecule in a liquid may be struck by two or more molecules in rapid succession and forced toward the surface of the liquid. If this molecule is traveling at a high enough velocity, it can overcome the force of attraction that tied it to the other molecules in the liquid.

Mind Expansions

States of matter are distinct forms of matter. The most common are gas, liquid, and solid. Another state of matter is plasma. This state occurs at extremely high temperatures, like in the Sun. Plasma looks almost like a cross between a liquid and a gas. It is made of an equal number of free protons and electrons, which gives it a neutral electric charge. The glowing part of a neon light is plasma. You may have seen the popular glass balls that contain miniature lightning bolts that are attracted to your fingers when you touch the glass; that's also plasma.

Solids

When a liquid is cooled enough, it freezes into its solid form. In a solid, the molecules move very slowly, and the force of attraction between them basically holds each one in place. The molecules are no longer free to move around and mix with the other molecules. All they can manage to do is vibrate.

Understanding the different molecular states of matter is important in order to follow the first theories that Albert Einstein developed prior to his relativity theories. He verified the molecular state of matter in one of his first papers.

The other reason for examining the different states of matter was so you could see the presence or absence of heat is responsible for any transformation of matter from one state into another. Now you can answer the question of what produces heat: The movement and collision of the molecules within a substance produces heat. No movement, no heat. To put this in terms of physics, heat is the kinetic energy released by the movement of molecules.

Then Chaos Set In

As you learned previously in this chapter, the first law of thermodynamics states that the change of internal energy in a system must be equal to the amount of energy added to the system during the change minus the work done by the system. This law is represented by the following formula:

$$\Delta U = Q + (-W)$$

In this formula, ΔU is the symbol for internal energy in a system. Q represents heat added to the system, and (-W) is the work done by the system.

To understand the distinction between internal energy, heat, and work, consider your bank account. You can add money to your account. You can also withdraw money from the account. The change in the amount of money in your account is equal to the sum of the deposits and withdrawals. The total balance shown in your monthly statement doesn't depend on which transactions were in cash or which were by check, or where you were when you did the transaction, or what the weather is—only the sum of the deposits and withdrawals is reflected in the balance. It's the same with internal energy. Only the sum of the added heat and subtracted work matters.

One of the concepts that's important in physics is the idea of a system. A *system* is a set or group of related things that is unified or forms an organic whole, such as the solar system or Earth's environment.

There are two types of systems:

➤ Closed or isolated systems

➤ Open or unbounded systems

In a closed system, the total amount of energy is always the same. The state of a closed system can be described in a number of ways. It can be described by calculating the content of its total internal energy or by measuring its temperature, size, volume, mass, internal pressure, or other conditions. Properties such as these are called the *parameters* of the system, and if any one of them changes, the entire system changes.

In an open or unbounded system, the total amount of energy is constantly changing. It's just the opposite of a

Mind Expansions

The need to standardize measurement was becoming very important by the middle of the eighteenth century. In 1790, the present-day metric system was adopted. It is an easy system to understand, because it is based on multiples of 10. The units of measure could be applied across the sciences and were very precise. The units for time, length, volume, temperature, and pressure were now used as the means to convey all forms of scientific quantities. The Scientific Revolution could not have occurred without this standardization.

Albert Says

It's important to understand what a system is. In many areas of physics, a system is a way of describing how certain events or conditions unfold. The rules governing how systems operate and the parameters that define systems are all part of what's called *general systems theory*.

closed system. It's very difficult to accurately calculate its contents unless the parameters that affect the system are very closely measured. Even so, open systems also have no limitations; therefore even unknown parameters could be affecting the total energy content, preventing any accurate estimation of it. Some physicists think that our universe is an open system.

Another parameter that can be used to describe a closed system is entropy. *Entropy* means chaos or disorder. Within a closed system, entropy can only increase or remain constant. An egg in your hand is an example of an ordered system; an egg smashed on the floor is an example of a disordered system. The entropy of the broken egg is higher than the entropy of the whole egg.

Now, you've never seen a broken egg re-form into the egg it was before it broke. Not even all the king's horses and all the king's men could perform that feat. How come? Because entropy increases as a function of time. The future is the direction in which entropy increases. It can't go the other way. The nature of entropy also lies at the core of the second law of thermodynamics, which states that the entropy in a closed system always increases. We'll explore the second law in more detail in a little while. And this brings us to one of the big paradoxes in physics.

Nuclear Meltdown

Science is a continuously changing system of thought. Discoveries continue to expand our understanding of the universe from year to year. Ideas and theories put forth today often change the long-held cherished theories of yesterday. It is not uncommon for two sets of ideas or theories to come into conflict. They will often remain so until a unifying theory comes along that shows their relationships rather than their differences. Sometimes such a theory never comes along. This process has occurred throughout the history of science and will continue in the future. It was one of the driving forces behind Albert Einstein's search for his unified field theory, which you'll get into later.

The discussion of Newton in the last chapter noted that that if you know the initial position of an object, and you know other things about it, such as its velocity and mass, you can predict its future position. This was Newton's second law. Using this law, you can predict the future behavior of an entire system. The interesting thing is that you can also do this in reverse. If you know where an object is now, given the same information as before, you can also find out where it was in the past. For example, astronomers use this law to show the position of the heavens at a certain time in the past. You can show the exact position of everything in the sky, for example, 3,000 years ago. Now that's pretty cool! All the equations in Newtonian mechanics work just as well projecting forward or backward. There is no distinction between the future and the past, which ultimately makes Newtonian mechanics capable of time reversal.

If the molecules and atoms of an egg or piece of wood obeyed Newtonian mechanics, they wouldn't distinguish between the future and the past. Yet did you ever see a broken egg become whole or a piece of burnt wood unburn? Of course not. Nature is not reversible, and that brings us to the paradox: You can't derive the second law of thermodynamics from Newton's laws. They are irreconcilable. Newton's mechanics are time reversible, but the second law of thermodynamics isn't. This difference may not seem like such a big deal, but Newtonian mechanics and the laws of thermodynamics are the two cornerstones of classical physics, and the fact that you can't derive one law from the other makes for an incomplete picture in what's basically a very complete system. It drives physicists crazy!

Don't Break the Law

We all know that heat flows from hotter objects to colder ones. If you drop an ice cube into a cup of hot tea, the ice cube gets warmer and melts into the tea, which becomes cooler than it was before. The heat flows out of the hot tea and into the ice cube. Remember Carnot's idea of the fluidity of heat? He thought that heat flowed downhill from higher temperature bodies to lower temperature bodies. In one sense he was right, but that idea doesn't sufficiently explain what happens. On a molecular level, heat has a lot to do with kinetic energy going. That energy, because of its activity, naturally spreads out by itself. This one-way flow of heat is the basis for the second law of thermodynamics.

Based on this knowledge, the second law goes something like this: In an isolated system, the flow of heat always goes from higher to lower. This process is not reversible. Heat can't flow uphill. This law leads to the understanding that heat is a form of energy that can't be fully converted to other forms of energy. Some of it is always lost or wasted. You can't get back what you put in.

For many years, scientists tried to design machines to show that this law could be broken. These machines were called *perpetual motion machines*. The ideas behind them were good, but in practice they could never be made. Let me describe how they worked.

Steam engines were the latest craze in the late 1700's, so they were considered the ideal machine. One idea was to minimize the friction between its moving parts. Friction generates heat, which is wasted and doesn't go toward producing work. The problem was that there was no way to get rid of the friction. (They didn't have Teflon yet.) So that idea fell through.

Nuclear Meltdown

Ludwig Bolzmann contributed a lot to the understanding of the second law of thermodynamics, but he ran into one small problem. He tried to show mathematically that you could derive the second law of thermodynamics from Newton's mechanical laws. He was mistaken, and his proofs were attacked. This mistake eventually drove him to commit suicide.

Mind Expansions

Perpetual motion machines have been a source of fantasy and research for hundreds of years. They're sort of the physicist's fountain of youth. They don't grow old, and they never stop running. The quest for unlimited energy, or at least the ability to get out what you put in, continues to this day. When nuclear fusion finally works on a practical level, we will have our first perpetual motion machine or at least something very close to it.

Another idea was to reheat the cool water in the condenser. A steam engine runs by extracting the heat from steam; the steam then cools and condenses into water. If the cool water could be immediately put back in the boiler, more steam would be produced, and in theory the engine would run forever. But cool water doesn't flow toward hot water; it's the other way around. And if you introduce an outside source to pump the water back to the boiler, where does this source get its energy from?

In any system in which heat is converted into another form of energy, such as mechanical or electrical energy, there is always a loss of heat. This loss lowers the efficiency of the system. This lower efficiency is entropy in action. And as you know, this running down can't be reversed, so entropy continues to increase. In all natural processes, the organized motion of molecules has a tendency to become disorganized or random.

Applying this concept to the world in which we live can account for our poor use of energy. Energy from the sun is stored temporarily in coal, oil, and other fossil fuels. We can release it as heat at a high temperature, and regardless of whether we use the energy for useful work, it always ends up at a lower temperature. All of the energy is conserved—no energy is ever lost. However, the value of the energy has been lost in the form of heat—it's become useless. That's the problem with the type of energy we use. It's not that we're using up energy, it's that we're transforming it into useless forms. Only some of it is used; the rest of it is wasted.

Relatively Speaking

Zeno's paradox were not named Daffy and Donald. A **paradox** is a self-contradictory event, idea, or situation. Zeno was a Greek who lived during the time of Aristotle. He put forth a number of theoretical paradoxes dealing with motion. His most famous paradox is often used in first-year philosophy courses to acquaint students with the concept of what a paradox is. Zeno used an example of Achilles in a race with a tortoise. Because Achilles runs so much faster than the tortoise, the animal is given a head start. But Achilles can never reach the tortoise, because he always has to travel half the distance to get to it—and the distance that separates them can be cut in half forever.

You Can't Get There from Here

The third law of thermodynamics arose from the question of how cold could an object get. Absolute zero is defined as a set temperature on the Kelvin thermometer scale. Temperatures on this particular scale are not measured in degrees, but in kelvins. There is a point on the Kelvin scale called absolute zero, but it refers to an actual temperature. This is not the absolute zero I'm talking about here. The absolute zero I'm discussing is the total absence of heat.

To get to absolute zero, you need a very, very good refrigerator. But the refrigerator gets less and less effective as the temperature goes down. It takes longer and longer to reach lower and lower temperatures. This is the essence of the third law of thermodynamics, which says that it isn't possible to reach absolute zero in a finite number of steps.

Translated into English, this statement means that it is possible to get extremely close to absolute zero, but you can always get just a little bit closer without getting there. You can eventually get there, but it will take an infinite amount of time.

The third law of thermodynamics is similar in concept to a philosophical concept known as *Zeno's paradox*. Zeno's paradox is the following: If you have a line that is a certain length, with point A at one end and point B at the other end, the distance between those two points is the length of the line. To get from point A to point B, you first have to cover half the distance. You're now halfway there. Now you have to travel half the remaining distance, from the midway point to the end. That distance is now cut in half, and once again, you have some distance to go. The remaining distance is cut in half again, but you're not there yet. Each time you move closer, you divide the remaining distance in half. The problem is that you never get there, because no matter how many times you halve the distance, even when the distance gets infinitesimal, it can still be cut in half. If you substitute temperature for distance in Zeno's paradox, then you have the third law of thermodynamics.

Dying from the Heat

One of the implications of the second law of thermodynamics, which states that heat is transferred from hotter to less hot substances, has to do with our universe as a whole. If the universe is a closed system and there is nothing else outside of it, then there is a fixed amount of energy within it. The stars represent high concentrations of energy and reflect an ordered system. But you know what happens to order in a closed system: Entropy increases. Eventually, after millions of years, all the stars will cool and burn out. There won't be enough heat energy left in the universe to sustain any life. This is known as the heat death of the universe. So why prolong the agony? To what end are we doing anything? If the end is chaos, what is the point of anything that seeks to make order? Well, that's one cheerful view. Let's look at the other side.

The assumption that the universe is a closed system is questionable. Albert Einstein once thought it was, and then changed his mind. Besides, who's to say that the second law of thermodynamics operates in the same way throughout the entire universe? Whatever you believe, it's important to note that these ideas have had an impact on many writers and thinkers of the nineteenth and twentieth centuries. The breakdown of people who live isolated lives, as well as groups or societies that isolate themselves from outside influence, has been a main theme for writers such as Herman Melville, Henry Adams, and Sigmund Freud. It has also been applied analogously to economic theory and ecology. So as you can see, when you look at things from a larger perspective, the connections that you make with other ideas can ultimately broaden your overall understanding.

The Least You Need to Know

➤ The first gas laws defined the relationship between heat, temperature, and pressure.

➤ Heat is a form of energy that does work.

➤ The transformation of the three states of matter (gas, fluid, and solid) is dependent on the amount of heat present.

➤ Energy can neither be created nor destroyed. It can only change form. This is the first law of thermodynamics.

➤ The entropy of an isolated system never decreases. This is the second law of thermodynamics.

➤ It is not possible to reach absolute zero in a finite number of steps. This is the third law of thermodynamics.

Chapter 6

Magnetism Marries Electricity

In This Chapter

➤ The discovery of magnetism

➤ Earth's magnetic field

➤ Volts, amps, and ohms

➤ Faraday's field theory

➤ Maxwell's mathematical contributions

Our ability to generate huge amounts of electricity has made our lives very comfortable. Television, medical diagnosis, satellite communications, and the heating and cooling of our homes have all been made possible by electricity and magnetism. Much of what they do is taken for granted until the computers, refrigerators, and other devices we have come to depend upon stop working.

This chapter will discuss two interrelated forms of energy. Both magnetism and electricity occurred naturally in the world before people discovered what each was and how they were connected to each other. The study of magnetism and electricity is called *electrodynamics*, and the theories in this area of study were instrumental to the development of Einstein's theory of relativity.

Don't Rub Me Out

Stories about discoveries in science make for great conversation at parties when you've run out of things to say about world problems and Hollywood gossip. Try this one about the discovery of magnetism at your next get-together: It all began in ancient Greece (where else?), with a shepherd boy by the name of Magnus (of course). One day, while tending his flock on Mount Ida, he touched the metal tip of his staff to a large stone. No matter how hard he tried, he couldn't pull his staff free from the stone. The stone became known as Magnus's stone, and to this day you can still see where his staff got stuck. Well, that's not true, but the legend about Magnus might be.

What is probably closer to the truth is that the word *magnet* comes from an ancient country of Asia Minor called Magnesia. (Bet you never knew that there was a connection between magnetism and laxatives.) In Magnesia, a number of unusual stones were discovered that had magnetic properties. Today, these stones are called *magnetite* and are known to be a form of iron ore.

Ancient peoples had many strange and unusual beliefs about these magnetic stones. If they were placed on your head, you were supposed to be able to hear the voices of the gods. They also were also used as cures for cramps, gout, and arthritis. Rubbing a paste of powdered magnetite and oil on your scalp was supposed to be a cure for hair loss. (Trust me, it doesn't work.) In ancient singles bars, wearing a charm or ring made of magnetite was supposed to find you a lover. (As in, "I find the magnet you're wearing strangely attractive.")

Remember Thales, from the first chapter? He was the Greek who thought that the primary substance of the universe was water. He also had some interesting ideas about magnetism. He observed that if he rubbed *amber* with a piece of wool, it could pick up light objects such as feathers and dried grass. He thought for sure that rubbing the amber made it magnetic, but the one thing it didn't attract was metal. He had some magnetite that attracted metal, but the amber didn't. What he had discovered was static electricity, not magnetism.

Static electricity is a common phenomenon that we encounter when we dry our clothes or walk across a nylon or wool carpet. It gives us a little shock when we touch a metal doorknob, another person, or our favorite pet, who then looks at us accusingly.

Relatively Speaking

Amber is a form of petrified, fossilized tree resin, usually of a translucent or clear yellow/gold color. If you've seen Jurassic Park, it's the stuff that the little mosquito was trapped in. That mosquito contained the dinosaur DNA.

The ability to understand and harness this kind of electricity wouldn't come until the seventeenth century.

In China, around 376 B.C.E., a general by the name of Huang Ti was the first to use another type of iron ore, lodestone, as a compass. Many military commanders during the Han dynasty used magnetic compasses to find their way over the land. About 900 years later, compasses were finally used on ships for navigation.

In the thirteenth century, because of trade with China, Arab sailors started using magnetic compasses as well. They eventually made their way to Europe, opening the door for the Age of Exploration. The use of compasses is about as far as the understanding of magnetism would progress until a man named William Gilbert came along.

The Love-Hate Relationship

In 1600, William Gilbert was appointed physician to Queen Elizabeth of England. He was highly regarded by his peers and often met with an active group of natural philosophers to discuss his ideas of magnetism. The Queen herself spoke and read Latin and was said to have attended many of his meetings.

Gilbert wrote the earliest work on magnetism, which was called *The Magnet.* (He probably spent all of five minutes coming up with that title.) In it, he discussed how the small spheres of magnetite that he experimented with were similar to the magnetic nature of Earth. He thought Earth was just a huge magnet. This idea was one of his greatest contributions to the understanding of Earth's magnetic field. It gave a rational explanation for the attraction of compass needles to the north and south poles. The publication of this work gave him such a great reputation that in his honor, the unit of magnetic strength is called, you guessed it, the gilbert. However, this name didn't last very long as a standard unit of measure, and was replaced with the current standard for representing magnetic field strength, the tesla. (Named after Nikola Tesla, who invented the means to apply alternating current in practical, large scale ways. It's what runs in our homes and businesses.)

Besides his theory of the magnetic nature of Earth, Gilbert also noted many other substances that, when rubbed, behaved just like amber. The word *electricity* comes from *elektron*, the Greek word for amber. Gilbert used the word *electrum*, which is the Latin word for amber, as the basis for a word that he invented, *electrica*. He used this word for all substances that behaved like amber. His studies ultimately showed that magnetism and static electricity were different.

Nuclear Meltdown

It's not uncommon in science to try to disprove new theories when they come out. But such refutations should be based on proof, not on conjecture. In 1629, Niccolo Cabeo published a book in which he tried to prove Gilbert wrong. Unfortunately for him, experimentation proved his beliefs wrong, and he became a footnote in a *Complete Idiot's Guide.*

Progress in understanding electricity was about to take off. Studies in magnetism, though, did not progress nearly as rapidly. Most of what is known about magnetism came as a kind of incidental result of studies and experiments on electricity. But today, magnetism is a branch of study all by itself. Its use in medicine, industry, and particle physics has led to an effort to conduct more research into its effects. Today, research is linking it to weather conditions, earthquakes, and other climatic phenomena.

Magnetism can be described as the physical phenomenon of a magnetic field. It is an invisible force of attraction that has power; it can do work, and people use it every day. Most of us think of a magnet as something that picks up nails or holds notes to the refrigerator. But magnetism is also something that you spend your entire life enveloped in and hardly ever think of it.

If you've played with a magnet, you know that it has two poles. These poles are called its north and south poles. If you take a piece of paper, put a bar magnet under it, and sprinkle some iron fillings on the paper, you will soon see a pattern take shape that shows the lines of force created by the field of the magnet. The magnetic field of Earth looks very similar. One of the earliest discoveries about magnets was that the like poles of two magnets repel each other, and the opposite poles attract. In that way, magnets have sort of a love-hate relationship.

The needle of a compass always points north. The needle has to be a small magnet, or it wouldn't be attracted to Earth's magnetic poles. Based on what you already know, which pole on the compass's needle do you think points north? That was an easy one. The south pole of the needle is attracted to the North Pole of Earth. The N on the needle is just there to confuse you. The magnetic North Pole of Earth is not the same as the true geographical north shown on maps or globes. Magnetic north is about 1,000 miles away from true geographical north and 70 miles below Earth's surface. The same is true of the magnetic South Pole. Thus, compasses never point to true north, but it's close enough that if you're lost, you can use a compass to find your way home. Hopefully, you won't get lost at the North Pole. If you do, here's a bit of advice: Go south. It's your only option.

Earth's magnetic field should not be confused with gravity. It used to be, until Newton showed that gravity was a force all its own. Gravity is concerned with weight and mass and affects everything. Magnetism, on the other hand, has a close relationship with electricity and is highly selective.

Mind Expansions

Did you know that birds use Earth's magnetic field for their seasonal migrations? A tiny crystal located between the brain and skull of pigeons was discovered, strongly suggesting that they use the magnetic field of Earth for orientation.

Some people believe that gravity and magnetism are related. For example, human brains are essentially electrical in nature, but they are affected by the phases of the Moon and gravity. Mental patients tend to become more disturbed during the full moon—I know I do. (Incidentally, this phenomenon is the basis of the word *lunatic*.) Someday, physicists will connect all this information together.

Shocking Discoveries

Quite a few individuals contributed to the discoveries in electricity, enough to write a whole book on. But I have only a part of a chapter, so let's take a look at the most significant individuals who advanced the study of electricity and electromagnetism.

Stephen Gray

An Englishman named Stephen Gray began conducting experiments with electricity in the early 1700s. He wanted to see if electricity could be transmitted from one point to another. This idea sounds simple, but don't forget that the battery hadn't been invented yet, and there was no sustainable source of any kind to produce electricity. The only way it could be produced was statically by rubbing a glass tube with silk or wool.

Gray ran a long piece of twine held up by silk threads between a glass tube and an ivory ball. When the glass tube was rubbed, the ivory ball connected to it via the twine attracted light objects. If Gray wet the twine, he got even better results. This experiment gave Gray the idea that electricity was fluid in nature. Sounds a bit like the first ideas of the nature of heat, doesn't it? He replaced the silk support threads with metal ones and found that the charge quickly dissipated. This experiment made him realize that conductors quickly lose their charge, but nonconductors such as glass and silk hold a charge. Gray had discovered insulators.

Albert Says

The idea that electricity is a fluid is very similar to the idea of heat being a fluid, which was covered in last chapter. It is common to associate new discoveries or ideas with objects that we are already familiar with to try to understand their nature. Such associations may prove to be false after further investigation, but they are a good place to start.

Charles DuFay

Charles was a Frenchman who replicated the experiments of Gray. In one of his experiments, he discovered that a glass rod rubbed with silk had a different type of charge from an amber rod rubbed with fur. After a series of experiments he concluded that charge created by friction was of two types: *vitreous,* like that produced on glass, and resinous, like that produced on resinous material such as amber or rubber. DuFay imagined two distinct electricities—weightless fluids that would flow from one body to another to electrify it, and he thought that all bodies contained mixtures of these two fluids. Today, these electricities are known as electric charge. Charles also established the general law that like charges repel and opposite charges attract, and that any force between charged objects depends on the type and amount of charge. Later, Benjamin Franklin would rename these two types of charges, but let's keep you in suspense until we get to him.

Leyden Jars

In the mid 1700s, Ewald Jurgen von Kliest, the son of a Prussian official, speculated about whether he could put electricity in some sort of container so that it would retain its charge longer. He took a glass jar half-filled with water, put a cork in it, and then drove a nail through the cork until the nail touched the water. Using a friction machine to generate a charge, he applied the charge to the head of the nail. He then touched the nail, and it gave him quite a shock. He had invented what's called a *capacitor.*

Relatively Speaking

A **capacitor** is anything that can store electric charge. A condenser is a very large capacitor. Smaller capacitors are mostly found in electronics, where small voltages and currents are used. The unit of measurement for capacitance is the **farad,** named after Michael Faraday.

Shortly thereafter, a group of scientists in Leyden, Netherlands built a much better apparatus that could hold a much larger charge. Their invention was known as the Leyden jar. These jars developed into today's condensers. Condensers consist of a large number of metal plates separated by thin layers of air, glass, or mica. They can store huge amounts of electricity and are commonly used in physics. The first atom smasher, built in 1930, had a battery of condensers that could be charged with up to one million volts. When the condensers were discharged through a glass tube filled with hydrogen, they produced projectiles of such high energy that the lithium target at the end of the tube was smashed in two.

Mind Expansions

With the ability to generate quantities of static electricity came larger friction-generating machines that improved this ability. These machines became popular carnival attractions, and many medicine shows, like the ones that toured the Old West, traveled through small towns. Uneducated professors of electricity guaranteed that their machines could cure all ills. But the only effect that these machines had on the patient was that the static electricity made his or her hair stand on end.

Conservation of Charge

The word *charge* keeps coming up in reference to the early experiments in electricity. When you rub a glass rod with silk, you have charged the rod. This term doesn't mean that you have *created* a charge; it just means that you have separated the positive and negative charges. If the rod becomes positively charged, the silk is negatively charged. No experiment has ever created a charge, nor is it likely that one will in the future. Benjamin Franklin stated this truth in 1747: Electric charge is neither created nor destroyed. The total amount of electric charge in the universe remains constant. This statement is known as the theory of conservation of charge.

Benjamin Franklin

Benjamin Franklin was a serious scientist who came to physics at the ripe old age of 40. Many of his experiments dealt with trying to understand the two forms of electricity that had been discovered. After quite a few experiments, he came to agree that these two types of electricity existed and were distinct. He named them positive and negative, but he got them backwards. Current flows in the opposite direction from what he thought, but this mistake is a minor one. However, today we use the definition of charge used by Franklin ... even though he got it a little backwards.

In 1752, Franklin performed his famous kite-in-the-storm experiment. He wanted to find out if lightning and electricity were the same thing. At the top of his kite, he attached a stiff wire pointing up. This wire was connected to the kite string with a key down at his end of it. In the storm, the string got wet, which made it a better conductor, lightning struck the wire and traveled down the string to the key, and sparks jumped from the key to Franklin's Leyden jar. Lightning and electricity were the same thing.

Charles Augustin de Coulomb

Around the second half of the eighteenth century, physicists were studying electricity more than any other area of inquiry. One such individual, a Frenchman by the name of Charles Augustin de Coulomb, discovered the electrical force of attraction. His law states that the force of attraction or repulsion between two electric charges is directly proportional to the product of their charges and inversely proportional to the square of the distance between them. It's virtually identical to Newton's law of gravitational attraction between two masses in that they are both inverse square laws.

Remember how inverse square laws work? When you double the distance between two charges, the force between them decreases by a factor of four. If you triple the distance, the force decreases by a factor of nine. Newton's law of universal gravity works the same way. When the distance between two objects doubles, the force of gravity between them decreases by a factor of four. There are two differences between the two equations. One is that Coulomb's law deals with charges, and Newton's law deals with mass. Mass is measured in kilograms, and electrical charge is now measured in coulombs. The other difference is that electrical force can be attractive or repulsive, while gravitational force is only attractive.

Although these two equations are identical in form, electrical force is much stronger than gravitational force. Looking at some numbers will make the difference clear. The electric force between an electron and proton (in an atom, the electron is negatively charged, and the proton is positively charged) is about 10^{39} times stronger than the gravitational force between them. That's a 1 with 39 zeros after it, which is an enormous number. How enormous? This number is larger than the total number of atoms in 100 billion people.

Coulomb's law is very important in physics and daily life for many reasons, including the following:

➤ The coulomb force keeps electrons in their orbits around the nuclei of atoms.

➤ The coulomb force holds atoms together in molecules.

➤ The coulomb force binds molecules together into liquids and solids.

➤ Coulomb's law, though not proven, strongly suggests that the total electrical charge of the universe is zero.

Albert Says

Just because a well-known name is associated with an idea or invention doesn't mean that person came up with it first. Many unsung heroes in history discovered inventions or theories before their more famous counterparts. Joseph Henry, an American, was one of these forgotten innovators. He invented the telegraph five years before Samuel Morse did. He invented the transformer and discovered magnetic inductance before Faraday. He also discovered radio waves before Heinrich Hertz. In most cases, he didn't know that these discoveries were being pursued in Europe, and therefore didn't publish his work in time. His claim to fame is in having the electromagnetic unit of inductance, the henry, named after him.

This last point may indicate that everything, including you, is electrically neutral. There are about 10^{29} protons and electrons in your body. If for some reason, one out of every 10,000 of your body's electrons migrated to the center of the earth, leaving you with a small excess of protons, you would be smashed to the ground by a coulomb force a million times the force of gravity. So don't take sides—stay neutral!

Electric Italians

The only kind of electricity that has been mentioned in this chapter so far is static, or nonmoving, electricity, whose charge can be expressed in coulombs. But the type of electricity that you are most familiar with is the kind you get by plugging an electrical device into a wall socket or using something that runs on batteries. That is moving electricity, not static. To understand this type of electricity, you need to know about a couple of Italian physicists who put on the moves.

Relatively Speaking

Luigi Galvani never had a unit of measure name after him, but the galvanometer, which is used to measure and detect small electric currents, and the word **galvanize,** which is a process used to plate metal with zinc, both use his name.

Luigi Galvani

The story goes that one day while sitting in a restaurant, Luigi Galvani noticed that the severed frog legs (frog legs are a favorite Italian delicacy) that were hanging from copper hooks jumped when they touched the iron balcony railing. He decided to conduct some experiments with frog legs. When he inserted a two-pronged fork made of one iron and one copper prong, he noticed that the frog leg moved or contracted. He called this animal electricity and thought that the nerve in the frog leg was responsible for producing this movement. He was wrong. In a moment, you'll see why.

Alessandro Volta

Alessandro Volta had been doing experiments using different types of metals. He thought that he could stimulate his sense of taste by placing two different metals on his tongue. When he placed a silver coin on the back of his tongue and touched the tip with a piece of tin, he perceived a strong acid taste. Touching other dissimilar metals to various parts of his body—don't ask where—caused a sensation of light in his eye. From these experiments, Volta concluded that the flow of electric charge was due to an electrical imbalance between the two metals. The key to this flow was the interposition of a moist conductor between two dissimilar metals. The nerve in Galvani's frog legs simply acted as the conductor for the electrical charge.

Alessandro constructed what is known as a voltaic pile using a large number of alternating disks made of copper and iron or zinc and separated by layers of cloth soaked in a salt solution. This pile was the prototype of today's modern electric batteries. In this process, he also discovered what is known today as EMF, or electromotive force, which is more commonly called volts. As you can see, the highest honor a scientist can receive is to have a unit of measure named after him or her.

Electromagnetism

Early investigators of electric and magnetic phenomena knew there was some sort of relationship between the two, but they couldn't quite get the connection. Magnets didn't influence electric charges, nor did charges affect magnets. The person who finally made the connection was a Danish physicist named Hans Christian Orstead.

Hans had built a voltaic pile to conduct experiments with. In 1820, while lecturing at the University of Copenhagen, he got an idea. Static electricity didn't seem to have any effect on magnets, but maybe the electricity moving through a wire connected to the two poles of his voltaic pile would. He connected a piece of wire across the two poles and placed a compass next to it. To his amazement, the compass needle pointed toward the wire, and no longer pointed to magnetic north. Orstead then reversed the wires and found that the compass needle reversed direction. He had found a relationship between electricity and magnetism and coined the term *electromagnetism*.

Nuclear Meltdown

Famous names don't necessarily mean famous lives. André-Marie Ampère's father was guillotined during the Reign of Terror in Paris. Ampère's first wife died after five years of marriage, and his second marriage ended in divorce. His daughter Albine married an insane, alcoholic officer in Napoleon's army. He was extremely absent-minded, never had any money, and died alone. Within this context, Ampère's achievements are even more amazing.

Monsieur Ampère

While in Paris, a French mathematician and physicist by the name of André-Marie Ampère heard of Hans Orstead's famous discovery. Within a few weeks, he performed a number of his own experiments and found that not only does current traveling through a wire affect a compass needle, but currents traveling through two separate wires affect each other. If you place two wires next to each other and both are carrying electric currents, one of two things happens:

➤ The wires are attracted to each other if their currents run in the same directions.

➤ The wires are repulsed from each other if their currents run in opposite directions.

In other words, the wires behave like magnets. Every time you replace a blown fuse, remember Ampère and the name given to quantities of current, the *amp*. (Yes, this is another unit of measure named after the person who discovered it.) If you blew a fuse, you tried to use too much current.

Being a mathematician, Ampère wanted to be able to quantify what he saw. He developed what's known as Ampère's law. This law states that the magnetic field at a given point in space is proportional to the current and inversely proportional to the distance. The important thing is that this law tells you exactly how much of a magnetic field is produced by a given current. So electric currents produce magnetic fields!

Relatively Speaking

An **ohm** is a unit of electrical resistance. **Resistance** is the property that impedes the flow of current in a material. The opposite of resistance is **conductance,** which is the property that allows current to flow in a material. Because conductance is the exact opposite of resistance, the unit of measure of conductance is called the **mho.** It sounds like a joke, but it's true.

Don't Resist Me

The last physicist in this section contributed the last of the three most common units of electricity. You've met Volta and his volts and Ampère and his amps; now it's time to meet Ohm and his *ohms.*

The German physicist George Simon Ohm was a schoolteacher in Cologne. He was interested in the effect different materials had on the flow of electricity. By using wire made of different substances and constructed in various thicknesses, he investigated whether the strength of the current depended upon these changes. He also wanted to know whether such changes influenced the amount of voltage needed.

He made two important discoveries:

➤ The strength of the current is directly proportional to the diameter of the wire, is inversely proportional to the wire's length, and is dependent upon the material of the wire.

➤ The strength of the current is directly proportional to the voltage and inversely proportional to the resistance of the wire.

What all this means is that Ohm had discovered and defined *resistance*, which is measured in ohms. He had also come up with Ohm's law, which establishes the relationship between voltage, current, and resistance. His law is at the heart of all electronics. Every technician knows Ohm's law by heart. Here's what the law looks like (in the following equations, V equals volts, I equals amps or current, and R equals ohms or resistance):

➤ Voltage is equal to current times resistance. The formula is $V = I \times R$.

➤ Current is equal to voltage divided by resistance. The formula is $I = V/R$.

➤ Resistance is equal to voltage divided by current. The formula is $R = V/I$.

There you have it. If you understand Ohm's law, quit your day job and go into electronics and computer repair.

All Without Mathematics

Michael Faraday was the son of a poor blacksmith and had very little formal education. But his accomplishments reflected a brilliant, innovative mind and a strong belief in the fundamental unity of nature. He invented the first electric motor, showed that magnetism could be converted into electricity, and designed and constructed a dynamo. He also made the first electrical transformer and showed that magnetism could affect polarized light (or light that vibrates in only one direction, not in all directions like ordinary light). Not bad for someone with an eighth-grade education.

Much of Faraday's early adult life was spent trying to catch up on the education he missed in a formal academic environment. He was an avid reader and fell in love with science in his late teens. He joined a scientific club and conducted a lot of experiments of his own design using the latest developments in chemistry and electricity. In 1812, he went to the Royal Institution to hear a lecture by one of its more popular members, Humphry Davy. Faraday was so impressed with Davy's lectures that he asked to become his laboratory assistant.

Over the next few years, Faraday made some important contributions to the study of electrolysis. These discoveries alone would have been enough to make him famous in the field of chemistry, but he is most famous for his discoveries in magnetism and electricity.

Electrolysis is the study of chemical reactions brought about by the passage of electricity through a substance. Through this process, Faraday showed that water is made up of molecules in which two atoms of hydrogen are combined with one atom of oxygen. (He didn't make this discovery; he just used it to show how electrolysis worked.) He placed two electrodes covered with tubes in water. When he ran an electrical current through them, they converted water into hydrogen and oxygen gas. Faraday noted that as the tubes filled with gas and the water diminished, there was twice as much hydrogen gas than there was oxygen gas. The next sections cover two of Faraday's most significant contributions to the study of electromagnetism.

Relatively Speaking

Electrolysis is not the removal of unwanted hair with an electric needle. Well, it is, but that's not what it means in chemistry. In chemistry, it refers to passing an electric current through a substance. Electrolysis was a very important discovery that was instrumental in breaking down compounds into their elements. Humphry Davy identified potassium, sodium, calcium, strontium, barium, magnesium, chlorine, and iodine using electrolysis.

Induction

Faraday already knew that electricity could produce magnetism, so he wondered if he could use magnetism to produce electricity. During his construction of an electric motor, he had built wire coils of different sizes. Through experimentation, he found that if he passed a magnet through a coil of wire, the meter hooked up to the ends of the coil showed that a current was created. Simply placing a magnet within the coil did nothing; the magnet had to be moving in and out of the coil to produce the current. The coils of wire broke the lines of force that were part of the magnetic field of the magnet. The more coils that there were, the more lines of force were broken, and more current was produced. The changing magnetic field induced a voltage in the coil of wire.

Mind Expansions

The nineteenth century saw great growth in the popularization of science among the general public. No one was more instrumental in popularizing science than Jules Verne. His novels captured the imaginations of thousands of readers and were responsible for the large public attendance at lecture halls throughout France and later other parts of Europe.

Field Theory

Many scientists by now felt that any description of the world needed to be expressed mathematically, but Faraday hardly knew any mathematics at all. This lack of mathematical knowledge didn't prevent him from developing a theoretical framework that he used as a guide to carry out his experiments. His belief in the unity of all types of physical interaction was at the core of his pursuit and theories. Einstein held this fundamental belief as well and spent the last half of his life trying to prove that it was true.

Physicists during Faraday's time thought that the interactions of electricity and magnetism were similar to that of gravity. Each object had its own amount of whatever was being measured, and forces that acted between objects entirely depended upon the distance between them. Faraday thought there was more to it than that. He put forth the idea that the space between electric or magnetic objects was filled with lines of force. Remember the metal filings scattered on the paper with the magnet under it? That's how Faraday pictured this space. The lines of force were invisible, but they were there. This concept came to be known as a field, and the transference of force within this space came to be known as Faraday's field theory. In the end, Faraday ended up doing for electromagnetism what Einstein would do for gravity.

Maxwell's Mathematics

Faraday's field theory was missing only one thing: the mathematics to show how and why it worked. This task would be left to James Clerk Maxwell (1831–1879). Maxwell was the opposite of Faraday when it came to mathematics. At 14 years of age, he won Edinburgh Academy's mathematics medal. For the rest of his life, he would devote himself to the application of mathematics to various problems of physics.

Maxwell's first contributions were to the kinetic theory of heat. (I see by the smile on your face that you remember that from the last chapter.) His most important work, though, would be the mathematical formulation of Faraday's ideas. He developed famous equations that bear his name to relate the rate of change of a magnetic field with the distribution in space of the electric field. In the same way that Newton's equations show the position of an object in space at any given time, Maxwell's equations do the same for the distribution of an electromagnetic field at any given time.

Maxwell's equations eventually led him to discover how he could apply mathematics to other theories and experiments that his fellow physicists were doing. What he did developed into a whole new branch of physics. He discovered that light is actually an electromagnetic wave of very short length. The branch of physics born at this time was the study of the electromagnetic theory of light. You'll find out more about this topic in a couple of chapters.

Take a moment now to try to put together the ideas that you have encountered in this chapter on magnetism and electricity. There's only three more chapters before you get to Einstein, and if you see the big picture of how physics has developed up to his time, you'll have better understanding of how he fits into the whole scheme.

Albert Says

Understanding concepts and ideas is as important to some as doing the mathematics behind the concepts is to others. If you ever think that one is more important than the other, think of Michael Faraday. Again and again, in one field after another, Faraday was the one who made the crucial discovery, performed the critical experiment, and had the deepest insight into the nature of things. But he couldn't do the math and wasn't particularly interested in it in the first place.

The Least You Need to Know

➤ Static electricity was the first kind of electricity to be discovered.

➤ The conservation of charge states that electric charge is neither created nor destroyed. The total amount of electric charge in the universe remains constant.

➤ Electromagnetism is the relationship between electricity and magnetism. Electric currents can produce magnetic fields, and magnetic fields can produce electric currents.

➤ Michael Faraday was the first to define the nature of a field.

➤ James Clerk Maxwell's equations determined how a field works and also defined the nature of electromagnetic waves.

The Name's Bond, Chemical Bond

The previous chapters covered the history and evolution of thought in astronomy and physics. This chapter sidesteps to look at the developments in chemistry. Tracing this branch of science from its earliest days to the end of the nineteenth century will provide you with the knowledge of how atoms and molecules were discovered. It was in chemistry, not in physics, that this important breakthrough took place.

Understanding atomic theory will help you when you begin exploring some of Einstein's contributions. The insights gained in this chapter will also aid you later in understanding quantum mechanics and the structure of the atomic and subatomic worlds. Because this part of the book is devoted to looking at different types of energy, this chapter also explains how energy is related to chemical interactions. By the time you're done reading, you may even want to become a practicing alchemist.

Laying the Foundation

You learned in Chapter 1 that many of the questions of the early Greek thinkers dealt with the fundamental building blocks of nature. Some thought that there was a primary substance out of which everything in the universe was made. Others believed that the substance wasn't important, because the universe was just a manifestation of a mathematical order. But one individual in particular thought that everything was made up of atoms.

Democritus's concept of atoms was right on the money, but his description of them was far from what we now know to be true. Nonetheless, his idea had great impact on our history. Remember the long conflict between the geocentric and heliocentric theories of the solar system? There was a similar conflict going on between atomic theory, philosophy, and religion. The arguments didn't have quite as severe consequences as the astronomical conflicts—nobody died—but they were still a source of strong disagreement.

This conflict brings up a very interesting point, if you look at the larger perspective. In the last few chapters, this book has covered many of the subjects of classical physics. Classical physics pertains to these areas of study:

➤ Newtonian mechanics

➤ Heat

➤ Electricity

➤ Magnetism

➤ Theories of light

➤ Electromagnetism

➤ Optics

➤ Sound

This list is basically what you find in the table of contents of any introductory physics course. You need to study some of these fundamental ideas before you can investigate relativity theory and quantum mechanics.

The study of physics covers a broad range of material. This book basically examines three main categories:

➤ Certain areas of classical physics

➤ Einstein's theories

➤ Certain aspects of quantum mechanics

All three of these broad areas are related to two larger topics: the *microcosm* and the *macrocosm*, or the very small and the very large. For example, atomic theory deals with the microcosm, the smallest particles of nature. But it also deals with the macrocosm, the universe as a whole, because as far as we know, everything in the universe is made of the same stuff that's discussed in atomic theory.

Each of the three categories covered in this book can be applied to the study of the microcosm. You learned that according to classical physics, heat is caused by the movement of molecules. In Chapter 6, "Magnetism Marries Electricity," you saw that Coulomb's law explained the forces that kept atoms and molecules bound together. Soon you'll learn about the first paper that Einstein published, which showed that molecules do indeed exist. Quantum mechanics deals specifically with the microcosm.

These three broad categories are also related to the macrocosm. In classical physics, you saw that Kepler's laws and much of Newtonian mechanics deals with the motion of the planets and the gravitational forces that keep the universe running like a big clock. Later, you'll learn that Einstein's theories of relativity pertain to much of the macrocosm. And although quantum mechanics focuses on the atomic and subatomic worlds, in the long run, these theories help to explain the dynamics of the basic structure of the entire universe.

Both the microcosm and the macrocosm are part of an even larger perspective on things, and that is the study of *cosmology*. In philosophy and physics, cosmology is the study of the origin of the universe. Can you see how both the microcosm and the macrocosm relate to that study?

Whether you approach cosmology from the very small or the very large, the questions remain the same:

➤ How was the universe created?

➤ What is the basic underlying structure of the universe?

➤ What is the smallest piece that can exist on its own?

➤ Is the universe infinite?

➤ Why was it all created in the first place?

Of course, these are only a few of the profound questions that can be asked. No doubt, you have some more of your own. The core question is this: Is the origin of the universe spiritual or material in nature? This question has been the source of much conflict and argument for individuals and societies. It's probably one of the oldest questions in the

Relatively Speaking

Microcosm and **macrocosm** are terms that are often used to describe relationships between small systems and large systems. They are used in many fields of science and even in economics. They pertain only to systems in general and are defined specifically within the context of how they are being compared.

Mind Expansions

The study of cosmology is a branch of philosophy that looks at where, how, and when the universe came into existence. It asks profound questions regarding our creation and forms the basis for many stories in mythology. It also lies at the heart of the evolution versus divine creation debate, which has been an issue in our society since Darwin published *The Origin of Species*. Religion places God at the center of creation; physics gives us the big bang theory. Some people are seeking to reconcile the two, but others take sides.

world, yet it is still rigorously debated today. In Part 6, this book addresses this question in much more detail, but it's worth bringing up here because chemistry and the discovery of atoms, elements, molecules, and compounds became a big part of that debate.

As you've seen, conflict has historically been a part of the lives of astronomers and physicists alike. Many scientists suffered for their beliefs if those beliefs appeared to threaten the paradigm held by the majority. The advances that took place in chemistry were no different. Fortunately, many of the disagreements weren't severe enough to warrant killing people, but things did get very unpleasant along the road.

Al the Chemist

Alchemy has existed in many cultures around the world for the last 3,000 years. It existed in India, China, Greece, Egypt, and the Middle East. The word *alchemy* comes from the Arabic *al-kimiya*, which is said to be derived from the Egyptian word *keme*, or black earth, which was a name for Egypt. The earliest Greek alchemical manuscripts are thought by some scholars to be copies of Egyptian papyri. The eastern traditions of alchemy in India and China are similar in essence, but they are nonetheless different from their western counterparts. This book is concerned with the western alchemical tradition only, and its impact on the development of chemistry.

Albert Says

Alchemy and other words associated with it, such as occult, astrology, and magic, are loaded words in our society. Our interpretation of these words has a lot to do with how they are colored by our belief systems. The true meaning of the word often has very little to do with what we think the word means.

To understand what alchemy was and how it changed over time, it's important to look at its two sides. From about the nineteenth century through the present day, alchemy has commonly been looked at as a primitive forerunner to chemistry. Many have adopted the rather simplistic view of alchemists as a bunch of greedy men trying to turn lead into gold. As in any pursuit, some seek only wealth and fame, and alchemy certainly did cater to such individuals, but that is only one side of the story.

The Great Work

Alchemy began as a rich cultural tradition that was passed on orally. It incorporated elements of mathematics, art, music, astrology, philosophy, and religion. The priests of ancient Egypt were at one time the only people who had access to this knowledge. Over time, Egypt passed its alchemical heritage to Greece, and from there it was passed to the Middle East. You already learned that one catalyst for the Renaissance was the interaction between Europe and the Middle East, which did a better job of preserving the ideas of the ancients. The alchemical traditions that the Islamic countries inherited were some of the first texts to be translated by the scholars of the Renaissance. From this point on, alchemy began to have a great influence.

The Scientific Revolution emphasized logic, rationality, and the examination of the outside or objective world. The alchemical tradition focused upon intuition, symbolism, and the exploration of the inner or subjective world. It was often referred to as the Great Work. Contained within its teachings was the mystical pursuit of *gnosis*, or divine knowledge. There was a blending of alchemical symbolism with Christianity, and many of the Gothic cathedrals that were built featured alchemical symbolism in the architecture.

Individuals such as Kepler, Newton, Boyle, and many members of the church were influenced by the teachings of alchemy. They weren't after gold. Instead they sought a deeper, more mystical understanding of the nature of the universe. Although this idea was at the core of the original doctrines of alchemy, many more ideas were involved.

The purifying process an individual had to go through to obtain divine knowledge was understood using the analogy of turning a base metal into gold. As I mentioned, gold was understood to be the embodiment of the sun, a life-giving presence, by the earliest civilizations, especially Egypt. Many of the first gold coins were minted with a symbol of the Sun on one side. As a divine metal, gold became very valuable. Having a gold coin was like carrying God in your pocket. Things really haven't changed that much; we still live in a society that worships money.

Mind Expansions

The first metals to be used for coinage were gold and silver. They were considered sacred metals because of their association with the Sun and Moon, once considered divine beings. Everything connected to processing these two metals, such as minting the coins and the metallurgy, was controlled by the priesthood of ancient cultures. Prayer and fasting were necessary before one was allowed to come in contact with sacred forces.

Part of the practice of alchemy also dealt with turning the inner process outward and vice versa. If you could transmute or change one type of substance in nature into a totally different type of substance, maybe that process could give you insight into how to do it to yourself. This outer process of transmutation eventually led to chemistry and the knowledge of atoms, molecules, and compounds. Along the way, the inner transmutation got lost in a world that was becoming more rational, analytical, and scientific.

However, the inner process didn't die out all together. Even though chemistry separated itself from the inner tradition, a rebirth of alchemy took place in the latter half of the nineteenth century. It became part of an occult renaissance that swept across Europe and other parts of the world. There was a resurgence of interest in the ancient teachings of the mystery schools, as well as a swell in membership in mystical groups such as the rosicrucians, theosophists, freemasons, and spiritualists. Alchemy was studied along with the Cabala, tarot, and astrology. It also influenced certain branches of psychology. The brilliant psychologist Carl Jung utilized the powerful tools of symbolism found in alchemy to help him unlock the unconscious patterns of the human mind.

Fortune and Glory

As I mentioned earlier, part of the practice of alchemy involved trying to apply the process of transmutation to substances in nature. These early attempts led to processes of distillation and to the invention and use of all types of specialized glassware: retorts, beakers, flasks, tubes, evaporating dishes, and mixing utensils. Chemists still use these tools of the trade today.

Alchemists discovered new alloys of metal, tinctures of various herbs, and compounds of different substances, both poisonous and beneficial to people. Some of the earliest medical practitioners adopted alchemical practices in their attempts to find cures for their ailing patients. This wasn't the best time in history to be a patient. More than a few doctors lost patients by inadvertently poisoning them. But not all alchemical work was done for altruistic purposes. The search for gold led many a person down the alchemist's path seeking unlimited wealth and fame.

At the heart of this search was the *philosopher's stone*. Determining the composition of this legendary stone was the goal of all alchemists, because this stone had the capability to change anything it touched into gold. But what was this stone made of? How many different substances combined to make it? What kind of substances were they? What were the substances themselves made of? Was there a primary substance from which all things were made? What was the smallest piece of anything that could exist? These are very familiar questions, aren't they? Philosophers asked these questions, physicists asked these questions, alchemists asked these questions, and soon after chemists wanted to know the answers, too.

Nuclear Meltdown

Mercury was one of the most common ingredients in an alchemist's laboratory. Unfortunately, it's also extremely poisonous to humans. Many alchemists died or went crazy because of overexposure to it. A little bit of knowledge can truly be dangerous if you don't know where it can lead. Atomic energy is another example of this principle.

Relatively Speaking

Phlogiston is a very strange word, aside from being a strange idea. Meaning flammable in Greek, it was part of a theory that helped account for observations people made while watching substances burn. Remember how long it took people to question anything that Aristotle said? What he stated as true jibed with what people saw. Phlogiston was also based on observations, and it's hard to go against the senses because we're so used to trusting them. That's why this theory persisted for such a long time.

Get the Gist, Floe?

The German chemist (alchemist) George Stahl thought that all substances were composed of water and three varieties of earth, one of which was a combustible material that he called *phlogiston*. Phlogiston was one of the first materials that alchemists came up with to show how chemicals interacted with one another. The theory of phlogiston became widely accepted, because it seemed to explain why substances burned.

Stahl's theory of phlogiston marked the first significant change from Aristotle's view that all matter was composed of the four elements: earth, air, fire, and water. The fact that Aristotle's ideas regarding astronomy and motion were shown to be incorrect helped open the door to questioning other areas of his thought. Stahl's theory wasn't such a tremendous change—Stahl still proposed that there were four elements, just different ones—but at least the influence of Aristotle regarding the composition of matter was beginning to wane.

According to Stahl's theory, substances burned not because they absorbed something from the air (which we know to be oxygen today), but because they lost phlogiston. In theory, this loss would make the substance lighter after it burned. But we know today that some substances gain weight when they burn, especially metals. But not all substances get heavier; gasoline disappears when it burns, and that's a lot of weight loss. This discovery poked the first holes in the phlogiston theory.

Ideas about our respiratory system had also advanced little from the time of Aristotle. He thought that the purpose of breathing was to cool and ventilate the blood. According to the phlogiston theory, people ceased to breathe for the same reason that a flame stopped burning in an enclosed space: because the air had become saturated with phlogiston. In the natural cycle, plants, which absorbed the phlogiston released into the air and prevented everyone on Earth from suffocating, purified the air. His idea is relatively close to how things work.

One of the main reasons why the phlogiston theory lasted for as long as it did (almost 100 years) was that it was consistent with observations. Most of the chemical reactions that took place in alchemists' laboratories were between liquids and solids. Air was just considered to be a waste bin for phlogiston fumes. Air was thought to be an element with certain properties, one of which was the ability to be put under pressure. The study of air pressure finally led to newer ideas about gases and their composition.

Albert Says

You should try to understand how each idea developed into another. The last chapter investigated the evolution of thought regarding electricity and magnetism. See if you can do the same for alchemy and chemistry, so you'll have a clear understanding of the ideas that led to our present-day atomic theory.

I've Got Gas Again

Investigation of the properties of air became a crucial turning point in the hunt for atoms. Galileo had an assistant working for him by the name of Evangelista Torricelli. He was interested in the nature of a vacuum. Aristotle (him again?) had said that nature abhors a vacuum, but suddenly it was kind of in vogue to question Aristotle. After all, he'd been proven wrong about so many things recently.

As the story goes, Torricelli filled long glass tubes, which were like big test tubes, with various liquids such as wine, water, and mercury. He inverted these tubes, standing them up in pans filled with the same liquid contained in the tubes, and lashed them to the masts of ships. The fluids dropped down quite a bit inside the tubes, leaving an empty space at the top. He noticed that lighter fluids didn't drop inside the tubes as much as heavier liquids did. For example, mercury, the heaviest liquid known, is four times heavier than water. Accordingly, a column of mercury inside such a tube would drop four times more than than water.

Was the empty space inside the tube a vacuum? Was that what was sucking up the liquid into the tube? After all, if you want to drink a bottle of soda with a straw, you have to form a vacuum in your mouth by sucking on the straw. No, thought Torricelli, that's no vacuum sucking up the liquid; it's the weight of the air in the atmosphere pushing down on the liquid in the pan, which in turn forces the liquid up the tube. The lighter the liquid, the more liquid could be supported in the tube. Hmmm, fascinating captain. Torricelli appears to be saying that air has weight. Very good, Mr. Spock.

Relatively Speaking

A **torr**, named after Evangelista Torricelli, is the unit of measure for atmospheric pressure. It is defined as the amount of pressure needed to support a column of mercury one millimeter high.

So Torricelli had discovered the idea of atmospheric pressure. Can you guess what the unit of atmospheric pressure is now called? That's right, the torr. Torricelli's theory was proven correct by Otto von Guericke, the inventor of the air pump. When von Guericke placed a barometer inside a sealed chamber and pumped out all the air, the mercury in the barometer fell. There was no more atmospheric pressure to hold it up.

Remember Robert Boyle from Chapter 5? He discovered the laws governing the pressure of gases. In addition to his work with pressure, many consider him to be the first rational chemist, meaning that he sought a more mechanical explanation for the interaction of chemicals. He wanted to do this for a number of reasons:

➤ His previous work with gases convinced him that chemistry also needed to be explained in more scientific terms.

➤ Newton had gone far in describing the workings of the universe according to a mechanical model, and Boyle thought that chemistry should also become part of this great clockwork mechanism.

➤ Boyle was an adherent of the ancient atomic theory of Democritus, but with some significant differences.

Boyle published a very influential work called *The Skeptical Chymist,* in which he convinced many chemists to look for explanations of chemical reactions other than the traditional alchemical notions. He attacked Aristotle's four-element theory and the alchemical concept of substances having sympathy or abhorrence as principles as an explanation of why chemicals would or would not combine with one another. He defined an element as a chemical substance that can't be broken down into any other substance and proposed that each element was made up of atoms.

There were a few more significant advances made with gases:

➤ Joseph Black isolated a gas that he discovered while heating up magnesium carbonate. This gas was heavier than ordinary air. He called this fixed air. Today we call it carbon dioxide.

➤ A Russian by the name of Mikhail Lomonosov, along with an English scientist Henry Cavendish, discovered a flammable gas 10 times lighter than air. We call this gas hydrogen. (Hardly seems fair, huh? What's wrong with calling it *lomonogen*?)

➤ Joseph Priestly isolated a gas that was purer than regular air. Candles burned much brighter when exposed to it. He called it dephlogisticated air. We call it oxygen.

➤ Daniel Rutherford isolated another gas about the same time as Priestly. He called it mephitic air, or air saturated with phlogiston. We call it nitrogen.

Gases were at the forefront of chemical investigation. Although numerous gases were discovered, they were all explained in relation to phlogiston. The notion of atoms was not catching on. It would take the work of one more chemist to destroy the phlogiston theory and usher in the chemical revolution.

Mind Expansions

A very interesting relationship exists between inventions and scientific discoveries. One of the reasons why new discoveries are made is that a new invention or new piece of equipment becomes available. Telescopes enabled Galileo to make his discoveries. Air pumps provided the equipment for Boyle to develop his gas laws and for Torricelli to show the existence of atmospheric pressure. Inventions and scientific discoveries have a symbiotic relationship.

It's Elementary

Antoine Lavoisier (1743–1794) is considered to be the father of modern chemistry. Through his experiments, he destroyed the theory of phlogiston and developed a new idea of the composition of matter. He showed that by burning various substances under controlled conditions, he could account for all of the substances in the process. He determined that burning a substance doesn't give off phlogiston, but instead combines the substance with oxygen.

When you burn a piece of wood, its carbon combines with oxygen and forms carbon dioxide, which dissipates into the air. The weight of the ashes is lower than the weight of the original piece of wood. That's why people thought that the phlogiston was released. Something with weight was thought to have escaped, leaving you with something much lighter. But Lavoisier showed that under controlled conditions, everything is accounted for. The total weight of a piece of wood and the air surrounding it before burning is equal to the total weight of ashes, smoke, and air after burning. Lavoisier published his findings in a book called *An Elementary Treatise on Chemistry*. This book would do for chemistry what Newton's *Principia* did for physics.

That's Heavy

Although Lavoisier did a lot to further chemistry, he never accepted the idea of atoms as the basic building blocks of elements. That honor goes to John Dalton (1766–1844). Dalton lived in rural England and was the professor of mathematics and physical sciences at New College, Manchester. He had been following Lavoisier's work and was convinced that a relationship existed between the weight of an element and the amount of that element that was needed to combine with another element to form a compound. He came up with the following conclusions:

➤ Elements consist of very small indivisible particles, which he called atoms in honor of Democritus.

➤ All of the atoms in each element are alike, but they are different from the atoms that form other elements.

➤ A chemical combination occurs when atoms from two or more elements form a fixed union. (He didn't quite have the word molecule yet.)

Out of these conclusions came the concept of atomic weight. Atoms themselves are too small and light to weigh individually, so you have to weigh a large amount of a pure element to see how it combines to form a molecule. Dalton found that $85^2/_3$ parts of oxygen by weight combined with $14^1/_2$ parts hydrogen by weight. He didn't know that a water molecule is H_2O, and the ratio he found wasn't two to one. So even though he was on the right track with the concept of atomic weight, the table of 20 atomic weights that he compiled was all wrong.

Please Set the Table

Over the next few decades, Dalton's atomic theory became more and more widely accepted by scientists. But corrections needed to be made to his atomic weights, and the definition of a molecule needed to be clarified. Through the work of two more scientists, Jakob Berselius and Amedro Avogadro, these corrections were made. Their conclusions can be stated as follows:

➤ Gases of equal volume contain an equal number of particles.

➤ These particles can either be single atoms or molecules. It depends on the gas.

➤ A molecule is the smallest combination of atoms of any substance that can occur naturally.

By 1869, there were 69 known elements. Only one thing remained to complete the atomic theory we have today: These elements had to be categorized into a standardized system.

Enter Dmitri Mendeleev. He prepared a set of cards for each element. On each card, he listed the atomic weight of the element and its other specific properties. He arranged his cards on the basis of increasing atomic weight and noticed that the properties of the elements recurred periodically. He arranged his cards into a table that we know today as the *periodic table of elements*. Within the table are groups of elements that have similar properties. For example, all the inert gases such as argon and neon are in the same group. Metals such as iron, nickel, and cobalt are grouped together. The great thing about the table was that Mendeleev was able to predict the existence of many yet-undiscovered elements. His table had many blank spaces, so he knew that other elements needed to be discovered to fill in these blanks.

Relatively Speaking

The **periodic table of elements** is one of the greatest achievements in the history of chemistry. It shows the beautiful mathematical order of all of the known elements in the universe. The discovery of each element is a story in itself. Some elements unraveled like a soap opera; others were found by accident. Today, there are 113 known elements. Some were discovered in nuclear reactions and exist for an incredibly short amount of time before they change or disappear from our universe.

So we've come full circle from Democritus's atom to the periodic table of elements. But what about the core question that was postulated before we began looking at the alchemists? I mentioned that the development of atomic theory had repercussions similar to those that occurred in relation to astronomy, but that they weren't nearly as intense. One reason was because the chemical advances came after the Scientific Revolution. Another reason was that these developments didn't necessitate as great a paradigm shift as some of the previous ones.

But the philosophical implications of the developments in chemistry still raised the same question. Is the origin of the universe spiritual or material in nature? If God created the world, how could we account for the seemingly dead, unconscious nature of atoms and molecules? Philosophers argued that if we are made of these things, they must be alive. Refusal to accept the atomic theory was common for decades. It wasn't even proven to be true until Einstein finally vindicated the existence of atoms. But the proof of their existence still doesn't answer this fundamental question. Maybe the next chapter will shed some light on this subject.

The Least You Need to Know

➤ Alchemy laid the foundation for chemistry.

➤ Alchemy addresses both inner transformation as well as outer transformation.

➤ The phlogiston theory was the first attempt after alchemy to try to explain chemical interactions.

➤ Robert Boyle is considered to be the forerunner of rational thought in chemistry.

➤ Antoine Lavoisier developed the principles of modern chemistry.

➤ The periodic table of elements lists the known elements and their relationships to each other.

Chapter 8

Let There Be Light!

In This Chapter

➤ Two forms of light

➤ The study of optics

➤ The development of perspective drawing

➤ Kepler and Newton's optics

➤ The electromagnetic spectrum

Albert Einstein once said that he spent his whole life trying to understand the nature of light. One day, when he was 16 years old, he was watching sunlight reflect off the surface of a lake and asked himself this question: What would it be like to ride a beam of light? This question is very simple, but like many simple questions, it is very difficult to answer. Within 10 years, Einstein would be able to answer that question; asking the question planted the seed. Underlying all of Einstein's brilliant achievements was the pursuit to understand the nature of light.

This chapter ends this book's coverage of the history and philosophy of physics with a topic that takes you right up to Einstein. This chapter looks at the many ways that people have thought of light and how these ideas provided Einstein with the tools he needed to advance his innovative theories.

In the Beginning ...

What is light? It depends on what kind of light you're talking about. The word *light* has many meanings. Light can be a metaphor, a symbol, an idea, a source, a color, or a

Relatively Speaking

Metaphysics is the branch of philosophy that examines the nature of reality. It is often associated with some of the most theoretical questions in philosophy, questions that haven't necessarily been proven to be valid in the first place and that can be answered based only on belief. Examples of these questions include: Is there life after death? What is the definition of reality? Does God exist? What is the soul? What is the mind? These questions deal with spirituality, consciousness, and other intangibles. The word *metaphysics* is a reference to Aristotle, in that it refers to the work that followed his writings on physics. *Meta* simply means after.

form of electromagnetic radiation. It also makes sight possible. Looking at much of what's associated with light provides a context to understand the nature of light.

The meaning of light can be divided into two fundamental categories. The discussion of cosmology in Chapter 7, "The Name's Bond, Chemical Bond," introduced the question of whether the source of the universe was spiritual or material in nature. You can apply that question to light as well. Is light spiritual or material (physical) in nature? Does this question ultimately imply that there are two kinds of light? It depends on how you look at it. From a scientific point of view, light is strictly a physical phenomenon. Within the context of spirituality, light takes on a *metaphysical* meaning. It's important to consider both approaches. The true nature of anything can best be seen in a larger perspective.

Good vs. Evil

In the beginning God created the heaven and the earth. And the Earth was without form and void; and darkness was upon the face of the deep. And the Spirit of God moved upon the face of the waters. And God said, Let there be light: and there was light. And God saw the light, that it was good: and God divided the light from the darkness. And God called the light Day, and the darkness he called Night. And the evening and the morning were the first day.

The beginning of the Old Testament in the King James Bible tells the Judeo-Christian story of creation and describes light as being invoked by God. In the New Testament, Jesus is referred to as the Light of the World. Throughout the western religious traditions, light is associated with what is good, pure, and right, and darkness is connected with evil and wrong.

Even in the eastern traditions, light is seen as the source of insight and guidance. In Buddhism, the goal is enlightenment, which manifests as compassion and wisdom. Buddha taught that each person should be a lamp unto themselves.

In all religions, light is understood as a power capable of transforming a person's life, consciousness, and being. In this context, it is not only a metaphor and a symbol, but also a source of power that can transform, enlighten, and infuse one with life-enhancing energy.

Light has a similar meaning in philosophy. Remember Plato's allegory of the cave, in which a person who has been chained up in a cave escapes and comes out into the light of day? The light represents that which illuminates the world of reality, as opposed to the world of shadows in the cave. For Plato, light is a metaphor for truth, beauty, and what he calls the Good.

Lighten Up!

Why has light come to be associated with all that is good? Think philosophically and see if you can answer this question. What comes to mind when you think of light? What role does light play in the world we live in?

> ➤ The most obvious source of light and warmth is the Sun.

> ➤ On a biological level, many forms of life require the light produced by the sun to live.

> ➤ The energy produced by the Sun is used for many purposes when converted to solar power.

> ➤ We also use artificial light, which allows us to extend our work and play time into the night.

> ➤ Light has been used extensively in medicine and technology, for example with lasers and fiber optics.

> ➤ Many artists use light in its various forms to create works of beauty.

It's hard to see anything bad or negative about these ways in which light is manifested in our world; this could be one reason why light's associations tend to be positive. But this list covers only the concrete, physical expression of light. What about light's symbolic and metaphoric aspects? Some ancient cultures saw things in a different light.

Ra the Sun God

Probably no other ancient culture made the Sun more of a major focus of worship than Egypt. The layout of the entire society reflected the passage of the Sun across the sky. The temple complexes of Luxor and Karnak show this relationship very clearly. The Nile runs south to north in Egypt. Thus,

Mind Expansions

Besides being a physical phenomenon that people respond to, light also has a psychological affect on some people. When the sun disappears for many days at a time, as it tends to do in the winter, many people become depressed from not having any sunshine. This condition is called seasonal affective disorder. People with this disorder may not even realize that the lack of sunshine is the cause of their depression.

Mind Expansions

During the reign of Pharaoh Amenhotep IV (1364–1347 B.C.E.), the worship of a single god, or monotheism, was introduced into Egypt. This fact is interesting because most people think that only Judaism, Christianity, and Islam are based on monotheism. Amenhotep IV changed his name to Akhenaton to reflect his association with the Sun god, Aton. Aton was seen as a universal god, and not merely a god of Egypt. After the death of Akhenaton, Egypt reverted back to its tradition of polytheism, or the worship of many gods.

it offers a dividing line for the passage of the Sun from east to west. On the eastern side of the Nile, where the Sun rises, was the land of the living. All the great temples, granaries, government offices and so on were all found on the eastern shore. In the west, where the sun sets, was the land of the dead. The Valley of the Kings and the Valley of the Queens, as well as all the mortuary temples, were situated here.

To the Egyptians, the Sun was the embodiment of life. Without it, no life would be possible; Earth would be a frozen hunk of rock. This view of the Sun as the source of life is why so many early cultures considered the Sun sacred.

Light My Fire

What are the main attributes of the sun? Light and warmth. Light comes first and foremost, and warmth is an important secondary quality. What other substance has the same qualities? Fire, of course. Fire was also considered sacred in many cultures and was seen as a manifestation of the Sun on Earth. Many cultures used symbols of fire and light in their religions:

➤ Greece had Apollo, the Sun god, riding across the sky in his chariot.

➤ Both the Incas and the Aztecs worshipped a Sun god.

➤ *Agni*, or fire, is one of the oldest and holiest objects of veneration in Hinduism. In the heavens, it is the sun; in the air, it is lightning; and on Earth, it is fire.

➤ The Zoroastrians of Persian origin worship *Athra*, or fire. Athra is universal energy, life, heat, and radiance. It is the flame of consciousness that burns in every heart. It is also the light of reason and the glow of love.

Albert Says

Light is one of the most common symbols associated with the religions of the world. It has many meanings and has been symbolically important to all cultures of the world at one time or another. Other symbols are seen in many cultures and religions as well. The cross in Christianity, the star of David in Judaism, the crescent moon of Islam, and the wheel of rebirth in Hinduism and Buddhism, among others, are not unique to the traditions that currently embrace them. They are cross-cultural symbols that have appeared in other cultures and other times with different associated meanings. None of these symbols is the property of any specific religion.

As you can see, light has been interpreted similarly by many cultures around the world at different times. These interpretations of light and the sun naturally extended into all early religions. The notion of what light represents lies at the very core of human consciousness. Regardless of how you interpret it, everyone seeks it in one way or another.

Optical Illusions

Up to this point, the discussion of light in this chapter has dealt mostly with the meaning of light in an abstract sense, that is, how people relate to it apart from its strictly material manifestation. We've considered the role it plays in our lives and how it has been interpreted through symbols, philosophy, metaphors, and religion. This section examines how the study of light developed through science.

Aristotle's Vision

The scientific study of light begins with the ancient Greeks (doesn't everything?). I've already mentioned Plato's views on light, but his allegory of the cave is more philosophical than scientific. Aristotle was one of the first Greeks to try to explain what light was in a scientific sense. He didn't have too much to say about light in particular; he was more concerned with vision and how the human eye perceived the outside world. Aristotle had the following thoughts about light:

➤ Light is not a substance, and it has no qualities that can define it.

➤ Light is the activity of what is transparent. Aristotle means that when you look at an object that's opaque, you can perceive what it is. It's defined in space by its shape, size, color, and other visual qualities. Light is what allows you to see and distinguish one object from another. Its function is to be transparent; otherwise, you wouldn't be able to distinguish one thing from another. Light has an active nature, and that nature is to be transparent (sounds like a typical Aristotelian assumption).

The study of the nature of light wouldn't progress much over the next 1,500 years, thanks once again to Aristotle. Light wasn't understood as a subject to be studied. Vision was the main thing that needed to be understood. People believed that the ability to see had nothing to do with light. The eye was thought to somehow produce the light that allowed perception to take place.

Nuclear Meltdown

Theories and ideas about how the eye produces or reflects light abounded. One of the more interesting ideas went something like this: The sun gives off light, and at night it appears as though everything in the sky also gives off its own light. We can perceive objects on Earth because they give off their own light as well. Like everything else, the human eye has its own inner light that helps it to perceive objects. When the sun sets, the reason we can't see as well is because the light given off by the objects and by the eye is too weak to be perceived without the aid of a stronger light, like a torch.

Relatively Speaking

Optics was originally a branch of natural philosophy that began with Aristotle and later became an area of study in physics. It essentially deals with the nature and properties of light and vision. Today, optics applies to many areas of study and research, such as lenses, lasers, fiber optic applications, eye surgery, and much more.

Refraction occurs when light passes through a medium such as water and gets bent such that an object appears closer or farther away from its actual position. For example, when you put a stick or pencil in a glass of water, the pencil looks like it has been broken, because the part sticking out of the water and the part in the water don't appear to continue in a straight line.

The Beginning of the Lens

You may be wondering at this point exactly what the connection is between light and vision. Why was everyone thinking that vision was somehow related to light? What do light and vision have in common? They both allow you to perceive the world around you. During the Renaissance, the studies of light and vision were combined into a science called *optics*.

One of the most significant developments in the study of optics was the invention of the lens. Many early natural philosophers like Euclid, Hero, and Ptolemy (the astronomy guy) studied an optical illusion known as *refraction*. The early study of refraction never went very far, but it was responsible for the invention of the lens.

A man named Robert Grosseteste (1170–1253) experimented with various glass flasks filled with water. By examining how sunlight was focused through the flasks and how this magnified objects, he supplied some of the best geometric explanations for the shapes of curved surfaces. The technology to make a good lens just hadn't been invented yet.

Eventually, with the advancement of medical theory and dissection, the structure of the human eye was understood. This knowledge, combined with experimentation with different lenses, led to the invention of eyeglasses and the telescope. Optics was about to become a serious study. But before we look at it, we need to discuss another important and interesting development.

It's Your Perspective, Not Mine

Chapter 2, "Great Minds of the Renaissance," highlighted some of the paradigm shifts that took place during the Renaissance. One of the most significant was the change in how humanity saw itself in relation to the world. People no longer saw themselves merely as creations of God. Now they considered themselves creators as well. This new concept manifested itself in the development of a new art form.

One of the most difficult techniques to master in painting is the portrayal of three-dimensional subjects on a two-dimensional canvas. Until the 1400s, most paintings portrayed flat, unrealistic subjects. How could an artist paint what he saw so that the work reflected reality? The method that evolved from solving this problem came to be called perspective drawing.

One of the first artists to develop this method was Filippo Brunelleschi. He wanted to paint an image the way it looked in a mirror. He decided to paint a Florentine building by using a mirror and painting what he saw in its reflection. What he ended up with was the reverse image of what one would normally see without the mirror. He had demonstrated that the point of view from which one sees an image defines how one ends up painting it.

Another artist, Leon Battista Alberti, formalized this technique into an exact method. He came up with what is called the vanishing point. This is a point to which lines appear to converge in the distance. If you look at the figure on this page, you'll see how it works.

The study of optics was instrumental in the development of this new technique. During this time of the Renaissance, it was not uncommon for artists to also have an interest in science. Artists such as Leonardo da Vinci and Michelangelo showed how the mastery of this method could convince viewers that they were looking into space rather than at a flat surface.

Albert Says

The development of perspective art was an important leap in human consciousness. Not only did it revolutionize the way artists portrayed what they saw, but it was also a direct reflection of the Renaissance ideal. Because people had more freedom of thought to question and examine the world around them, from their own perspective, the ability to portray the world in a truer form was a natural development. Perspective drawing was an artistic technique that altered the way people perceived the world.

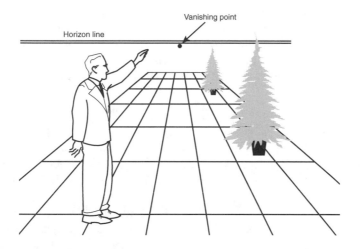

Construction of a perspective drawing.

Locked in a Prism

Galileo, Kepler, and Newton all made important contributions to science. Galileo didn't contribute much to the study of light or optics, but something he said regarding light is worth mentioning. Galileo started modern scientific experimentation, so you would expect him to apply that method to anything he wanted to investigate. When it came to light, he set up an experiment to show that it moved very fast. Beyond this experiment, he didn't do a heck of a lot in the area of optics. In a letter that he wrote near the end of his life, he admitted he was still in the dark about what light was. But he made an interesting statement about it. He believed that light had to do something with the world of atoms. (Remember that the notion of atoms goes back to Democritus, and most scientists of the Renaissance knew the writings of the Greek philosophers.) He didn't know how close to the truth he was.

Kepler's Optics

Although Kepler is best known for his work on the planetary laws of motion, he also published a book on optics. It was mainly concerned with the relationship between optics and astronomy. All of his brilliant work was done without the aid of a telescope, so much of what he wrote had to do with his observations on the nature of light as seen with the naked eye. He developed a set of propositions based on experience and imagination. His first few propositions are as follows:

➤ The flow or projection of light conveys it from its source toward a distant place.

➤ Light from any point flows along an infinite number of straight lines.

➤ Light can flow infinitely far.

➤ The lines along which light flows are called rays.

➤ Light doesn't move from place to place in time; it is instantaneous.

Mind Expansions

Most scientists during the Renaissance tried to study all of the known areas of natural philosophy, and many of them studied the works of Aristotle. Although some scientists broke with Aristotle's tradition and developed their own theories and ideas, many scientists developed their ideas based on what Aristotle already said. The influence of Aristotle's ideas was difficult to overcome, because so much of what he said was based on common sense and observation. It's hard to go against what your senses tell you is true. The development of experimentation was important in helping to verify information gathered from the senses.

These propositions are simple enough to understand and reflect a common-sense approach to the nature of light. However, the last point that he made was taken from Aristotle's ideas concerning motion. Aristotle thought that everything is moved by something. (Remember his concept of the prime mover, which set the universe in motion?) If there is nothing pushing, there can be no motion.

It also takes more force to throw a heavier object than a light one because a heavy object has more resistance than a lighter object. Kepler thought that light, which has no weight (no mass), offers no resistance to its movement, and consequently its speed is limitless. While this idea may be false in one respect, in that light does have a speed (186,000 miles per second), it is true that light has no mass.

Kepler went on to explain his ideas about vision, which are too lengthy to go into in detail here. But he did develop two more interesting ideas in relation to heat and color:

➤ Color and light have the same nature, but they are not the same thing. Color is potential light and becomes activated in an object when struck by light from the sun. He wasn't too far off with this idea. You see color in an object because that color is reflected back to your eyes, and all the other colors are absorbed. In other words, when you see green, green is reflected back, and all of the other colors of the spectrum are absorbed or scattered from the line of sight. If you think about it, in a way something that looks green is everything *except* green.

➤ Light is a form of heat. A heated object gives off light and color. He was pretty close with this idea, too. We'll explore this idea in more detail when we discuss light as radiant energy.

Color Me Newton

Chapter 4, "Newton's Clockwork Universe," already covered some of Newton's contributions to the study of light. With the publication of his book *Opticks*, Newton advanced the study of light to a whole new level. This was almost expected of him because of his previous brilliance in mathematics and mechanics.

Trying to cover even the most important theories that Newton put forth in his book would require at least two or three chapters, so I'll just summarize his most important ideas:

➤ White light, or sunlight, is the mixture of all the colors of the rainbow. Prisms act to separate these colors.

➤ Light consists of particles (he called them corpuscles) that move through space in straight lines.

➤ When these corpuscles run into a surface, they set up vibrations in the particles of the surface, and we see the color produced by these vibrations. The introduction of the notion of vibration was very important.

➤ The particles that make up light move at a constant speed.

➤ Light was created at the beginning of the world. It is a unique substance, and it plays a special role in maintaining nature.

➤ Ether is the medium that carries light particles from one place to another. This concept played a big part in the early study of light.

In contrast to Newton's idea that light was made of particles, another scientist, Christiaan Huygens, advanced the idea that light was a wave. He thought that if light were made of particles, these

Albert Says

It's always hard to convince others of a new idea or theory that goes against established and accepted ideas, especially if the previous work placed its author at the forefront of society. Even if the old theory is wrong and the new theory is shown to provide a better or alternative explanation, it's difficult to change people's minds. This happened with the established teachings of Aristotle and Newton and has been a common occurrence in physics, biology, medicine, philosophy, mathematics, psychology, and even religion.

particles would collide when two rays of light crossed. He argued that this doesn't happen with sound waves, which are all over the place, so light must also be a wave phenomenon. He went on to explain the mechanics of how light operates as a wave and presented some good evidence. His arguments were much better than Newton's, but because of Newton's reputation, Huygens' theories were not widely accepted. When you get to the quantum nature of light later in the book, you'll see that both Newton and Huygens were right.

Lights, Camera, Action

Toward the end of Chapter 6, "Magnetism Marries Electricity," you were introduced to James Clerk Maxwell. He developed the mathematical equations that showed the relationship between electricity and magnetism. The brilliant conclusion that he reached explained that light was a combination of electric fields and magnetic fields traveling together as a wave. These electromagnetic waves travel through space like waves that travel on water. Whenever a source produces an electric charge or a moving magnetic field (and remember, either of these can produce the other), waves containing these combined electromagnetic forces move out in all directions.

Nuclear Meltdown

It was believed that space wasn't empty, but was filled with a substance called ether. This theory helped to explain how waves moved through space. When you look at waves on water, the water is the medium that the wave moves in. The wave is carried by the water. With this idea in mind, how did electromagnetic waves get from one place to another? Something had to carry them. Ether was the invisible medium that carried these waves. This widely accepted theory would not be disproved until shortly before Einstein came along. Einstein would help explain how these waves traveled from one place to another.

How did Maxwell figure out that light is an electromagnetic wave? In addition to his mathematics, he made a relatively easy connection. The approximate speed of light had already been determined to be about 193,000 miles per second by a few scientists (which is a little faster than the actual speed of 186,000 mps), and Maxwell's calculations showed that electromagnetic waves travel at approximately the same speed. Because light and electromagnetic waves both travel at the same speed, Maxwell reasoned that light must be a form of an electromagnetic wave. Although Maxwell's equations are still considered to be correct, he never performed any experiments to verify his findings. Two other individuals would do that for him.

A Man Before His Time

Joseph Henry was probably one of the most unrecognized scientists of his time, although he did manage to have a unit of measurement named after him (the henry). He never published anything in time to be recognized for his contributions. For example, he detected electromagnetic waves when Maxwell was only 11 years old. While working in his lab and studying the electrical discharge from a Leyden jar, he found that the jar didn't just release its charge and become empty. Instead, the charge oscillated, or surged back and forth, changing direction thousands and thousands of times before it died out.

He also found that while performing his experiment on the second floor of a building, the charge that was released was sufficient to magnetize needles he placed in the basement. The experiment and the needles were separated by 30 feet as well as two floors and two ceilings, each 14 inches thick. He had produced and detected electromagnetic waves.

Ouch, That Hertz

Heinrich Rudolph Hertz was more famous than Henry, not only because he had a more familiar unit of measure named after him, but also because his work helped develop electromagnetism into an industry. Hertz picked up where Henry left off by designing circuitry to improve upon the strength of the electromagnetic waves that radiated out from the source. He was able to find both the *frequency* and the *wavelength* of the waves, and from that he deduced the speed at which electromagnetic waves traveled. They traveled at the same speed as light. As a matter of fact, all forms of electromagnetic radiation travel at the same speed.

Shortly thereafter, the idea of using electromagnetic radiation to transmit messages was turned into one of the most significant inventions all time: the radio. In 1896, Guglielmo Marconi produced the first wireless telegraph system. Five years later, in 1901, he transmitted the first wireless signal across the Atlantic.

Over the next 20 years, more and more discoveries would be made about electromagnetic radiation. The spectrum of frequencies would be discovered and made available for use. The graph on the next page shows the entire known electromagnetic spectrum. The range of visible light is only a very small portion of the entire graph. Our eyes are limited to seeing a wavelength range of about 400 to 700 nanometers. (A nanometer is equal to one billionth of a meter.)

Our study has brought us to the end of the nineteenth century. The last eight chapters have followed the evolution of ideas in physics and, in many cases, the underlying philosophy that accompanied these ideas. Each chapter has examined how humanity's desire to understand the world in which we live has provided a piece of the larger puzzle of human existence. Astronomy, physics, chemistry, and philosophy, though each its own separate field of knowledge, are interconnected. One of the main purposes of this book is to show that a greater understanding of each is possible when you relate them to each other.

Relatively Speaking

Frequency pertains to the number of cycles per second that electrical current or an electromagnetic wave oscillates. It is expressed in hertz. For example, 60 hertz is 60 cycles per second.

Wavelength pertains to the physical length of the wave, as measured in meters. Low-frequency signals have a wavelength from crest to crest of 10,000 meters, or about 6 miles. Microwaves are about 10 centimeters long, or about 4 inches. Shorter wavelengths are measured in angstroms. An **angstrom** is equal to one hundred millionth of a centimeter. X-rays are one angstrom long, and gamma rays are 0.001 angstroms long. Visible light has wavelengths between 4000 and 7000 angstroms. The higher the frequency, the shorter the wavelength.

109

The electromagnetic spectrum.

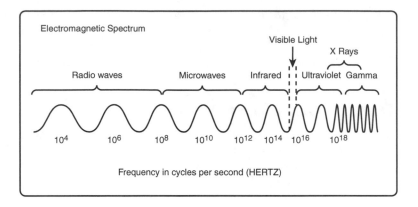

All of the ideas, inventions, mistakes, innovations, and beliefs go into forming our unique perspective, which has developed over the centuries. Everything continues to change and evolve into a greater understanding over time. This continual change in understanding is a dynamic process that has reached a high velocity in the late twentieth century. Where it will go is what the adventure of human existence is all about.

At the end of the nineteenth century, many physicists believed that they had discovered almost all that could be known about how the universe operated. They were all in for a nice big surprise. Albert Einstein came on the scene to clarify, disprove, disrupt, and take physics to the next level.

The Least You Need to Know

➤ Light can be understood symbolically or scientifically.

➤ The Renaissance ideal was reflected in the development of the perspective technique in art.

➤ Newton theorized that light was made up of particles called corpuscles.

➤ Christiaan Huygens theorized that light was made of waves, not particles.

➤ Heinrich Hertz showed by experimentation that light is an electromagnetic wave, just like radio waves and all other types of waves like x-rays and microwaves.

➤ Visible light waves are a very small portion of the entire electromagnetic spectrum.

Dark Clouds on the Newtonian Horizon

This chapter examines a few discoveries that are relevant to Einstein's work. Most of these discoveries occurred just a few years prior to the development of Einstein's famous theories. At the close of the nineteenth century, some powerful existing theories, ideas, and beliefs needed to be resolved or discarded. With the advancement of science, not all phenomena could be explained by the existing laws of physics. New discoveries led to new questions that required explanation.

Waves of Light

In Chapter 8, "Let There Be Light!" you learned about the two dominant theories of what light was made of:

➤ Newton's idea held that light was composed of particles, which he called corpuscles.

➤ Christaan Huygens believed that light was made of waves.

The followers of Newton were called *atomists*. They believed that these tiny particles that comprised light were the smallest possible particles. However, some phenomena made the particle theory hard to maintain. Newton even studied a few of them. With his usual humility, he called them Newton's rings. You've probably seen them yourself.

Have you ever noticed the way a soap bubble looks in the sunlight? You can see a rainbow on the curve of its surface. When you see oil floating on water or a patch of oil on the road, muted rainbow-like colors move around the edges of the oil. These rainbow rings are caused by the reflection and refraction of light. Newton had a hard time explaining these rings with his particle theory. However, this problem wasn't big enough to shake his followers, who continued to insist that light was made up of particles. After all, light moved in straight lines. It didn't bend around corners like sound waves did, or at least that's what they thought.

Huygens believed that light traveled as a wave. When light waves meet, one of two things can happen. They can cancel each other out, or they can add to each other's height. You can see this happen when you throw rocks in a pond. When ripples collide, some of them combine to make bigger ripples; others cancel each other out. Huygens hoped to show why light refracts in water using this idea of wave interference. Remember, an example of refraction is when a pencil appears bent in a glass of water.

Relatively Speaking

To **propagate** means to transmit or extend something through space. This term is used a lot in physics in connection with different types of waves. It has other meanings in other sciences, such as biology, in which it refers to the reproduction of a species.

Diffraction refers to the phenomenon that occurs when rays of light interfere with one another. It's essentially the breaking up of light into light and dark bands.

As was mentioned earlier in the book, James Clerk Maxwell developed a mathematical theory to show that light was an electromagnetic wave. Hertz's experiments also provided strong evidence that light was a wave phenomenon. But years before Maxwell developed his theories and decades before Hertz's experiments, Thomas Young performed one of the most famous experiments in physics, which showed that light was *propagated* as a wave.

For his experiment, Young made two holes in a screen a few millimeters apart and let light from a distant source shine on the screen. He then put a target screen behind the first screen so that the target screen would be illuminated by the light that was passing through the two holes in the first screen. Two patches of light appeared on the target screen when the holes in the first screen were large.

When he made the holes in the first screen smaller, he expected the patches of light on the target screen would grow smaller, too. The particle theory of light predicted exactly how much smaller the patches of light would become. This experiment worked as he predicted, until he made the two holes extremely small and faint rings appeared around around each patch. Instead of becoming correspondingly smaller, the patches of light on the target screen became larger. Whoa! This couldn't happen if light were particles, because particles move only in straight lines.

Young found that if he made the holes even smaller still, the patches of light on the target screen began to overlap and became crossed with fine dark lines. These lines were caused by the waves of light interfering with one another, just like the waves created by rocks thrown in a pond.

You can see this for yourself by doing a simple experiment. Hold two fingers in front of your eye, as if you were going take a close look at a grain of sand between them, and then look at a light source behind your fingers—but not the sun! The retina of your eye will take the place of the screen used by Young. Just look through your fingers as you slowly squeeze them together. Right before you extinguish the light coming through, you will see a series of alternating light and dark bands. This phenomenon is *diffraction*.

These light and dark bands are the result of the electromagnetic waves interacting with one another; some amplify each other, and others cancel each other. Where they are amplified, you see a light band. A dark band is caused when the crest of a light wave meets a trough of another wave, and they cancel each other out. The result is darkness, because the light in that very small area has been canceled out. Normally we don't see interference patterns, because the crests and troughs of light waves are very, very small. But when light waves are forced to travel through a small space, such as the space between your two fingers or the two tiny holes in Young's experiment, the waves bend into each other and form visible interference patterns.

Young received nothing but ridicule and contempt from his fellow scientists for his discovery, because like Huygen's ideas, it didn't agree with what Newton had said. He was eventually exonerated by Maxwell and Hertz.

Frames of Reference

Eventually the wave theory caught on, and everybody was happy, at least for a while. Everybody knows that waves need something to wave in. Water waves are waves in water, sounds waves are waves in the air, at a baseball game "the wave" occurs in people—so what did light waves wave in? Nineteenth-century scientists were convinced that waves couldn't travel in a vacuum. They knew this because when they placed a ticking clock inside of a glass jar and slowly sucked out all

Relatively Speaking

Luminiferous referred to the quality of ether to carry or transmit light. This term came from the Latin word for light, *lumen*. Any word that contains a variation of the word *lumen* always refers to some aspect or quality of light. So words such as luminescence, luminous, and luminosity all pertain to having light.

Albert Says

Nothing is more detrimental to the development of understanding than rigid, fixed belief systems. This problem occurs throughout the history of humanity. You've seen examples of this more than once in this book and will run into it again. Science as well as religion uses belief to establish the validity of systems of thought. The problem with this occurs when beliefs become so rigid that they don't allow any room for growth or deeper understanding. To use an appropriate phrase, "You have to be able to look outside of the box."

the air to create a vacuum, the clock kept running but they could no longer hear the ticking. There were no sound waves, because there was no air for them to wave in.

At the end of the nineteenth century, physicists were convinced that space was filled with a kind of fixed, invisible substance that was called *luminiferous* ether, in reference to the light waves that it carried. It was everywhere; it pervaded and penetrated all materials. In the experiment with the ticking clock, even when all the air was pumped out, the jar was still thought to be filled with ether. If it weren't, the light waves wouldn't be able to travel through the vacuum created in the jar. The area inside the jar would be dark. But it wasn't dark, so the logical conclusion was that ether must pass through the walls of the jar to keep it filled, even when there is no air inside.

This ether was invisible, had no smell, no taste, couldn't be heard, and was a nonmaterial substance. Nonetheless everybody believed it had to exist, because a light wave needed some medium to travel in. All types of known waves traveled in something. This was, after all, Newton's mechanical universe. This system of thought was founded in part on these basic ideas:

➤ Objects moved in a continuous manner, as described by the laws of motion. This law applied to all objects in the universe.

➤ Objects moved for reasons, governed by cause and effect. It was therefore possible to predict all motion, because it was determined by a previous observable action.

➤ All motion could be analyzed and broken down into its component parts. Each of these parts was seen as playing a role in the mechanism of the universe as a whole. This complex clockwork-like machine could be understood as the simple movement of its various parts, including those parts beyond our perception, like ether.

➤ The universe was a determined, mechanical system that could be completely understood. Space existed as an absolute quantity and so did motion.

Mind Expansions

Lord Kelvin, William Thompson (1824–1907), was a famous British mathematician and physicist. He was the first to show that Faraday's fields of force could be expressed mathematically. He just never went through the legwork of coming up with the formulas—James Clerk Maxwell did. Kelvin also formulated one of the first objections to Darwin's theory of evolution, stating that the earth couldn't be more than 20 million years old, thereby questioning the length of time it took for the whole evolution process. He was also the first to formulate the second law of thermodynamics.

During the mid-nineteenth century, a well-known scientist by the name of Lord Kelvin saw a few dark clouds on the horizon of this perfect Newtonian world. He predicted that there would be trouble when people tried to come up with a mechanical explanation of light and heat.

All objects, including planets, suns, baseballs, you, and this book, were all thought to move in relation to this fixed, motionless ether. Objects were like sieves moving through water. So the question was asked, how can one measure the movement of the earth with respect to this stuff that we believe is there, but can't perceive? The answer was to compare Earth's motion to the speed of a beam of light. To understand this comparison, you need to do your first *thought experiment.*

Relative Motion

Imagine that you're riding on a train and feel the sudden urge to get a drink of water, so you get up and begin walking toward the back of the train to the dining car. The train is traveling north at 75 miles per hour. Suppose that you are walking at 3 miles per hour. In what direction are you moving and at what speed? You can't answer that question without choosing a frame of reference. If your frame of reference is the train, you're walking south at 3 miles per hour. If your frame of reference is the ground, you're moving north at 72 miles per hour (75 for the train minus 3 for your walking).

Relatively Speaking

Thought experiments are useful tools for understanding concepts in physics. Einstein made them famous. He formulated some of his theories based on thought experiments that he developed. Thought experiments are good techniques to train your mind to visualize scenarios, and they require you to utilize your imagination. The only difference between a thought experiment and a real experiment, is that one takes place inside of you, in your mind, while the other takes place outside of you, in the world.

Is your ground speed your true, absolute speed? No, there are other frames of reference to take into consideration. You have to look at the bigger picture. Earth is rotating on its axis and is in orbit around the Sun. The Sun and the entire solar system is moving though the galaxy. The galaxy moves and rotates relative to other galaxies. These sets of galaxies form galactic clusters that move relative to each other. It's not known how far this continues—infinity is a hard concept to grasp. But as you can see, there is no fixed, absolute motion, because there is no frame of reference from which all motions can be measured. Is your head spinning a little? (Relative to what?)

Measurement is relative. Up and down, left and right, large and small, slow and fast, motion and rest, all need some frame of reference to be able to compare them to. There is no way to measure the motion of one object unless you compare it to the motion of another object. However, if measuring motion were this simple, Einstein would never have needed to develop his theory of relativity. As it turns out, Einstein would show why relative motion wasn't so easy to explain; that's coming up in a few chapters. For now, because you understand a little more about frames of reference and relative motion, it's time to return to measuring the movement of Earth through the ether by comparing it to the motion of a beam of light.

Constant Speed

Why was light used to measure the speed at which Earth was traveling through the ether? To answer this question, try your second thought experiment. Imagine that you're flying in the cockpit of a plane. You pull out a gun, stick your arm out the window, and fire a bullet straight ahead of you in the direction you're traveling. The bullet shot straight ahead from the moving plane has a ground speed greater than a bullet shot from a gun by someone standing on the ground. Suppose that the plane is flying at 200 mph, and the bullet fired from the plane travels at 400 mph. The combined speed of the plane and the bullet gives the bullet a ground speed of 600 mph, but the bullet shot by the person on the ground is only traveling at a ground speed of 400 mph, because it doesn't have the speed of the moving plane added to it. Get it?

Light presents a unique condition. The velocity of a beam of light is not affected by the speed of its source. In other words, if you repeated the preceding experiment with a flashlight instead of gun, the speed of the beam coming from the flashlight on the plane would be the same as that of the beam from the flashlight on the ground. The speed of the plane would not add to the velocity of the light, because the speed of light is its maximum velocity. It isn't possible for the beam to travel any faster from its source. Because the speed of light is fixed, it can be used as a point of reference to determine the velocity of an object as it moves through the ether.

Albert Says

The idea of relative motion is a crucial concept to understand. Einstein's theory of relativity is derived from this primary idea, and it will also come into play in the discussion of quantum mechanics. Relative motion is one of the most important concepts in this book. If you don't feel that you completely grasp it, take a little more time and think about it. You might be sitting still as you read this book, but what about what you're sitting on? Even your apparently stationary house is flying along on the surface of our planet as it rotates and orbits the Sun—so are you moving or not? The answer depends on what your reference point is.

Just Ether in the Wind

How could light work as a measuring tool? Say you were moving in the same direction as a beam of light; you would see the light beam move past you with an apparent speed less than its fixed speed. (Of course, this is assuming that light moves slow enough to see at any attainable speed, which it doesn't. This is just an example of how light can be used as a measuring tool.) If you were moving toward a beam of light, it would appear to be moving toward you faster than its fixed speed. Your movement is relative to the fixed velocity of the light beam, and you could calculate the apparently different speeds of light to find out how fast you're moving. The speed of light appears to change because you are moving. In fact, the speed of light is always constant.

To take this idea a step further, imagine you're back on that train, walking back to the dining car at 3 mph. The speed at which you are walking is always the same relative to the train. It doesn't matter if you walk to the front or back of the train; your speed is always 3 mph relative to the frame of reference, which is the train.

The same is true for the speed of sound waves in the railroad car. If you're sitting in the middle of the car and shout, a person sitting at the front of the car will hear you at exactly the same time as someone sitting at the back of the car. The air inside the car is carried along with the car as it moves, so the sound waves move through the air in both directions at the same time.

But what happens if instead of sitting in an enclosed car, you want to get some fresh air and go out to the open flatcar next door? Picture it. The sun's shining, the air's fresh, the birds are singing, and you're holding on for dear life because the wind is blowing in your face at 75 mph. (The speed of the train is 75 mph, so the air is now blowing across the open flatcar at that speed.) You shout for help to the person you see standing in the doorway of the car in front of you, but instead of that person responding, another passenger who was in the doorway of the car behind you hears you and pulls you in to safety. The person in the car in front of you, the one you originally yelled to, turns around but doesn't see anyone there, because you've already been pulled to safety by the passenger behind you who heard you first. The other person thinks they're hearing things and goes back to the dining car for another drink. (This story was exaggerated a bit to illustrate the movement of sound in air. It would happen much quicker in reality.)

So why did the person behind you hear you first, instead of the passenger in front of you? Because of the wind blowing toward you at 75 mph. The wind slowed down the speed of the sound waves moving forward in the air and accelerated the sound waves moving behind you.

Mind Expansions

A test was made in 1955 by Russian astronomers to confirm that the speed of light is independent from its source. They did this by examining the light that comes from the opposite sides of the rotating Sun. One edge of the Sun is always moving toward us, and the other edge is moving away from us. Hold up a ball and spin it and you'll see how this works. It was found that light from both edges traveled to Earth at the same velocity. This is one of a number of experiments that led to the same conclusion. You would think that one would have been enough, but part of the fun of scientific experimentation is coming up with all the various ways you can approach a situation.

This is how nineteenth-century physicists viewed the movement of Earth through the ether. The ether must behave like the wind rushing over the flatcar. In motionless ether, an object moving through it, like Earth, would encounter an ether wind blowing in the opposite direction. Earth moves through space as it orbits the Sun. Physicists thought that ether wind was blowing past Earth at the same speed that Earth was traveling in its orbit. So how do you measure this ether wind? With a light beam, because ether is the medium that carries the light waves, just like air is the medium that carries sound waves.

The idea was to measure the motion of Earth with respect to the fixed ether by measuring the speed of light as it moved back and forth in different directions on the surface of Earth. The ether wind would affect the speed of light waves, just like wind affects the speed of the sound waves. According to this theory, the speed of light would

117

increase in some directions and decrease in other directions. When the various speeds were compared, it would be easy to calculate the speed of Earth. Remember, the speed of light is a fixed velocity coming from its source. Any fluctuations in its speed would be accounted for by the ether wind.

The First Dark Cloud

Two gentlemen, Albert Michelson and Edward Morley, got together to check out ether wind and the movement of Earth. This famous experiment is therefore called the Michelson-Morley experiment.

To set up the experiment, Michelson and Morley floated a stone slab five-feet square and one-foot thick on liquid mercury. This setup eliminated any vibrations, kept the slab evenly horizontal, and allowed them to rotate the slab around a central axis. They then arranged mirrors on the slab, such that a beam of light introduced into the system would bounce back and forth between them until it made eight round trips. Then they arranged another set of mirrors in the same way, but at 90-degree angles to the first mirror setup.

Nuclear Meltdown

The Michelson–Morley experiment was one of the biggest failures in the history of physics. Their experiment was brilliant. It should have demonstrated the presence of the ether that everyone believed to exist. No one could understand why it didn't work, because the experiment was sound. But they didn't give up. The belief that ether existed was so strong, the most incredible theories were devised to save the belief system. This failed experiment ended up changing the face of physics. In the end, it had a more positive effect than negative, because it forced people to rethink their ideas about the structure of the universe.

The mirrors were arranged such that one beam of light would travel parallel to the ether wind, and the other would travel at a right angle to it. When Michelson and Morley introduced beams of light to each set of mirrors, they expected the following to happen: The parallel beam of light, traveling back and forth in the same direction as the ether wind, was expected to speed up as it moved with the ether wind and slow down on its way back against the either wind. But, you say, wouldn't this acceleration and slowing down just cancel each other out? No, my friend. According to their calculations, the light moving back and forth with and against the wind would take longer to complete its circuit. The beam traveling at right angles would also be affected similarly, but not as much. The perpendicular beams would just give more information about the speed of the wind. If there was an ether wind, and Earth was moving through it, this experiment would detect it.

Guess what they found? Nothing. Zip. Zilch. Nada. They tried it many times, but they never detected a change in the velocity of light. News of their failure spread like wildfire throughout the scientific community. Physicists couldn't believe it. Now what?! They continued to come up with all sorts of explanations to save the ether wind theory.

Light from Heat

The second dark cloud in Newton's mechanical landscape is called the *ultraviolet catastrophe*. This concept will take a few pages to explain, but it is not quite as complicated as the last dark cloud. This second problem has to do with the light that's produced when elements are heated. The known mechanical theories couldn't explain this light, and the solution to this problem would give birth to quantum mechanics and would earn Einstein a Nobel prize. (His prize was awarded not for solving the ultraviolet catastrophe, but for the theory that was produced as a result of finding the solution to it.)

All material objects glow or become luminous when heated to a high enough temperature. This is how old-fashioned gas lamps on the street produced light. This is also how the filament in light bulbs produces light. The Sun and all the stars in the galaxies give off light because their surfaces are very hot. At lower temperatures, you can get heat without visible light, for example, from your radiator.

Every physical object gives off electromagnetic radiation of all wavelengths, including visible light, all the time. Bet you didn't know that you glowed! Have you ever seen an infrared picture of yourself or a CAT scan of your brain? (If you're a psychic, you've probably seen people's electromagnetic auras—but that's a topic for *The Complete Idiot's Guide to Being Psychic*.) These types of pictures are possible because of the infrared wavelengths that the heat in your body emits.

The hotter a body is, the more it radiates. If it gets hot enough, it radiates enough visible light to be seen. The end of Chapter 8, "Let There Be Light!" introduced the electromagnetic spectrum. The spectrum comprises a wide range of wavelengths and frequencies, and visible light is only a small portion of the whole. The relationship of heat to the spectrum can be explained like this: As the temperature of a physical body goes up, the electromagnetic waves it emits rapidly become more intense and richer in the short wavelength range. To put this concept a little more plainly, as you heat up an object, it changes color. First you see red, then orange, then yellow, and then blue. The wavelength of red is longer than that of orange, and the wavelength and frequency of each color in the progression gets shorter and shorter. So far, so good. Physicists explained the mechanical

Mind Expansions

A German physicist named Gustav Kirchoff discovered an important law that states that all substances absorb the same light frequencies that they emit. This theory is used a lot in physics, chemistry, and astronomy. It led to the development of something called spectral analysis and the invention of a machine called the spectrometer. These devices are used to find the chemical composition of the sun and stars, as well as any physical substance. Each known element, when heated, gives off its own unique signature of color frequencies. It's as though each element has a unique fingerprint. So if you view an object with a spectrometer, you can see what elements it's composed of.

Relatively Speaking

Thermodynamic equilibrium is a term that applies to the study of thermodynamics, or the study of heat. This term means that a body absorbs as much energy in the form of heat as it radiates back out in the form of electromagnetic waves. This idea was brought into play with the question about blackbody radiation. Using classical Newtonian theory, no one could explain how blackbody radiators attained thermal equilibrium. Thermodynamic equilibrium was happening, and no one could explain it.

Nuclear Meltdown

The ultraviolet catastrophe was the second major failure of classical physics to explain a physical phenomenon. The first was the inability to explain how light waves traveled through space without something to carry them. This so-called catastrophe challenged the concept of the perfectly mechanical and predictable world of Newtonian physics and Maxwell's electrodynamics. These two areas are the cornerstones of classical physics, but by the end of the nineteenth century, these established concepts were beginning to wobble.

and mathematical relationships between the amount of heat applied, the energy emitted, and the corresponding colors of the spectrum.

This brings us to something called the *blackbody problem*. A *blackbody* absorbs all the radiation that it encounters, in contrast to a perfect mirror, which reflects everything it encounters. A surface that can absorb all frequencies (colors) of light can also emit all frequencies. You know what it's like to wear black in the sun—you get hot very quickly, and the material quickly becomes warm or hot to the touch. Wearing white has the opposite effect.

When an object is in *thermodynamic equilibrium*, it absorbs as much as it radiates. It does this at each wavelength, so it doesn't get hotter, or cooler, or change in any other way. That's what the term equilibrium means. A blackbody is said to be in thermodynamic equilibrium if it absorbs the same amount of light that it radiates back.

The important point about blackbody radiation is that its properties depend only on its temperature. The hotter an object becomes, the more electromagnetic waves it radiates. The frequency of the waves increases with the temperature, with a corresponding shift in color. This shift can be seen by looking at what happens to an iron poker placed in a fire. When the poker is warm, you can feel the heat in the form of infrared radiation, but you can't see it. As the poker gets hotter, it first glows red, then orange, and eventually white as the intensity of the blackbody radiation it emits shifts across the visible spectrum.

The problem is that if radiation is explained in terms of waves, using the same theory which describes sound waves, only the brightness of the light should change with temperature. The color should remain the same! Classical theory could not explain why the color of a blackbody changed as it got hotter.

This explanation still doesn't help you understand why it is called the ultraviolet catastrophe. The reason for the name ultraviolet catastrophe is mathematical. The classical wave theory predicted that most of the radiation emitted from a blackbody was high frequency (or short wavelength) in the ultraviolet, x-ray, and gamma-ray regions of the spectrum. (Ultraviolet rays are the culprits behind sunburns.) In cases where classical theory predicted an excess of ultraviolet rays emitted from a blackbody, there were none! This discovery was a catastrophe for classical physics theory. It wasn't until Max Planck developed the first quantum theory of radiation at the end of the nineteenth century, that this could be explained.

In other words, the amount of heat energy that went in didn't correspond with the mathematical prediction of what should be coming out. According to classical theory, when you turn on your oven and heat it up, and then after a while open the oven door, you should be bombarded with a lethal dose of radiation, in the form of high frequency X-rays, and gamma rays. Of course, this doesn't happen. The real world didn't work the way classical Newtonian mechanics and Maxwell's electromagnetic mathematics said it should.

There you have it. Two dark clouds are obscuring the once completely mechanical and predictable universe. Dark clouds have gathered, so it's time for Einstein to bring the rain.

Mind Expansions

Ironically, the sun and other stars are blackbodies. Like all blackbodies, the Sun glows because it is hot; it's hot enough that it glows yellow/orange. If you look at other stars in the night sky, you can see different colored stars. Go out some night (or early morning) and find the constellation Orion; it's the one that has three stars in a row to form Orion's belt. There are two bright stars in this constellation, one near the head of Orion and one near the foot. With the naked eye, on a clear and dark night, you can see that one star is kind of bluish and the other is kind of reddish. The blue star is hotter than the Sun, and the red one is colder.

The Least You Need to Know

➤ By the middle of the nineteenth century, all physicists accepted that light was made up of waves.

➤ All motion is relative to a frame of reference. Absolute motion is nearly impossible to calculate because everything in the universe is moving in relation to something else.

➤ The speed of light is not affected by the velocity of its source.

➤ Classical physics wasn't able to explain two problems. One was known as the ultraviolet catastrophe and the other was the lack of a medium to carry light waves. This inability brought the mechanical structure of the universe into question.

Part 3
Into the Heart of Einstein's Mind

One of my favorite quotes by Einstein is: "The most difficult part about understanding something is that we understand it at all." The title of this book goes right to the heart of that statement. How do we understand Einstein? We have a hard enough time trying to know ourselves, let alone someone else. I know that there have been people in my life whom I thought I knew, but then I came to the realization that I didn't know them at all. There are others whom I sometimes think I know better than I know myself.

There have also been times when I have met strangers for the first time and after a brief visit, felt like I knew them well. What happened? What was the key to this sense of connection I felt? Do you know? If so, then Einstein's life will unfold to you in a way that will give you an immediate connection. If not, we will get to know him together. Let's become familiar with Einstein, his ideas, beliefs, and theories, and see whether we can figure out who this fellow is.

Chapter 10

Albert's Love of Life

In This Chapter

➤ Einstein teaches himself math

➤ Two events that changed Einstein's life

➤ Einstein's views on life and love

➤ Einstein's job at the patent office

If you want to understand something, you should go to its foundation to see what it's built on. From there, you can explore, analyze, compare, and eventually appreciate everything that has gone into making it what it is today. This is true for architecture, ideas, and especially people.

This chapter takes a look at Einstein's early life. It's important to have a clear understanding of Einstein's life before he developed his theories. Although his theories are extremely significant to physics and the advancement of technology, they don't explain who he was as a person. The theories are a product of a man whose passion, wisdom, and humor set him apart from his contemporaries. Although other physicists understood the importance of his ideas and even made important contributions, few affected the world on the scale that Einstein did.

How Albert Saw the World

Albert Einstein wrote the following when he was 52 years old, in 1931:

> *What an extraordinary situation is that of us mortals! Each of us is here for a brief sojourn; for what purpose he knows not, though he sometimes thinks he feels it. But from the point of view of daily life, without going deeper, we exist for our fellow men— in the first place for those on whose smiles and welfare all our happiness depends, and*

next for all those unknown to us personally with whose destinies we are bound up by the tie of sympathy. A hundred times every day I remind myself that my inner and outer life depends on the labors of other men, living and dead, and that I must exert myself in order to give in the same measure as I have received and am still receiving. I am strongly drawn to the simple life and am often oppressed by the feeling that I am engrossing an unnecessary amount of the labor of my fellow men. I regard class differences as contrary to justice and, in the last resort, based on force. I also consider that plain living is good for everybody, physically and mentally.

You can tell by his approach to life that he was as concerned about the well-being and welfare of others as he was about himself, perhaps even more so. How did he develop this compassion for his fellow man? Was he born with it? Did he get it from his family and environment? To find an answer to these questions, you must go back to the beginning of his life.

Albert Einstein was born on March 14, 1879, in Ulm, Germany. His father Hermann had struggled to make a living for many years. About a year after his son was born, Hermann went into business with his brother Jakob and established an electrical and engineering firm in Munich. Albert's mother loved music, and under her influence, Albert started to play the violin when he was six. He played the rest of his life.

Mind Expansions

Ulm, Germany, where Einstein was born, had a curious town motto: "The people of Ulm are mathematicians." The town adopted this slogan during medieval times during a period of prosperity. Presumably, Ulmers had a lot of calculating to do to keep track of their income. Little did anyone know how appropriate this motto would be for Einstein.

School Daze

Albert and his family were Jewish, which wasn't the safest group to belong to in those days. Up until eight years before Albert's birth, Jews weren't full citizens of Germany, and they weren't allowed the rights and privileges of other Germans. Despite their religion, Albert's parents decided to send him to a nearby Catholic school because they felt he could get a good education there.

Albert was the only Jew in a class of 70 students, and he felt like an outsider. He was liked by his teachers and earned pretty good grades overall. One of the things he hated—he still had strong negative recollections of it when he was older—was the disciplined, military-style authority that pervaded the school. His distaste for the military was in sharp contrast to his classmates, who couldn't wait to grow up and serve their king and country. While most of the other boys played with toy soldiers, Albert was engrossed with his metal construction set and a toy steam engine his uncle had given him. Even when he was very young he spent a lot of time working on puzzles, building elaborate structures with his building blocks, and using playing cards to construct houses of incredible heights.

Albert and Maja Einstein as children.

From a very young age, Albert seems to have had a kind of natural mechanical ingenuity. Once when he was asked where he thought his abilities came from, he answered that he inherited his love of math from his father and his love of life from his mother. But he wasn't a boy genius like many others who became famous mathematicians. His teachers called him a dreamer and felt that he didn't apply himself as much as he could. He was also prone to temper tantrums until he was almost seven years old.

The quality that governed his innate curiosity was the sense of wonder that he had as far back as he could remember. Almost all youngsters exhibit this child-like quality, but it seems to disappear in most people with age. Einstein attributed much of his love for physics throughout his life to this feeling of wonder, which he never lost. Einstein reflected on his childlike fascination with the world in a letter he wrote to a colleague who was also a Nobel prize winner:

Nuclear Meltdown

One of the more common stories that's told about Einstein's life is that he wasn't good at math when he was young and that his teachers gave up trying to teach it to him. That story was probably made up to encourage young students who were having problems learning math. It just isn't true. Einstein was a child prodigy when it came to math and taught himself more in this area by the age of 14 than he could learn in school.

127

When I ask myself why it should have been me, rather than anyone else, who discovered the relativity theory, I think that this was due to the following circumstance: An adult does not reflect on the space-time problems. Anything that needs reflection of this matter he believes he did in his early childhood. I, on the other hand, developed so slowly that I only began to reflect about space and time when I was a grown-up. Naturally I then penetrated more deeply into these problems than an ordinary child would.

What Triggered It All

Two significant events in the life of young Albert put him on the road to becoming a physicist. When he was four or five years old, while sick at home from school, his father brought home a compass to help him occupy his time. Albert was intrigued by the magnetic compass. He felt that this event left one of the deepest and most lasting impressions on him in his life. What made the compass needle swing? There had to be some hidden force behind it, making it move. Albert's sense of wonder kicked in.

The second notable occurrence came when his uncle Jakob showed him a book about Greek mathematics. In it, Albert found the Pythagorean theorem (in a right triangle, the sum of the squares of the perpendicular sides equals the square of the hypotenuse). He was struck by its simple elegance and decided that he wanted to prove the theorem to himself. So for the next three weeks he worked exclusively on proving the theorem to himself. Albert came up with a unique proof of the Pythagorean theorem. There aren't too many people who would feel the need to prove a theorem that's already proven, and there certainly aren't a heck of a lot of people who would be able to even if they had the desire. But this is a key to understanding Einstein. He felt that people can learn a lot by studying concepts and ideas, but to really understand them, people have to work out the concepts for themselves.

Relatively Speaking

A **bar mitzvah** is a coming-of-age ceremony for boys in Judaism. It is celebrated at the thirteenth birthday. It literally means "son of the commandment." Girls celebrate a bat mitzvah. The counterpart to this ceremony in Christianity is confirmation.

The **Talmud** is the collection of writings that constitutes Jewish civil and religious law. It consists of two parts, the *Mishnah* (text) and the *Gemara* (commentaries). It was compiled in the sixth century C.E.

High School Dropout

Although Albert's parents were Jewish, they were very liberal about their beliefs. They followed some of the Jewish customs, but they were lax compared to the strict orthodox standards. Albert's sense of wonder spread to his religious roots when he was about 12. He studied the Old Testament ferociously and even composed his own hymns to the glory of God. He was preparing for his *bar mitzvah*, to officially join the Jewish community as a man, but his religious life was about to change.

It was a common practice in those days to invite a poor *Talmudic* scholar to lunch on the Sabbath to listen to his teachings. However, one young man the Einstein family invited into its home had no real desire to become a rabbi. He was a medical student who brought many books on the natural sciences for young Albert to read. Most Jewish families wouldn't have liked this action, but because Albert's parents were liberal thinkers, they supported anything that might help expand his knowledge about the world.

Albert devoured the 20-volume series on science for young people, and by the time he was ready for his bar mitzvah, his views on religion had changed drastically. He was convinced that most of the stories in the Bible couldn't possibly be true and believed that young people were being deliberately lied to by the state. This conviction laid the groundwork for Albert to question the truth behind many widely held beliefs concerning religion, government, politics, and of course science. He became a free thinker in the strongest sense, a quality that would make him many friends and some enemies.

Upon turning his back on religion, Albert devoted himself to mathematics and music. He loved to play the violin, and in typical fashion, he gave up on the regimented mastery of scales and boring musical pieces in favor of teaching himself works that he liked. He also began an in-depth study of geometry. It's interesting to note that another individual who we've already studied, Galileo, discovered his passion for geometry at about the same age that Albert did.

Albert Says

Regardless of your personal beliefs about the stories in the Old Testament, Einstein found many of them to be inconsistent with science and reason. He was only 13 years old at this time—an early start for questioning religious traditions. Many people consider the stories to be more symbolic or allegorical in nature, rather than literal history. We all have our own views on religious matters based on faith, personal experience, reason, and other factors. The important thing is to respect other people's beliefs. No one holds a monopoly on truth.

The Move to Milan

After Albert completed his first year of high school, his father ran into financial difficulties. He made a bid for a contract to furnish the power and materials to supply Munich with electricity. Unfortunately, he lost the contract to a competitor with much more political power. This loss left him without enough work to keep his company going, so he decided to move his business to Milan, Italy. This move posed some problems for Albert. He still had three years to go before graduation, and he didn't want to spend them by himself in an antagonistic atmosphere by staying behind. Living in the dormitories at school would bring him closer to all the things he didn't want anything to do with: anti-Semitic comments and overly authoritarian teachers. So he did what many of us would do in a similar situation. He bailed.

Albert got a letter from a doctor who was a friend of the family to obtain a certificate of release from school. The reason was explained as nervous exhaustion. (I'll have to remember that one for future use.) He also asked his math teacher to write a letter of recommendation on his behalf stating that he had the equivalent knowledge of mathematics of a graduating senior, which was true. The school accepted these documents, and by Christmas he was in Milan with his parents.

Albert as a teenager.

Albert recalled this period of time in Milan fondly:

> *The happy months of my sojourn in Italy are my most beautiful memories … Days and weeks without anxiety and without worries.*

Mind Expansions

One thing that Albert considered a weakness was his inability to remember words. Terms eluded him, which was the main reason he did so poorly in Latin and other foreign languages. Although he liked French and loved Italian, he never became very good at speaking these languages, although he liked to hear them spoken.

Soon Albert resumed his academic studies. He wanted to enter the famous Polytechnic Institute in Zurich, Switzerland. One of the requirements for entrance was that the applicant had to be 18 years old. Albert was only 16. Fortunately, another friend of the family (it's good to be connected!) knew the head of the Polytechnic. Upon his recommendation, Albert was allowed to take the entrance exams. If you think that today's SAT tests are difficult entrance exams, forget it. The exam that Albert had to take lasted several days.

This test proved to be a rude awakening for young Einstein. Although he was recognized as a prodigy in mathematics, the rest of his scores in liberal arts were quite deficient. He learned the value of a well-rounded education and returned to another school nearby to fill in the gaps in his formal education.

Albert's First Love

At the age of 17, while attending the liberal arts school, Albert met and fell in love with the daughter of the family friend who had recommended him to the Polytechnic Institute. Her name was Marie Winteler. The members of her family were the closest friends the Einsteins had in Zurich, and Albert lived in their home during his time there. She was a year older than he and had just graduated with a teaching credential. They became friends quickly and fell in love soon after.

His feelings for her are clear in this excerpt from a letter he wrote to her during the Easter vacation:

> *My beloved darling: I have now, my angel, had to learn the full meaning of nostalgia and longing. But love gives much more happiness than longing gives pain. I only now realize how indispensable my dear little sunshine has become to my happiness.*

Everyone thought they would soon be married, but within six months, Albert broke off their relationship. He had begun his studies at the Polytechnic in Zurich and had a new love: physics. In a letter he wrote to Marie's mother, he explained himself:

> *So as not to continue fighting a mental conflict whose outcome to me is unshakable: I cannot come to you for Pentecost. It would be unworthy of me to buy a few days' pleasure with new pain—already I have inflicted too much on the dear child through my own fault. It fills me with a strange kind of satisfaction to have to savor myself some of the pain that my frivolousness and my ignorance of such a delicate nature has caused the dear girl. Strenuous intellectual work and observation of God's nature are the angels that will guide me, reconciling, strengthening, and yet implacably severe, through all the troubles of this life.*

Albert wouldn't get seriously involved with another woman again until he met his first wife, Mileva. But he had many women friends, and more than one woman was attracted to his good looks or enchanted with his violin playing.

Nuclear Meltdown

Women probably found it very hard to live with Einstein. For him, physics would always come first, and his relationships with women and even some friends would come second. It's not that he didn't care about people—he was very concerned with humanity as a whole. But on a personal level, his passion and conscious focus was on the hidden workings of nature and not with individual relationships.

Einstein as a young man.

Religion and Politics

Much of what is known about Albert Einstein comes from three sources. His collected writings cover a tremendous amount of information, much of which has nothing to do with his theories. His papers on relativity comprise only a small portion of the many topics he wrote about for various journals and conferences and in his correspondences. From Judaism to Christianity, politics to pacifism, and education to nuclear war, Einstein had ideas and opinions on many of the issues that were important in his world.

A second source of information comes from his family members, especially his sister. She provided unique insights into his early life, providing a lot of information that Albert either forgot or didn't consider important enough to record. Other family members also helped to fill in the blanks.

The third source is the very recent discovery of letters written to Mileva during the years 1897–1903. No one knew of these letters until the 1950s, when they turned up in the hands of Hans Albert Einstein, the oldest son of Albert and Mileva. The contents weren't made public

Albert Says

Albert Einstein has been interpreted by quite a few people, and this book represents another effort. Hopefully I've been able to present a clear and accurate picture of him. The challenge is being as objective as possible, letting Albert speak for himself whenever possible, and clarifying only simple points of interest.

until 1987, when a new series of volumes containing Albert's previously unpublished work was released.

Changing Points of View

Wisdom is one of those elusive things that many of us seek, some never acquire, and others use to make the world a better place to live. If we acquire it at all, it's usually later in life, because wisdom is most often gained through life experiences. In looking at Albert's life, we can see a big change in some of his views. One example is the change in his view of his religious heritage. Albert wrote the following when he was around 20 years old:

> *The religion of the fathers, as I encountered it in Munich during religious instruction and the synagogue, repelled rather than attracted me. Nor did I feel anything like national community or community of destiny. The Jewish bourgeois circles, which I came to know in my younger years, with affluence and lack of a sense of community, offered me nothing that seemed to be of value. Loneliness, at first painful, then productive and strengthening, was the result.*

He wrote the following a year before he died:

> *At that time I would not even have understood what leaving Judaism could possibly mean. Traditional religion had no place at all in my consciousness. But I was fully aware of my Jewish origin, even though the full significance of belonging to Jewry was not realized by me until later.*

His earlier comments reflect disatisfaction with aspects of a religious tradition that got caught up in the pomp and circumstance of society. Other religious traditions could be criticized for this very thing as well. People and cultures are not necessarily the best representatives of what a religion stands for. Some of the worst instances of man's inhumanity to man have occurred in the name of God.

Yet wisdom is the key with which we can see beyond our earlier misunderstandings. What we perceive to be true when we are young does not necessarily become less true, but we gain the ability to see it within a larger context. This certainly appears to have been the case in Albert's life. He renounced his Jewish heritage for two decades before embracing it later in a whole new way. The elements that caused him pain and loneliness were forgiven and were replaced by the essential qualities of Judaism and the needs of others who called upon him for help.

Germany, No Thanks!

Around the same time that Albert was renouncing his Jewish heritage, he decided to give up his German citizenship and become Swiss. He always disliked anything that had to do with military-style authority, and Germany was moving more and more in that direction. All young men in Germany were required to serve in the military at the age of 20. Anyone who was a German citizen at the age of 17 was expected to serve

133

when he became 20. Albert was 16 when he decided to give up his German citizenship, because if he had waited until he was 17, it would have been too late to get out of military service.

Mind Expansions

The choice to renounce his citizenship to avoid serving in the German military was not due to any fear of having to serve, but to a complete disdain for the use of force in resolving problems. Albert's anti-authoritarian views stemmed from his dislike of people who believed that their ways should be imposed upon the lifestyles of others. He always believed that a democratic approach to politics and religion would best serve all concerned.

Relatively Speaking

Ernst Mach's *Science of Mechanics* was as much a teacher to Albert as any of the instructors he had at the Polytechnic, perhaps even more so. He considered Mach a breath of fresh air in an overly conservative curriculum. Mach questioned many of the Newtonian laws, because they weren't verifiable by experience—they were just theories that worked. Albert enjoyed this skepticism, because it reflected his own distrust of entrenched scientific beliefs.

His decision to give up his German identity was not an easy choice, but his friend and political mentor Jost Winteler offered him many rational reasons for doing so. Jost Winteler was the father of Marie, Albert's first love, and he and Einstein remained close friends until Winteler died in 1929. It seems likely that Papa Winteler, as Albert often called him, probably influenced Albert's decision to become Swiss. Winteler also warned of Germany's desire to expand militarily. Almost prophetically, he saw that Germany would be the focal point of the world heading towards war. In a letter that Albert wrote to Winteler's son-in-law in 1936, he stated:

> *… human affairs in our age are less than agreeable, not to mention the clowns in Germany. Now it is obvious what a prophetic mind Prof. Winteler had when he perceived this grave danger so early in its full magnitude.*

Albert and Mileva

In October 1896, Albert enrolled at the Polytechnic in Zurich. He didn't care too much for the social life of the typical student and considered himself a loner. He did make a few friends. Among them were Marcel Grossman, a mathematics major a year older than Albert, Michele Besso, a mechanical engineer six years Albert's senior, and Mileva Maric, who would become his first wife. All three of these people would play significant roles in Einstein's life.

Michele Besso encouraged Albert to read certain books on physics that influenced the development of his theories. One of the most important books that Einstein read was *Science of Mechanics* by Ernst Mach. Einstein and Besso also spent hours on end discussing the philosophical foundations of physics. In the years to come, Albert always discussed his ideas and theories with Besso, relying on him to give good critical feedback.

Marcel Grossman was responsible for getting Albert his first permanent job, which was at the patent office. Marcel's father knew the director of the Swiss patent office (there's those connections again) and recommended Albert to him. Within a year he was working as a technical expert, third class. While working in this office, Einstein published his famous theories of relativity. Grossman also helped Albert later with the mathematical calculations for his general theory of relativity.

Physics or Mileva

Albert met Mileva in his first year at the Polytechnic. They had enrolled in the same year. She was of Serbian heritage and more than three years older than Einstein. Although she started out in the medical school at the University of Zurich, after a year she switched to studying mathematics and physics at the Polytechnic. Because Albert and Mileva were in the same curriculum, they became friends quickly, and soon they fell in love.

In many ways, Mileva was a lot like Einstein. Something of a loner and a free thinker, she had already accomplished a lot just by being accepted in the curriculum for mathematics and physics. The traditional role of women in European cultures at the time was not particularly compatible with a career in science. One of Albert's greatest pleasures in his relationship with Mileva was the opportunity he had to discuss his ideas about physics with her.

A few years passed before Albert and Mileva were married. A few obstacles stood in their way:

➤ Einstein's mother was opposed to their relationship. She didn't want anything getting in the way of her son's promising career. The fact that Mileva wasn't Jewish didn't help.

➤ Mileva failed her first final examination. Although her scores in physics were second only to Albert's, she didn't pass the mathematics section. She was very bummed.

➤ Albert had been unable to find a job, so he didn't have the income he needed to get married. His father was again in a financial crisis, and this weighed heavily on Albert.

Finally, in 1901, Albert got the job at the Swiss patent office. Things were looking good, at least for him. Mileva still had some problems to deal with. She had failed her second attempt at the final examination and left Zurich to return to her family in Serbia. Plus, she was pregnant. Albert's reaction to the pregnancy was mixed, and it's not really known what his true feelings were. But in a letter to Mileva in which he expressed his excitement about a paper that he wrote about the photoelectric effect, he said:

I am filled with such happiness and longing that you must also have some of it un-masked ... You'll soon see that you don't lie badly in my arms, even if I have begun a bit stupidly. How are you then, darling? How's the boy? Can you imagine how fine it will be when we can once again create together completely undisturbed, with no one allowed to interfere?

135

This letter is certainly an interesting response from a soon-to-be father. Mileva gave birth to a daughter named Lieserl in January of 1902. When Mileva rejoined Albert a few months later, she no longer had the child. Lieserl was either put up for adoption or died of a childhood disease. There is no record to provide the answer.

Mind Expansions

Six years after the birth of their first son, Albert and Mileva had a second son, Eduard. Very little is known about him, because in his teenage years he was diagnosed with severe schizophrenia and lived out his 50 years in a sanitarium.

Albert the scholar.

The Patent Office

In 1902, Albert's father died of a heart ailment. He was only 55 years old. But one thing that eased the relationship between Albert, Mileva, and his family was that his father gave his blessing to Albert and Mileva's union. Because Hermann had given his approval to his son's marriage, Albert's mother eased up on her disapproval of Mileva. Although she personally still didn't accept Mileva, she stopped outwardly opposing their marriage.

On January 6, 1903, Albert married Mileva. Within 16 months, Mileva gave birth to their first son, Hans Albert Einstein. This event put an end to Mileva's desire to pursue her career. She had already failed her final exams twice. She had also already given birth to a child she had most likely put up for adoption. She now assumed the role of a traditional housewife.

Meanwhile, Albert started getting into his work at the patent office. He could not have landed a better job to help bring his mental powers into focus. Albert had always been a visual person, that is, he was much more comfortable thinking in images rather than in verbal concepts. Most patents have blueprints that go along with the paperwork. Blueprints convey the workings of an idea or invention through images rather than verbal concepts. With all the different patents coming through the office, Albert had the opportunity to not only examine the latest inventions, but he also was able to understand how they worked through the images in the blueprints.

This important quality was the key to the development of Albert's theories. Imagery was Albert's great gift. By being able to visualize his ideas in the form of pictures, Albert was able to work them out and conquer the mathematics later. By putting existing theories into images, he was able to see any inconsistencies in the images that they evoked. Some of his famous thought experiments are perfect examples of how he figured things out. We'll take a look at some of these experiments later on.

The seclusion at the patent office allowed him time to think about the questions he had concerning physics. His greatest theories were all developed while working in the patent office. Years later, when he was teaching, he looked back upon this time nostalgically, as in this letter to Michele Besso:

> ... *those days in that temporal monastery, where I hatched my most beautiful ideas and where we spent such pleasant time together.*

In those recently rediscovered letters that Albert wrote to Mileva between the years of 1897 and 1903, much of what was not previously known about his personality has come to light. Overall, it's difficult to piece together what their true relationship was. In his later years, Einstein recalled that he married her out of a sense of duty, feeling very guilty about the fate of their little daughter. As in other matters, his views changed as he got older, and comparing quotes from when he was first married and after his divorce from Mileva show two very different views. In a letter to Michele Besso, he said:

> *So I'm a married man now, and lead a very pleasant comfortable life with my wife. She looks after everything splendidly, is a good cook, and is always cheerful.*

Later, when he was living in Princeton, he was asked by a group of Jewish students whether marriage outside the tradition was acceptable. He answered:

> *That is dangerous ... but then, any marriage is dangerous.*

Do you feel as though you're beginning to understand Einstein? Do you identify with any of his traits? Are you disappointed in him, or do you understand some of what he was going through? Whatever your first impression of him is, hold off on forming your final opinion. There is much more to come. Chapter 11, "A Warm-Up to Relativity" looks at the discoveries in physics that led to the publication of Einstein's first work.

The Least You Need to Know

➤ The most important quality Albert maintained throughout life was his childlike wonder at the world.

➤ Albert renounced both his religious tradition and his German citizenship because they went against some of his basic principles of rationality, free thought, and opposition to authoritarianism.

➤ Albert loved his first wife for many reasons, but one of the most important was her mind. They discussed his new ideas in physics, and this made them both very happy.

➤ One of Einstein's greatest abilities was to visualize his theories as images. He devised many of his famous thought experiments this way, so others could understand his theories.

Chapter 11

A Warm-Up to Relativity

In This Chapter

➤ Albert's early search for a unifying force

➤ Einstein's miraculous year

➤ The discovery of the electron

➤ Radioactivity and Marie Curie

➤ Max Planck gives birth to the quanta

Before Albert Einstein published his famous paper on the theory of relativity, he worked on a number of other ideas that laid the foundation for his most famous work. Between the years 1900 and 1905, Albert published six little papers. None of them were significant contributions to physics, but they did attract some attention. These papers dealt with his ideas about electromagnetism and the interaction of atoms and molecules.

In 1905, Einstein published four of his most famous papers. The first two represented further investigation of the ideas he had written about in his previous six little papers. Any one of these four great papers would have secured him a spot as a renowned physicist. Previous chapters covered some ideas that will be fundamental to under-standing Einstein's four important papers. This chapter looks at the work and ideas that led up to Einstein's special theory of relativity, as well as Max Planck's introduction of the quanta.

Underlying Forces

Einstein published his first paper four years after graduating from the Swiss Polytechnic. It was called *Deductions from the Phenomena of Capillarity*. Basically, the paper attempted to explain the forces that interact among the molecules in a liquid. He used statistical analysis to accomplish this. Einstein wanted to approximate the size of an atom. This idea was important because the very existence of molecules and atoms was still being debated by many physicists at this time.

Chapter 7, "The Name's Bond, Chemical Bond," covered the development of John Dalton's atomic theory. Out of his theory came the idea of the molecule, which is the smallest form in which an element can naturally occur. Molecular theory explained experimental results pretty well, so it seemed to be a good tool for interpreting data, but scientists still didn't know if molecules existed. After all, no one had ever seen an atom or a molecule. (No one had ever seen ether before either, but that didn't stop them from believing that it existed.) But molecular theory worked well, so why not use it until something better comes along? Einstein used statistical analysis as a mathematical tool in molecular theory, because this way molecules and atoms could be treated in large quantities. They were too small to be considered individually.

Einstein published his first paper for one main reason: to get a job. He didn't expect that the paper would convince anyone of the existence of atoms or molecules. He wouldn't get his job as a patent clerk for a few years yet, so he was hoping to land a teaching job at the Polytechnic from which he had graduated four years earlier. But Einstein, although a bright student, hadn't left a good impression on his teachers. His attendance in class was never very good. He preferred to spend his time thinking and pursuing his own ideas. He used his friends' class notes so he would know what the lectures had contained, but for the most part he found the instructors boring.

In his class on electrodynamics, he had suggested a number of unique and innovative experiments to test some of his professor's ideas, but the instructor refused. Einstein's attitude tended to alienate his instructors. He presented his new ideas in a very self-confident manner that looked like arrogance to the professors. (There probably was some arrogance mixed in.) His teachers' lack of willingness even to consider his experiments frustrated Einstein immensely, so he often preferred the company of his own solitary thoughts.

Even though Einstein's first paper was a good attempt at defining the size and interaction of molecules, its underlying theme may have been more significant.

Mind Expansions

Einstein spent a lot of time job hunting when he first graduated. His father's business in Italy had serious financial difficulties, and his family needed money. He wrote letters to every physics department in Europe inquiring about a position. He didn't get one response. His father even wrote a few letters, almost begging his former instructors to give his son a position. No luck. Einstein was off to a rough start, but he did finally manage to land a few substitute teaching positions that helped out a bit. He really didn't get settled until he landed the job at the patent office.

Einstein thought that a universal force governed the interaction of molecules. This force was very similar to gravity, but it controlled the interaction of the planets in our solar system. Einstein felt that there must be some sort of correlation between the universal force and gravity, so he tried to find a unifying factor that would combine them into one natural law. He was wrong. In a few years, he declared that this first paper was worthless.

He wrote a second paper pursuing the same ideas of molecular interaction and a universal governing force. This idea or theme defined his entire scientific life. If there was one dominating philosophical idea that acted as a major catalyst in his pursuit of understanding the hidden structure of the universe, this was it. He spent the latter half of his life searching for a unified field theory. (I'll explain what this means in Chapter 22, "Who's Got the Bomb?".)

Friends and the Academy

Two factors strongly influenced the further development of Einstein's thought:

➤ His study of the philosophy of science

➤ The discussions he had with friends

Both of these factors helped him develop the reasoning and insight that led him to re-examine the works of Newton and other physicists.

In 1902, the year that his father died, Albert helped to found the Olympic Academy in Bern, Switzerland. This academy was jokingly named after the academy founded by Plato in Greece (circa 400 B.C.E.). It began with three members and ended with three members. They were Maurice Solovine, Conrad Habricht, and Albert Einstein. Occasionally, other friends would join their talks. Einstein's friend Michele Besso, who you met in Chapter 9, "Dark Clouds on the Newtonian Horizon," often joined them. Solovine had met Einstein through an ad in the newspaper. Einstein had placed an ad advertising private lessons in physics, and Solovine had answered it. They discovered that they had a mutual interest in certain philosophers such as David Hume, Gottfreid Liebnitz, and Immanuel Kant. They shared an even greater interest in the philosophy of science. With this strong common interest, they decided to meet regularly. Habricht, who was studying to become a mathematics teacher, joined the club.

There's a great story about one of these meetings that occurred on Einstein's birthday. Solovine, who came from a well-to-do Romanian family,

Relatively Speaking

Plato's Academy was a famous school in ancient Greece where young men could study philosophy, mathematics, and music. It was the forerunner of Aristotle's Lyceum, where he taught and lectured on the profuse subjects he had written about. Both of these academies were responsible for later centers of learning, especially the great museum and library in Alexandria, Egypt.

loved to eat caviar. Einstein had never had any, so the other members of the club decided to get him some for his birthday. They brought it to his home on the night of their meeting. Einstein was giving a talk that night on Galileo. He became so interested and focused on the material he was presenting that he ate all of the caviar without paying any attention to it. When asked afterward how he liked the caviar, he said, "What caviar?"

Outside of the academy, Albert kept in touch with his friends via letters. Two of his closest correspondences were with Michele Besso and Marcel Grossman. They wrote to each other about their careers, family events, and latest ideas in physics. Here's a short excerpt from a postcard Einstein sent to Grossman in response to a note telling Einstein about the joys of parenthood and Grossman's latest mathematical treatise:

> There is a strange similarity between us. We too will have a child next month. And you too shall receive a paper which I sent to Wiedemann's Annalen a week ago. You deal with geometry without the axiom on parallels, and I with the atomistic theory of heat without the kinetic hypothesis.

In another letter to Grossman describing his ideas on molecular forces, Einstein referred to his idea of a unifying force:

> What a magnificent feeling to recognize the unity of a complex of phenomena which appear to be things quite apart from the direct visible truth.

Albert Says

Einstein was not the first or last physicist to pursue the idea of unifying force in nature. In many ways, this idea was indirectly connected to personal beliefs about the role of some divine force in the universe. Newton believed in such a force, as did Galileo and other scientists of the eighteenth and nineteenth centuries. Discovering this force is still a goal in today's physics, but it's called the TOE, or the theory of everything. In some circles, it's devoid of any spiritual connection; in others, the theory is not exclusive of a possible spiritual aspect. It all depends on your beliefs and perceptions.

Herr Doctor Einstein

After these two papers were published, Einstein wrote four more between the years 1900 and 1905. Three of the papers were on statistical analysis, and the fourth was on thermodynamics. None of these papers contained anything new. They covered topics that other physicists had already written about. Einstein had been out of touch with much of the academic community, especially when he worked at the patent office, so he didn't know what ideas and methods were being developed. But these papers still were of benefit to Einstein. Remember when he proved to himself the correctness of the Pythagorean theorem? He loved to understand concepts and ideas for himself. So with these last four papers, he essentially reinvented the wheel in statistical analysis and thermodynamics. It gave him a deep understanding of a powerful tool that would be the hallmark of his mathematical approach and also taught him the foundations of Maxwell's equations on thermodynamics.

Einstein had given up on pursuing a Ph.D. He thought that it was a waste of time and didn't want to play the academic games involved. But he changed his mind when he realized how important it would be if he wanted to get a teaching position and further his career at the patent office. So he applied to the University of Zurich.

One of the main requirements for graduation is the submission of a doctoral thesis, usually a topic picked by the candidate and the department head. But true to form, Einstein did what he wanted to and submitted his recently completed paper on the theory of relativity. The panel of professors who read the paper felt it was a little too "out there" (in other words, they didn't understand it) and rejected it. Little did they know what they had in their hands. So Einstein submitted another paper that he felt would be much more to their liking, not too radical, on a nice conservative topic. In this paper he laid the groundwork for his second paper of 1905, which dealt with determining the size of molecules in liquids.

A quote from Professor Kleiner, who read his paper, shows the degree to which Einstein had mastered statistical analysis:

> *The arguments and calculations are among the most difficult in hydrodynamics and could only be approached by someone who possesses understanding and talent for the treatment of mathematical and physical problems, and it seems to me that Herr Einstein has provided evidence that he is capable of occupying himself successfully with scientific problems. I have examined the most important part of the calculations and have found them to be correct in every respect, and the manner of the treatment testifies to mastery of the mathematical methods concerned.*

With the acceptance of his paper, he was now Dr. Einstein. Later, he dedicated this paper to his friend, Dr. Marcel Grossman.

The Miracle Year

The year 1905 is considered to be one of the greatest years in the history of physics. Within the span of nine months, Einstein published the four papers that would make him famous. He was only 26 years old. This year is called the *annus miribilis*, which is Latin for "miracle year." The four papers were published in the *Annalen der Physik*, the foremost physics journal in Germany. Einstein had burst on the scene as a virtual unknown. It took a few years for these papers to begin to have an impact on the physics world. These are the titles of the four papers, in chronological order:

Relatively Speaking

Heuristic is a term commonly used today in computer programming, referring to new ways for computer systems to learn. The meaning of the word comes from German and simply means "helping to discover or learn."

➤ On a Heuristic Viewpoint of the Generation and Conversion of Light

➤ A New Measurement of Molecular Dimensions & On the Motion of Small Particles Suspended in a Stationary Liquid

➤ On the Electrodynamics of Moving Bodies

➤ Does an Object's Inertia Depend on Its Energy Content?

Let's take a look at these papers so we can understand what they're about. His first paper was a contribution to the new field of quantum mechanics. The paper discussed Albert's revolutionary idea of photons and something called the *photoelectric effect*. I know, you're so excited you can't sit still, right?

Relatively Speaking

The **photoelectric effect** is something that most of us are already familiar with. Photocells, photo emmissive cells, and photoconductive cells operate elevators to ensure the correct level is attained when the door opens. They read film soundtracks in a projector, turn street lights on when it gets dark, and convert optical signals to electric signals in television cameras. A beam of light acts as part of an electronic circuit. When the beam is interrupted by an object, the circuit is broken, and this either stops or starts something.

The second paper covered his ideas about the interaction of molecules, a topic that he addressed in a few of his earlier papers. This paper was an extension of his doctoral dissertation. Albert published two separate papers a few weeks apart—that's why there's an ampersand between what look like two titles for this paper. Although separate, they're usually considered as one paper, because they deal with the same topic.

The third paper is considered to be the greatest of the four. It introduced the topic most commonly associated with his name—the special theory of relativity. This paper established a new mental approach to understanding the relationship between space and time. (There's that theme to the Twilight Zone playing in the background again.)

His fourth paper was a further advancement of his third work, again on relativity. In it, he showed that mass and energy are interconvertible. This paper is where he published his famous equation $E = mc^2$. Chapters 13, "The Special Theory of Relativity," and 14, "It's Still Relative," are devoted to these last two papers, so by the time you finish this book, you'll have a thorough understanding of what this theory means. But now, let's take a look at the discovery of the electron and the birth of quantum mechanics.

Electrons and Radioactivity

On a day in November 1895, nuclear physics was born. A physics professor in Wurzberg, Germany, was studying how certain mineral crystals glowed under ultraviolet light. His name was Wilhelm Conrad Roentgen. He was using a new apparatus called a cathode ray tube to direct ultraviolet light at the fluorescent minerals.

In the process of performing his experiments, Roentgen became aware of the fact that a number of crystals in various parts of the room began to glow. Some of these crystals were behind other objects on shelves several feet away. Whatever was emanating from his tube was passing right through the objects blocking the crystals. After many weeks of trial and error, he eliminated other possible explanations and concluded that these rays had the capability to pass through most solid objects.

In one experiment, he had his wife place her hand in front of a photographic plate and directed his rays at her hand. Upon developing the negative, he found that he had a photograph of the skeletal structure of her hand. He named these rays x-rays, because X was the scientific symbol for an unknown. In 1901, Roentgen received the first Nobel prize in physics for the discovery of x-rays.

Mind Expansions

A cathode ray tube, or CRT, is a device for producing a visual display using electronic circuitry. The earliest types of tubes were coated with a fluorescent material, which caused the material to glow when it was hit with electrons. Without the invention of this device, we wouldn't have TV today. A TV is essentially a CRT, as are radar displays and oscilloscopes.

About two months after the discovery of x-rays, a French physicist, Antoine Henri Becquerel, was sitting in an audience listening to a lecture on this latest discovery. He had been pursuing his own experiments with fluorescent minerals and was fascinated with these new developments. He found that some of the minerals even left images on photographic plates after they were exposed to the ultraviolet rays in sunlight. One mineral in particular gave him a new insight entirely by accident. He found a photographic plate with an image of a mineral on it in a dark drawer. This mineral had never been exposed to sunlight, yet it still left an image on the plate. Whatever was radiating was coming from inside the mineral! The mineral was uranium.

So the discovery of x-rays led to the awareness that certain minerals somehow gave off their own form of energy. This discovery was very confusing to physicists at the time, because it went against the law of the conservation of energy. You probably remember that this law states that energy can never be created or destroyed—it can only be transformed. It appeared as though this energy being radiated by the minerals, which was given the name of alpha rays or alpha particles, didn't come from anywhere at all! How could it come from inside of a rock?

Elect Ron for President

In 1897, a British physicist by the name of J. J. Thomson was working on identifying the particles that moved within a cathode ray tube from the negatively charged terminal (cathode) to the positively charged terminal (anode) and lit up the tube where they hit. Thomson discovered that it was possible to bend these particles in their journey using a magnetic field, which meant that they had an electrical charge. He found that the charge was negative. So at least one type of radiation coming from a cathode ray tube was not a ray at all, but a stream of negatively charged particles that came to be called electrons.

Albert Says

Although science prides itself on its description of the natural world and uses a method that is a powerful tool for evaluating observations and experimentation, it's surprising how many scientific discoveries have been made by accident. For example, the discovery of x-rays was accidental, as was the discovery of the radioactive qualities of uranium. Many inventions and ideas have been developed because of an intuition or even a dream. For example, the chemist Friedrich Kekule, who discovered the basis of the carbon chain for benzene, had a dream in which he saw a snake with its tail in its mouth, forming a circle. This image gave him the idea that the structure of the carbon chain was a ring or circle, instead of a straight link. So never underestimate your ability to find answers outside of the domain you're used to.

Thomson caused an uproar in the scientific community, because he claimed that he had found a particle smaller than an atom. No way, shouted the outraged scientists. The atom is the smallest possible form of matter. Here we go again: Those good old beliefs are getting in the way of new ideas and sound experimental data. Thompson even suggested that Becquerel's alpha particles emitted from uranium might be electrons, but he didn't have any valid evidence to back up this idea.

No one had any idea what an atom looked like. Some physicists thought that if there were a negatively charged particle in an atom, there might also be a positively charged particle. But how did it look? Some felt it probably looked like plum pudding, with the electrons stuck in the center. Others thought it was a spongy, gooey blob, or even something like a beehive with the electrons buzzing around it. They eventually came up with a good idea of what it looked like. We'll take a closer look at the atom in Chapter 16, "What Does an Atom Look Like?" when we enter the world of subatomic particles.

Try Not to Be So Radioactive

The first famous woman scientist in this book, Marie Curie, was a native of Poland. She went to Paris to study math and science in 1889. She didn't have a lot of money, so she worked as an assistant to a chemistry professor to help pay for her tuition at the Sorbonne. In 1896, she received her first degree and decided to pursue a Ph.D. in science. The topic of her dissertation was, *What was the source of the energy that darkened Becquerel's photographic plates? What were the rays emanating from uranium?*

Marie chose this topic for a number of reasons. First, it was a very new area and could only be explored by experimentation, rather than spending hours in a library doing research. It also was a very narrow field that had a clearly defined objective. In the years to come, Curie would develop a friendly rivalry with Becquerel and another man by the name of Ernest Rutherford. In the end, her devotion to finding the answers she was looking for would dominate her life and also take it from her. She would die of cancer due to overexposure to radioactive materials.

Curie's work led her to conclude that the rays that were being emitted definitely came from within the atom of uranium. She had discovered a new type of invisible ray that was different form the x-rays discovered by Roentgen. She adopted the term *radioactiv-ity* in reference to the rays that were being emitted. The word comes from the Latin term *radius*, meaning ray. In the years to come, she and her husband would discover two previously unknown elements: polonium, named after her home country of Poland, and radium. Radium turned out to be a million times more radioactive than the original piece of uranium that started it all.

In 1903, Becquerel, Marie, and her husband Pierre were awarded the Nobel prize in physics for the discovery of radioactivity. Her discoveries opened the door for others to follow, and by the early part of the twentieth century, the structure of the atom began to reveal itself. All of these developments are directly connected to Einstein's first two papers. The only other area we need to examine before we move on to a close look at Einstein's papers is the ad-vancement of the concept of the quanta.

How Much Is That Quanta in the Window?

As was explained in Chapter 10, "Albert's Love of Life," classical physics ran into a problem trying to explain something called blackbody radiation, which led to the ultraviolet catastrophe. Let's review blackbody radiation so that you have a clear understanding of how it relates to the ultraviolet catastrophe. With that freshly in your mind, you'll see how Max Planck avoided the problem.

Nuclear Meltdown

Marie Curie was the first woman in Europe to receive a Ph.D. in physics. She had an uphill struggle the entire way, because she was going against centuries of traditional values. Even in today's western technological society, many people still don't accept the intellectual or spiritual equality of women. This book has covered many developments in science throughout the ages, with little reference to the contributions of women. Unfortu-nately, most of western history has been dominated by masculine perspectives that have not allowed women to participate. Fortunately, this is changing. Women have made many contributions throughout the ages in many areas of science, but their voices were rarely heard.

A blackbody is an object that absorbs all light, or electromagnetic radiation, that hits it. When the light is absorbed, it heats the blackbody. The heat is then re-released as light, but often at a different color than the light it absorbed. The light emitted from a blackbody has a very specific spectrum that only depends on its temperature. Anything giving off heat in the form of light is radiating with the spectrum of a blackbody. This includes you, a plant, a cold glass of water, the TV, the sun, whatever. Most things are not hot enough for you to see the blackbody radiation with your eyes. When some-thing gets hotter, its color changes from infrared (which you can't see) to reddish and then to bluish. Really hot things, like the heaters in a toaster or the sun, give off blackbody radiation that we can see.

Rockin' on Your Sitar

Let's look at what the problem is concerning blackbody radiation in the form of an analogy. Make believe you know how to play the sitar, which is a multistringed instrument from India (you know, that cool instrument in the background of some of the Indian-sounding Beatles songs). The unique thing about a sitar is that when you play on the larger strings, those strings set into vibration, through a process called *resonance*, another group of finer strings below the main strings. If you've ever heard a sitar played, you know the unusual sound that this vibration produces. As you pluck the main strings, energy is sent to the smaller strings via sound waves, and the smaller strings vibrate and give off energy in the form of sounds that you can hear.

Imagine that this sitar has a huge amount of strings on it, of all different lengths and thicknesses. The long, thick strings give off lower tones and frequencies, and the other strings, which are shorter and finer, produce higher tones and frequencies. You're sitting in a cubicle room and begin to play one of your favorite Ravi Shankar tunes. To really get the room rocking, you want to fill it with as many standing waves as possible.

Do you know what a standing wave is, or what it looks like? Here's two ways you might have seen one:

➤ On the screen of an oscilloscope, you can see pictures of waves moving across the screen. You can adjust the frequency of the wave, so that it no longer looks like it is moving, but is standing still.

➤ An easier way is to take a long jump rope and have a friend hold one end. As you begin to wave your arm up and down with your end of the rope, you can make big waves in the rope. When you move your arm fast enough, you can create wave movements in the rope that look like they're standing still. That's a standing wave.

Only certain wavelengths produce standing waves. The distance between you and your friend holding the other end of the jump rope will only contain standing waves that fit perfectly in that distance. The same is true of the sound waves produced by the sitar in your room. There is only enough space in your room to fit a standing

Relatively Speaking

Resonance is a term that refers to the ability of one vibrating body to set into motion or amplify another body close by. It's also been called sympathetic vibration. Acoustic rooms are designed to eliminate resonance, as in a recording studio. If you hum low tones to yourself, you can feel parts of your body vibrate in resonance to that sound. If you hum the right note in an elevator, you can hear the walls vibrate. Try it, it feels pretty cool.

Relatively Speaking

Quantum, or the plural of the word, **quanta,** comes from the Latin word meaning "how much." It's an appropriate word when you don't know how much you're looking for. Instead of evoking images of complex numbers and ideas, the term quantum mechanics may just mean that you want to know how many repairmen it takes to fix your car (and how much it costs).

wave of a certain size of wavelength. You can always make smaller standing waves, with smaller wavelengths, but you cannot make one larger than the longest wavelength that will fit in the room. As it turns out, you can fit many more shorter waves into this room than long waves.

As you know, when you pluck the main strings, this energy sets all the other smaller strings vibrating, and then these strings evenly distribute the energy among themselves, happily vibrating away. This is called *equipartition*. If you continue to play the finer strings, more and more higher frequencies fill your room, so most of the energy gets concentrated in the higher frequencies. This is the problem.

Blackbodies contain a wide variety of charged particles that can produce all sorts of possible frequencies, just like all the strings on your sitar. When the particles begin to vibrate, they accelerate and produce higher frequencies, radiating more energy like the smaller strings. So if you heat a poker over a fire, the infrared energy given off gets evenly distributed over all the radiating frequencies, and just like in your sitar room, a lot more of the higher frequencies can exist than lower frequencies. This energy should set higher and higher frequencies vibrating, until you are vaporized by all the x-rays, and gamma rays given off by the poker. But you won't ever be vaporized. That's the crux of the problem. No one knows how the energy gets distributed to prevent this from happening. Enter Max Planck.

> ### Albert Says
>
> Physicists generally fall into two main categories: those who experiment and those who theorize. This division is the source of many disagreements, because sometimes theory says something should work that experimentation contradicts. Blackbody radiation is a perfect example. Classical theory couldn't account for what happened in reality. But experimentation got no closer to providing an explanation. With the introduction of Planck's constant, theory now explained reality, but nobody could experimentally verify it without more theories. That was one of the early problems with the acceptance of Einstein's theories. They explained how things worked, but it would be a while until technology was sufficiently advanced to show that they really worked.

Chunks of Energy

What did Max Planck discover? He found out that matter absorbed heat and radiated light discontinuously, or in chunks. These chunks were called *quanta* or in Latin, *how much*. It wasn't continuous or smooth. Who would have thought? This is the reason why we don't get fried in front of a fireplace or heater.

To understand the quanta, imagine that you have a candy bar. This candy bar can represent energy at any frequency. You have to follow these two rules:

➤ When you cut up the candy bar, it has to be divided into equal pieces with nothing left over.

➤ The higher you make the energy or frequency, the fewer chunks you can cut the bar into.

149

Following these rules, if the frequency is made higher, the energy candy bar would have to be cut up into larger pieces. That means that fewer chunks are available. This means that the higher frequency waves are not produced as much as the lower frequency waves. The lower waves are produced in more abundance, because it takes less energy to produce them for any given amount of energy. That's why there are more light waves produced in the lower frequencies.

Max Planck developed a mathematical constant, called Planck's constant (predictably), which allowed for the calculation and explanation for this unusual behavior of light. But he didn't like it. For years afterward, he tried to disprove it, but it always worked. It was just a mathematical construction. There was no way to see it or connect it to any other formulas like it. However, young Albert Einstein thought it was a great idea, and the next chapter will show how he put it to work.

The Least You Need to Know

➤ The theory of everything is the search for way to unify all known forces under one elegant theory. Einstein would spend his life in search of a unified field theory.

➤ J. J. Thompson used a cathode ray tube to show the existence of a particle smaller than an atom, the electron.

➤ Radioactivity was discovered to emanate from inside a substance, rather than being caused by something outside of it. It was discovered by Marie Curie.

➤ Max Planck's theory of the quantum solved the blackbody problem and overcame the ultraviolet catastrophe.

Chapter 12

Albert's World of Friends, Photons, and Molecular Motion

In This Chapter

➤ Max Planck's effect on Einstein

➤ The photoelectric effect explained

➤ The relationship between quanta and photons

➤ An introduction to the dual nature of light

➤ Brownian motion

A century has gone by since the discovery of the quantum. In that time, physics has changed tremendously. Although classical physics still explains much about the interplay of forces in the everyday world, the fundamental structure of the universe can be explained only by using the theories found in quantum mechanics.

The science of quantum mechanics didn't fully develop until the 1920s, when individuals such as Werner Heisenberg, Neils Bohr, Erwin Schrodinger and Max Born developed theories and concepts that helped explain the dynamics of the atomic and subatomic universe. Before then, Einstein's theories and Max Planck's quantum turned the world of physics upside down.

This chapter focuses on the first two papers that Einstein published in 1905. Chapter 11 summarized the content of these papers, but this chapter explores them in detail so you can understand how they made such a significant contribution to physics.

The Friendship of Max and Albert

Max Planck's incredible leap of insight in 1900 ushered in a break with classical physics and gave birth to the quantum revolution. He hit upon the idea that energy, which underlies all of nature, is not continuous, but comes in bursts or packets of energy, called *quanta*. He is regarded as the father of modern quantum theory. In 1918, Planck was awarded the Nobel prize in physics for this discovery.

Albert Says

Coming up with a theory or idea that goes against what you believe to be true must feel very strange. More than one scientist has had to wrestle with this problem. Experiments and mathematical analysis are difficult evidence to question, yet somehow a little voice inside says, "No, that's not the way things work." Many great ideas or advances in theory have brought into question what people believe. The trick in knowing whether the new or old theory is closer to the truth is in determining which one supplies you with a greater or more comprehensive understanding.

As I mentioned in the last chapter, Planck was convinced for many years that there was a flaw in his method. He thought that if he would just work a little harder, he could make the quanta disappear and still come out with a valid mathematical formula. He spent the next 10 years of life trying to do this, but he didn't succeed. Late in his life, he wrote:

> *My futile attempts to fit the elementary quantum of action somehow into the classical theory continued for a number of years, and they cost me a great deal of effort. Many of my colleagues saw in this something bordering on a tragedy. But I feel differently about it. For the thorough enlightenment I thus received was all the more valuable. I now knew for a fact that the elementary quantum of action played a far more significant part in physics than I had originally been inclined to suspect.*

Planck was deeply respected and much loved by all of his associates, as is evident in the following quote by Einstein:

The longing to behold harmony is the source of the inexhaustible patience and perseverance with which Planck has devoted himself to the most general problems of our science, refusing to let himself be diverted to more grateful and more easily attained ends. I have often heard colleagues try to attribute this attitude of his to extraordinary willpower and discipline—wrongly, in my opinion. The state of mind which enables a man to do work of this kind is akin to that of the religious worshipper or the lover; the daily effort comes from no deliberate intention or program, but straight from the heart. There he sits, our beloved Planck, and smiles inside himself at my childish playing about with the lantern of Diogenes. Our affection for him needs no threadbare explanation. May the love of science continue to illumine his path in the future and lead him to the solution of the most important problems in present-day physics, which he has himself posed and done so much to solve.

As you can see, Albert was very fond of Max. He felt that he owed a lot to Plank's discovery of the quanta. The Nobel prize in physics that Einstein was awarded in 1921 was not for his theories on relativity—he never received a Nobel prize for that. He received it for his first paper on the photoelectric effect. The revolutionary ideas put forth in this paper were influenced by Max Planck's concepts and mathematical formulas. In fact, Planck and Einstein spent many hours over the years discussing and comparing each other's ideas. Understanding Einstein's relationship with Max Planck is important to understanding Einstein's groundbreaking work in 1905.

The Berlin Years

In the early 1920s, four of the greatest minds in physics attended and presented lectures at the Berlin physics colloquium. This group consisted of:

➤ Albert Einstein

➤ Max Planck

➤ Erwin Schrodinger

➤ Max Born

Berlin was then the world center for theoretical physics and would remain so until the mid 1920s. The Berlin colloquium brought together outstanding scientists, chemists, and physicists on a weekly basis. Gatherings such as these, involving such tremendous scientific luminaries, have rarely been witnessed in history. The sheer brain power and brilliance of these individuals together would have been a sight to behold.

In his obituary for Planck, Max Born described the relationship that formed between Einstein and Planck during the years they met in Berlin:

> *Planck and Einstein met at regular intervals at the Berlin Academy, and a friendship developed which went far beyond the exchange of scientific ideas. Yet it is difficult to imagine two men of more different attitudes to life: Einstein a citizen of the whole world, little attached to the people around him, independent of the emotional background of the society in which he lived— Planck deeply rooted in the traditions of his family and nation, an ardent patriot, proud of the greatness of German history and consciously Prussian in his attitude to the state. Yet what did*

Mind Expansions

One great quality that Einstein and Planck shared was their ability to focus on what they were doing, exclusive of what was going on around them. Remember the story of Einstein eating the caviar and not being aware that he even ate it? This quality can seem like absentmindedness to some people. To a certain degree, it is. But this state of mind is much more than that. What you think about depends on what your mental priorities are. If you're concerned about whether your socks match, you aren't thinking about the secrets of the universe, and vice versa. This is one reason why Einstein's socks seldom matched. Well, they did when he was young, but they didn't as he got older.

all these differences matter in view of what they had in common—the fascinating interest in the secrets of nature, similar philosophical convictions, and a deep love of music. They often played chamber music together, Planck at the piano and Einstein fiddling, both perfectly absorbed and happy. Planck was an excellent pianist and could play on demand almost any piece of classical music, a great many by heart. He also liked to improvise either on a theme given to him or on old German folk songs which he dearly loved.

Nuclear Meltdown

It's a sad reflection of the twentieth century that many of the greatest advances in science were put to use not for the benefit of humanity, but for its destruction. That's not to say that science is a negative aspect of our world. The choices that people or groups in power make determine how science is utilized. The best intentions can often have negative consequences if the important choices are controlled by an outside source. Einstein would face this problem in his later years, as you will see when we get to the controversy surrounding nuclear warfare.

Planck's Last Days

Einstein was aware of the dangerous situation that had been arising in his homeland of Germany since the early 1920s. Nazism was already organizing outbreaks against Jews. However, Max disregarded these as a passing phase, either because he couldn't believe that Germany could be influenced by the Nazis or because he was getting on in years and wasn't aware of the threat and influence the Nazis already had. Planck even went to see Hitler around 1933 to ask him to spare the life of a Jewish friend, but Hitler flew into a rage and screamed at Planck. After this episode, Planck resigned himself to living out his life with as little to do with the outside world as possible. During World War II, his house and library were destroyed by Allied bombing, and his son Erwin, who was involved in the July 1944 assassination plot against Hitler, was executed by the Nazis. Max lived out his life in Gottingen, where he died in 1947 at the age of 88.

The Photon Heard Around the World

The concept of the photon was considered by many to be a revolutionary idea, but Einstein saw himself as anything but radical. He never saw science as a sequence of scientific revolutions. To him, it was more like a chain of connecting ideas. He was always very careful about describing his theories and deliberately avoided using words like revolutionary when they were published. In all of his writings, there is just one use of this word, in a letter he wrote to his friend Conrad Habricht, one of the original members of the Olympic Academy. Einstein had just finished writing his paper and was preparing it for publication:

… I'll send you four papers, the first of which I could send off soon, as I am to receive my free copies very shortly. It deals with radiation and the energetic properties of light and is very revolutionary, as you will see provided you send me your paper first. The second paper is a determination of the true size of atoms …

Before taking a look at this revolutionary idea, you need to understand something called the photoelectric effect.

Photoelectric Effects

The photoelectric effect was first discovered by Heinrich Hertz in 1887. (Remember him? He discovered radio waves.) It works something like this: When you connect a battery to two metal plates enclosed in a vacuum, you can send an electric current through the empty space between the two plates without the use of a wire, simply by shining light on one of the plates. One of the plates is connected to the negative terminal of the battery, and the other plate is connected to the positive terminal. If you shine a beam of light only on the negative plate, a meter hooked to the circuit will show that an electrical current is flowing.

A number of different kinds of experiments have shown that the current between the two plates is made up of tiny negatively charged particles—electrons, of course. You read about J. J. Thompson's discovery of electrons in Chapter 11. In this case, they're called *photoelectrons* because of their association with the photoelectric effect. But they're still plain old electrons.

So how can a beam of light activate a negatively charged metal plate and induce a current to pass between the two plates? It was first explained using classical electromagnetic theory. Remember that a light wave is essentially made up of a combination of moving electric and magnetic fields. The electric field in the light wave forces the electrons in the metal plate to begin vibrating and gives them enough energy to break their bond with the surface of the metal plate. These free electrons are attracted to the positive plate. When the battery is connected and voltage is applied to both plates, the negatively charged electrons fly across the empty space between the two plates because they are attracted to the positive plate. Remember that unlike charges attract. This continuous flow of negative electrons occurs as long as the light beam shines on the negative plate. Shut off the light, and the current stops.

Relatively Speaking

The discussion of CRTs, or cathode ray tubes, in the last chapter mentioned that the negative and positive terminals inside the tube are called the cathode and anode, respectively. In photoelectricity, the terminals have the same names, but with a prefix added on. So the cathode is now called the photocathode, and the anode is called the photoanode. Electrons are called **photoelectrons.** Later, this name was shortened to just plain **photons.**

This explanation of the photoelectric effect using the classical wave theory of light seems reasonable, but some unusual things couldn't be accounted for with this traditional paradigm. Let's look at some predictions one could make using the classical theory, and then look at the results of experiments that seemed to disagree. In this context, Einstein's theory will be easier to understand.

➤ **Prediction 1** Making the light brighter, or increasing its intensity, will cause the same number of electrons to fly off the plate, but these electrons will have more energy.

➤ **Result 1** Just the opposite happens. When you make the light brighter, more electrons fly off the plate instead of the same amount. Furthermore, the electrons have the same amount of energy, not more.

➤ **Prediction 2** Changing the color of the light, or changing its frequency, will cause more electrons to fly off the plate, but will not affect their energy.

➤ **Result 2** The opposite happens again. When you change the color of the light, the same amount of electrons fly off the plate, but now the electrons have either more or less energy, depending on the color of the light.

If light were just waves, as the electromagnetic wave theory states, the results should have coincided with the predictions, but they didn't. In most cases, the results were the opposite of what was predicted based on classical theory.

Relatively Speaking

A **frequency threshold** is the way a plate material reacts to light that is shone on it. If the frequency of the light isn't high enough, no electrons will be emitted when it shines on the plate. As soon as the light frequency goes above the threshold, the plate begins to emit electrons. If you shine white light on a zinc plate, nothing happens. But if you shine ultraviolet light on a zinc plate, electrons are emitted. The frequency of UV (ultraviolet) light is higher than that of white light. The UV light exceeds the frequency threshold of zinc and causes it to release electrons.

Scientists discovered some colors of light, regardless of how bright or intense the beam was, didn't cause plates made from certain metals to emit electrons. But they also found that various colors did cause other metals to release electrons, sometimes even at brightnesses lower than that originally used. For example, a very intense yellow light had no effect on copper plates, but an ultraviolet light affected copper even when the beam was extremely weak. This effect was called the metal's *frequency threshold*. This term means that each metal had a certain frequency of light beyond which it would release its electrons.

Imagine a beach with a lot of waves breaking on it. As the waves come in, they deposit sand, pebbles, and other things that were floating in the water. The waves wash these things up on the beach. On certain days, the waves are spaced very far apart from crest to crest. This space is so large that they aren't able to wash anything up on the beach. However, on other days there are huge amounts of waves, and the distance between each crest is very close. If the distance is small enough, just a ripple will wash things up on the beach.

What happens in the case of the photoelectric effect is basically the same problem as the one encountered with the ultraviolet catastrophe. It's not possible to explain the photoelectric effect using classical electromagnetic wave theory. Max Planck needed to develop the idea of the quanta to explain why the theoretical ultraviolet catastrophe wasn't such a catastrophe in practice. Einstein would do the same for the photoelectric effect by using Planck's idea of the quantum.

Quanta of Light

In 1905, Albert published his first paper, in which he developed the theory of the photoelectric effect. He called this theory *heuristic*, because it helped to uncover the nature of light in connection with the photoelectric effect. He couldn't base his theory on accepted electromagnetic principles, because they didn't work. He needed to discover a new means of explaining this phenomenon.

What Einstein proposed was that light energy came in chunks, or quanta of energy. The wave acted as the carrier of these chunks or quanta of energy. These special quanta of light came to be called *photons*. The energy contained in each photon depended on its color, or frequency. The mathematical number that Planck determined to calculate the amount of energy in each quantum was now used by Einstein to calculate the energy of a photon.

Einstein said that the energy of a photon (E) is equal to Planck's constant (h) times the frequency (f). As a formula, this idea looks like this:

$$E = hf$$

This formula is pretty simple. One of the great things about almost all of Einstein's formulas is that they're relatively easy to understand. The brilliance of his mathematics was that he was able to derive simple equations from a lot of work.

So what happens when a photon strikes a metal plate? Remember that a photon is a quantum of energy, and that the rules that apply to Planck's quanta also apply to photons. Photons come only in whole units, not in pieces. Thus, if a metal surface absorbs light energy, it must accept the whole photon, not just a piece of it. Actually, the electrons in the surface of the metal plate accept the energy of the photon.

Mind Expansions

Planck's constant is a mathematical constant that relates the energy of a quantum of electromagnetic radiation to its frequency. To put this another way, it relates energy's particle nature to its wave nature. It's an extremely small number: 6.626 divided by 10 million, divided by one billion, divided by one billion, and again divided by one billion. Another way to express this number is 6.626×10 to the negative 34th power. This number is so small that you wouldn't expect it to have a noticeable effect, and in most cases the effect isn't noticeable, because we aren't aware of the particle-wave duality of electromagnetic radiation.

Relatively Speaking

A **photon** is a particle of light. Many people think that Einstein came up with the name, but Gilbert Lewis, a chemist, named the particle in 1926. Einstein called his particle a point-like quantum. Did you know that about a thousand billion photons of sunlight hit a pinhead each second? When you look into the sky at night, every distant star that you see hits your eye with a few hundred photons each second.

If a single photon can give an electron enough energy to break the energy that binds it to the surface of the plate, the electron can escape. If the photon doesn't give the electron enough energy, the electron will clatter around inside the metal plate and dissipate the energy that it received from the photon. So one of two things can happen: either the electron escapes, or it doesn't. If the frequency or color of light is high enough, the photon has sufficient energy to knock an electron loose. If the frequency is too low, the photon doesn't have enough energy to free the electrons. Different metals have different types of chemical bonds that vary in strength. That's why the same frequency of light will release electrons in one metal plate, but not in another. This explains the concept of the frequency threshold.

Let's go back to the metal plate experiment, only this time we'll make predictions based on the photon theory. How does the theory hold up?

➤ **Prediction 1** When we make the light brighter, more photons will hit the metal plate. They won't have more energy; there'll just be more photons. There will be more collisions of photons with electrons, so more electrons will escape, each with the same amount of energy.

➤ **Result 1** Turns out exactly as predicted.

➤ **Prediction 2** When we change the color of light, which means we change the light's frequency, we're not changing the light's brightness. So there aren't more photons, but the photons hitting the plate have more energy.

➤ **Result 2** This experiment turns out exactly as predicted, too.

Einstein's revolutionary idea was that light waves were streams of individual particles. This idea did a good job of explaining phenomena that wave theory by itself couldn't handle. But the photon theory was hardly welcomed at first within the physics community. Everyone was convinced that light was a wave. Experiment after experiment had shown this to be true. Yes, the photon theory explained the photoelectric effect well and provided an accurate formula for doing the mathematical calculations, but how could light be made of these small particles? Light was a wave, and waves aren't made up of tiny particles. So now what?

Einstein Wrestles with the Nature of Light

Einstein was in a situation similar to that of Planck after he developed his mathematical formula. It worked, but it didn't fit with the accepted idea of reality. Even Einstein's equation contained elements of both qualities of light. Frequency (f) is associated with waves, and Planck's constant (h) relates to particles. So the formula itself contained aspects of both the wave and particle natures of light.

Philipp Frank, a close friend of Einstein's, wrote this anecdote concerning Einstein's struggle with the nature of light:

> *About this time, Einstein began to be much troubled over the paradoxes arising from the dual nature of light … His state of mind over this problem can be described by this*

incident: Einstein's office at the university overlooked a park with beautiful gardens and shady trees. He noticed that there were only women walking about in the morning and men in the afternoon, and that some walked alone sunk in deep meditation and others gathered in groups and engaged in vehement discussions. On inquiring what this strange garden was, he learned that it was a park belonging to an insane asylum of the province. The people walking in the garden were inmates of this institution, harmless patients who did not have to be confused. When I first went to Prague, Einstein showed me this view, explained to me, and said playfully: Those are the madmen who do not occupy themselves with the quantum theory.

In an excerpt from a letter Einstein wrote to a collaborator named J. J. Laub, he said:

… This quantum question is so incredibly important and difficult that everyone should busy himself on it. I have already succeeded in working out something which may be related to it, but I have serious reasons for thinking it is still all rubbish.

Albert Says

In 1951, Einstein wrote a letter to his long time friend Michele Besso. In it, he said:

All these 50 years of conscious brooding have brought me no nearer to the question: What are light quanta? Nowadays every Tom, Dick, and Harry thinks he knows it, but he is mistaken. That still is very true now as it was then. No one really knows exactly what light is! All we know is that it can be both a particle and wave. And that's the paradox. How many things do you know of that can be two different things at exactly the same time?

Albert spent the next 20 years trying to resolve this dual nature of light. Much of his work on this subject was never published. This unpublished work was unsuccessful, but it reflected his drive to find a single theory that would encompass both the wave and particle natures of light.

As you begin exploring quantum mechanics in more detail in a few chapters, you'll see that the nature of light becomes even more of a paradox. Under certain conditions, light acts like a wave, and under other conditions, light acts like a particle. Light is still thought to have a dual nature, and this subject remains open to debate even today. This duality has many interesting philosophical ramifications, which will be covered in later chapters. For now, let's take a look at Einstein's second paper, which deals with molecular motion.

Molecules in Motion

Einstein's second paper of 1905 is probably one of the easiest of his papers to understand conceptually. As Chapter 11 explained, this paper was a combination of two papers that were published very close together. It was also an extension of the doctoral dissertation that he wrote for his Ph.D. at the University of Zurich. His paper *On the Motion of Small Particles Suspended in a Stationary Liquid* was able to predict a phenomenon called *Brownian motion*.

In 1827, Robert Brown, who was a Scottish botanist, conducted some experiments that bear his name today. He used a microscope to study the behavior of pollen grains while they were suspended in a liquid. Brown discovered that the particles, which were only 1/5000 of an inch in size, were constantly jiggling around. He said:

> *These motions were such as to satisfy me, after frequently repeated observations, that they arose neither from currents in the water, nor from its gradual evaporation, but belonged to the particles themselves.*

He mistakenly assumed at first that the particles were some new state of matter that he called active molecules. (The other states of matter are solids, liquids, and gases, but you already knew that, right?) He then examined other forms of matter in suspension, because he thought that maybe the pollen was alive, so he tried gum resin, coal tar, nickel, iron, and others. He found that anything he suspended in a liquid exhibited this random, dancing behavior.

Relatively Speaking

Brownian motion, named after Robert Brown, refers to the motion of particles suspended in a liquid. These particles, which appear to just jiggle around randomly, are being moved by a huge number of molecules that make up the liquid and collide with the particles in suspension. Einstein developed his theory of molecular motion based on the early discoveries of Brown.

This motion, called Brownian motion, is caused by the thermal motion of the water molecules. Remember the discussion of the movement of molecules in gases and other states of matter? Whenever heat is introduced, it causes the molecules to move faster and faster. Water at room temperature has enough heat to cause the water molecules to move. The jiggling of the suspended particles is not due to colliding with just one molecule of water—the particles are way too big for that. They are pushed around by a large number of molecules that they collide with. These collisions are random in nature, so they kick the particle one way and then another way. These collisions give a particle the jiggling motion that is observed under a microscope.

Many people still didn't accept the idea of molecules as real things, aside from being useful concepts that could be used in chemistry or to explain thermal motion. But Einstein's paper conclusively proved that molecules exist. He worked from the assumption that molecules existed, and by combining that with what was known of Brownian motion, he added the powerful statistical mathematical methods he had mastered in his earlier papers. He showed that he could estimate the number of molecules per cubic centimeter of a liquid, as well as the size of the individual molecules.

At the beginning of this 1905 paper, he wrote:

> *In this paper it will be shown that according to the molecular-kinetic theory of heat, bodies of microscopically visible size suspended in a liquid will perform movements of such magnitude that they can be easily observed in a microscope, on account of the molecular motions of heat. It is possible that the movements to be discussed here are identical with the so-called Brownian molecular motion; however, the information available to me regarding the latter is so lacking in precision that I can form no judgment in the matter.*

You can see from this excerpt that Einstein knew of Brownian motion before he wrote his paper, but that's about as far as it went. His work was very original in its approach and takes its concepts and proofs way beyond anything that Brown ever accomplished.

Einstein's mathematical description of Brownian motion is known today as "the random walk," because his formula works not only for molecular collisions, but can also be used to describe the path of a drunk as he staggers down the street. I bet you never knew that after you've had one too many drinks, your movement down the street as you bounce off lampposts, parking meters, and fellow pedestrians can be accurately described with a scientific formula.

And that, my friends, brings Einstein's first two papers to a close. His first paper established the quantum theory of light and placed Albert Einstein alongside Max Planck as one of the two founding fathers of quantum mechanics. His second paper firmly convinced others of the existence of molecules. This knowledge, added to what had already been discovered about the electron and radioactivity, gave atomic and molecular theory a solid foundation. Upcoming chapters will show how both of these theories would usher in a more complete view of the structure of the atom and also aid in defining the dynamics of quantum mechanics.

Mind Expansions

You can find a drunk's average distance from where he began using Einstein's formula. Multiply the length of each step by the square root of the number of steps taken. Suppose a randomly staggering drunk whose paces measure 3 feet in length takes 49 steps altogether. The square root of 49 is 7. This number multiplied by 3 equals 21 feet from where he started. If he were sober and walking in a straight line, he would have covered 147 feet or 3 times 49.

The Least You Need to Know

➤ Photoelectricity is the phenomenon in which photons hit electrons in a metal plate, thereby releasing them and creating a flow of current in an electric circuit.

➤ Einstein's theory of light quanta shows that light is both a wave and a particle. Light has a dual nature.

➤ The formula that expresses the relationship between the amount of energy a photon has and its frequency is expressed by the formula $E = hf$, where E is the energy of the photon, h is Planck's constant, and f is the frequency or color of the light.

➤ Einstein was the first to determine the number of molecules in a given amount of a liquid, as well as the size of the molecules.

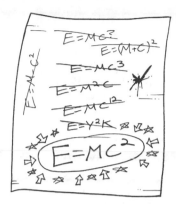

The Special Theory of Relativity

In This Chapter

➤ Early ideas of relativity revisited

➤ The relationship between time and speed

➤ Einstein's thought experiments on relative time

➤ How to understand difficult concepts

I was 16 when the image first came to me. What would it be like to ride a beam of light? At 16 I had no idea, but the question stayed with me for the next 10 years. The simple questions are always the hardest. But if I have one gift, it is that I am as stubborn as a mule.

These are Einstein's own words describing the concept that set him on a path that eventually led to his theory of relativity. I referred to this image back in Chapter 8, "Let There Be Light!" when I introduced some ideas and theories about the nature of light. I'm mentioning it here again because we've finally come to the point in Einstein's life when he discovered the answer to that question.

Einstein developed two theories of relativity: the special theory of relativity, and the general theory of relativity. The first was published during his wonder year of 1905 and is the focus of this chapter. The general theory came a few years later and will be the subject of the next chapter.

The special theory of relativity is considered one of the most significant ideas that has ever been conceived. The simplicity of its brilliance makes it accessible for all to understand. The application of this idea has led to many innovations in science and will continue to do so for many years. As we go deeper and deeper into the secrets of the universe, one of the brightest lamps that lights the way is Einstein's concept of relativity.

It's All Relative

Two sailors were stranded on a deserted island for 20 years. One day, while walking along the shore, one of them found a 32-ounce empty Coke bottle. Never having seen the new larger bottles before, one sailor exclaims to the other, "We've been out here in the sun for so long that we've shrunk!"

This story illustrates a point you became acquainted with back in Chapter 10, "Albert's Love of Life." You have to have some frame of reference to which you can compare a measurement. For the sailors, their erroneous frame of reference was the size of the Coke bottle. Assuming that it must be the same size that they remembered from years ago, they saw themselves as smaller in relation to the bottle.

A French mathematician by the name of Jules Poincare devised an interesting thought experiment. Suppose you go to bed one night, and while you're sleeping, everything in the universe increases in size a thousand times. Absolutely everything is bigger: you, the bed, Earth, the solar system, and all the atoms that make everything up. When you woke up in the morning, would you know that anything had changed? Would you be able to perform any kind of experiment to verify that everything had changed? In the first place, why would you even think anything was different? You wouldn't know that anything had changed. Besides, what could you compare everything to? Everything has grown as proportionately larger as everything else. How do you know this doesn't happen all the time? (Cue *The Twilight Zone* soundtrack.)

Size and all the measurements related to it, such as length, width, and height, are all relative. They have no meaning unless they can be compared to something else. Our standardized forms of measurement—inches, feet, yards, miles, pints, quarts, and so on—give us frames of reference to measure things. But they are still arbitrary units that have merely been defined so we can relate to and interact with our world. None of them are fixed or absolute.

Periods of time are like that, too. Time is just another form of measurement, and time has two frames of reference:

Mind Expansions

Research into the structure of the human brain has shown that our experience of time is directly related to how the brain functions physiologically. Our synapses fire in a linear fashion. A synapse is a point of contact between adjacent neurons, where nerve impulses are transmitted from one to the other. We experience time as a linear event, because that's the way information is processed in our brains. If our brains worked differently, we would experience time differently.

➤ Subjective or psychological time is how you experience time internally. When you are a child, time seems to pass much slower than it does as you get older. Even within the span of a day, you may experience time at different speeds. When you daydream or are intensely focused on something, aren't you sometimes amazed at the amount of time that has gone by?

➤ Objective or external time is the frame of reference we use to compare our subjective time to, but it is only a measurement. We use clocks as our frame of reference for external time, but seconds and minutes are arbitrary, standardized forms of measurement. They relate to our year, which is the amount of time it takes for Earth to complete one orbit around the Sun. This amount is also relative.

The idea of time being relative lies at the very heart of Einstein's special theory of relativity. You may think that time is fixed, that it can't slow down or speed up—especially external time. But Einstein would disagree.

Back to Galileo

The concept of relativity that Einstein advanced was based on ideas developed earlier by Newton and Galileo. Newton wrote in his *Principia*, "Absolute, true, and mathematical time, of itself and from its own nature, flows equably, without relation to anything external." We now know that Newton's statement isn't true. Einstein intuitively knew that, which is why he developed his theory.

There's nothing essentially wrong with Newton's physics. Einstein just realized that his principles needed to be extended. For Newton, distance was always expressed in relation to time. Whenever you move through space, whether you're walking or driving, or even if you're just reaching for a glass, time passes. Before Einstein, no one ever thought that time was not a fixed measurement.

Time passes differently for a person at rest than it does for someone moving at a great speed in relation to the person who's standing still. This difference is called relative motion. But for both people, their experience of time is real time. Neither one is aware that time is passing by differently for them. Chapter 10 covered a few examples of relative motion. Much of what was written about this topic came from Galileo.

Albert Says

At the core of it, there's nothing wrong with classical physics. Newton's principles, Galileo's relativity, and Maxwell's electromagnetic equations define most of the operations of the universe that we deal with on a daily basis. Einstein's theories were just needed to account for extreme conditions at very high speeds. Einstein's gravitational theory, however, did eventually replace Newton's law of universal gravitation. But it's important to realize that classical physics is just a set of connecting principles in the long chain of ideas that continues to unfold as our understanding grows.

The following examples should provide you with a better understanding of relativity. If a ship is moving in a harbor, and a ball is dropped from the top of the mast, the ball will drop straight to the deck—at least, that is how it appears on the ship. To someone standing on the shore, the ball appears to fall along a curve, not straight down. The ball maintains its forward motion due to the forward motion of the ship and simultaneously falls down to the deck of the ship. This combination is what causes the curve. On the ship, the object still hits the deck straight below it, because the ship is keeping pace with the forward motion of the ball. That's why, to the person on deck, the ball appears to fall straight down. Only by looking at the shore do the people on deck know that there is a relative motion between the ship and the shore. On the ship, it appears that the shore is moving, and the ship is standing still, even though the people on the ship know that they're the ones moving.

Suppose you walk to the front of the ship from the rear. Your speed relative to the ground is equal to the ship's speed plus your walking speed. Your speed relative to the ship is just your walking speed. In both this case and the case where the ball was dropped from the mast, there are only two frames of reference relative to each other. Let's add a third frame and see what happens.

You're standing at the rear of the ship and throw a ball towards the front of the ship at 60 mph. The ship is traveling north at 20 mph. Relative to the ground, the ball's speed is 80 mph. Relative to the ship, the ball's speed is 60 mph. You look up and see a small plane flying at a speed of 80 mph in the same direction the ship is going. Relative to the plane, the ball has zero speed—it's not moving.

Albert Says

Einstein wasn't the only physicist pondering these issues at the end of the nineteenth century. Others sought to find answers to questions dealing with the nature of light, the lack of ether, the structure of the atom, and relativity theory. But it took a gifted individual like Albert Einstein to make sense of it all. He did it first, and who knows if anyone else could have at the time.

As you can see, speed all depends on what your frame of reference is. In all of these measurements, Earth is the fixed frame that the speeds are all relative to, but Earth is also rotating and moving in an orbit around the Sun. So relative to the Sun, any object that moves on Earth has to have the speed of Earth factored in to figure out its speed. Of course, the Sun is moving in our galaxy, so relative to any other stars … As you can see, this can go on and on.

At the end of the nineteenth century, figuring out the absolute motion of Earth became very important. The concept of the ether was very important in finding a solution to this problem. If the ether was a fixed, nonmoving frame of reference, it would be possible to calculate the speed of Earth through it. But physicists eventually found out that the ether didn't exist. The Michelson-Morley experiment showed that there was no such thing.

Constant Light, Relative Time

A very significant discovery was made as a by-product of the search for the ether. Do you remember what it was? It was the realization that light travels at a constant speed, regardless of the speed of its source. Let's briefly go back to the ship moving in the harbor. One physicist sets up an experiment with a very accurate recording device to measure the speed of light on shore, and another physicist sets up the exact same experiment to measure the speed of light on the boat. Based on Galilean relativity and Newtonian physics, there should be a difference between the two recorded speeds of light. According to classical theory, the speed of light relative to the ground should be different from that relative to the ship. But it's not! To a person on the shore, the speed of light on the boat doesn't go any faster than the speed of light on the shore, even though it's coming from a moving source. Speed is nothing more than the relationship between distance and time. So if the speed doesn't vary, something else must!

Einstein asked himself: What if the speed of light is constant? He concluded that this would mean that time is *not* constant. That's the only thing that could account for the speed of light being the same regardless of whether its source is moving. Time must be relative to the speed of light. Light is the only fixed frame of reference, because it never changes. It is always the same speed. To everyone but Einstein, time was absolute, unchanging. It was the steady beat of the universe. The idea that a tick of time could waver was difficult for people to accept, including Einstein. The idea was *counterintuitive.*

Relatively Speaking

Many of the theories in this book are **counter-intuitive,** meaning that they are contrary to what would normally make sense to us. These ideas are some of the most difficult to imagine, because you must make quite a conceptual stretch to grasp their meaning. Quantum mechanics is so hard to understand because most of it doesn't make sense, and many physicists will be the first to tell you so. However, it works. And for now, that's all that *is* important.

How Did I Get Here?

How was it that Einstein zeroed in on the relativity problem in the first place? After all, many concepts in physics could have occupied his time. Everyone else was happy with the current understanding of the universe. Why bother with trying to figure out such an obscure topic? Some biographers contend that one of Einstein's greatest gifts was his ability to sift through large amounts of knowledge and come up with the most important questions that needed answering. In his own words, there was another key to his personality and his awareness of these problems:

> *The essence of a man of my type lies in what I think, not in what I feel ... And I have come to understand that I have to disengage from the momentary, the merely personal.*

This belief can account for some of his relationship problems as well as his realization that for him, his true passion was in the realm of thought, not in the fulfillment that comes from personal relationships. Underlying all of this was the one motivating idea that began at an early age. What was the nature of light?

I opened this chapter with a question Einstein asked himself when he was 16, "What would it be like to ride a beam of light?" To expand the question a little further, he later asked:

> *If light were a wave, then no matter how fast it traveled, it should be possible to catch up to its peaks and valleys. But then what would I see? Would light stand still? Would time stand still? Would I ride this wave forever?*

You can imagine what riding a wave of light would be like and why Einstein originally thought it was possible to catch up to a wave of light. If you've ever gone body surfing or water skiing, or even if you've just been in a boat, you know that it's possible to catch up to a wave in the water and ride on its crest. From that frame of reference, it appears as though the waves have stopped moving, because your body is now keeping pace with the moving wave and no other waves are passing you by. This is what Einstein thought riding a wave of light would be like.

Lightning Strikes

After much thought, Einstein knew that he had to find a solution to this problem. It came to him in a flash one day while walking with his friend Michele Besso:

> *I was doing most of the talking as usual, and I had been full of ideas about time and motion … I was just about to ask Besso that I needed his help in solving a problem, and then it came to me, before I could even finish asking the question, I had the solution. I ran home, and the next day I ran into him and said thank you, I have completely solved the problem.*

This was his big insight: Time must be relative if the speed of light is constant. This idea led him to ask the following question:

> *Can we all agree that two events are simultaneous if they occur at exactly the same time? When I say, for example, "The train arrives here at seven," that means that the passage of the little hand of my watch at the place marked seven and the arrival of the train are simultaneous events.*

One of Einstein's first thought experiments brings the concept of simultaneous events into question. Suppose you measure the distance between two poles alongside a railroad track, and then find the midpoint between the two poles. You have a right-angle mirror that allows you to see both poles when you stand at the midway point. Now imagine two bolts of lightning striking the two poles at exactly the same moment, in other words, simultaneously. You observe in your mirror that indeed, the two bolts have struck the two poles at the exact same time.

Coming down the track on a train is a friend of yours who also has a right angle mirror. When your friend gets to the midway point, lightning strikes the two poles again, and in his mirror your friend observes that the lightning bolts do *not* strike simultaneously. He sees the bolt hit the pole he's moving toward first, and then sees the other bolt hit the pole behind him. Why? Because the light has a shorter distance to travel from the pole that the train is moving toward than it does from the pole that the train is moving away from. Yet from your stationary position, you see them strike at the same time. You and your friend cannot agree as to when the lightning bolts hit. Time is relative.

This experiment verified for Einstein that time flows differently for each of us. It certainly does for you and your friend in the preceding example. This idea literally means that two clocks keep slower or faster time relative to each other.

Mind Expansions

A Polish physicist by the name of Leopold Infeld was one of Einstein's assistants in the 1930s. He wrote the following comment about Einstein's definition of simultaneous events:

This is the simplest sentence I have ever encountered in a scientific paper. So you have in Einstein, the ability to say the simplest and sometimes the most difficult concepts to comprehend.

There's Just Not Enough Time in the Day

In our daily lives, the relativity of time doesn't really affect us. We're moving too slowly to be aware of any of these changes in time. That's why Newtonian physics and Galilean relativity were never questioned. Our normal everyday world moves at rates that these principles can explain and account for. But things begin to change when you start approaching the speed of light. The closer you approach the speed of light, the more time slows down. As any object approaches half the speed of light, there is a noticeable effect on the speed of *time intervals*, which is the rate at which time passes. This rate change is also called *time dilation*.

Einstein mathematically derived a formula involving different speeds, the speed of light, and how they relate to each other. He called it his relativity factor. If you're on board a spaceship traveling at half the speed of light, according to the relativity factor your time interval between events would be 15 percent shorter than the time interval between events for someone at rest on Earth. For one hour of time that passes on board your ship, one hour and nine minutes pass by on Earth. That's because 115 percent of 60 minutes is 69 minutes. It's important to realize that this is not a perceived

Relatively Speaking

Time dilation is what happens to time intervals as an object approaches the speed of light. **Time intervals** are just periods of time between events. **Dilation** means to make wider or larger, so time dilation means that time intervals become larger as time slows down.

change in time. This is actual time for both individuals. Time passes slower at speeds approaching the speed of light.

The relativity factor is the essence of the third paper Einstein wrote in 1905. He developed his relativity factor to compensate or adjust for the difference in time between the two frames of reference. It wouldn't be possible to account for what happens and keep a constant speed of light without the relativity factor. This theory is known as the special theory of relativity. It's special because it deals with just two frames of reference. Einstein didn't give the theory this name; other scientists came up with this title. His paper was called *On the Electrodynamics of Moving Bodies*. He never referred to it as the special theory.

In answer to his original question of what it would be like to ride a beam of light, Einstein realized that he could never catch up to it. To understand why it is impossible to travel at the speed of light, you have to take a look at his other theories of gravity and general relativity, which are coming up in the next chapter. For now, let's take a look at two of the implications of the ideas explored so far:

➤ Time stops at the speed of light.

➤ The length of an object shrinks to zero at the speed of light. I haven't talked about shrinking length yet, but I'll get to it soon.

Neither of those concepts is easy to imagine. That is sometimes the frustrating thing about trying to understand material like this. There's a lot of information out there explaining Einstein's theory of relativity. Most of it can show you how he arrived at his theories and can also provide you with the thought experiments that help to explain the ideas behind the theories. The mathematics underlying the theory aren't difficult to follow either. All you need is high school algebra to understand how his theories work. However, we're not here to do the math; we're here to grasp the ideas themselves. But when you get right down to it, we're in the same boat that many of his contemporaries were. How do you comprehend the concepts that are the logical conclusions of his theories?

Mind Expansions

Time travel is probably one of the most popular topics in science fiction. H.G. Wells' book *The Time Machine* started it all. Most of us have thought about what it would be like to travel into the past or future. To know what tomorrow holds, to confirm an event from the past, or even just to live in another time has a strong psychological pull. Science says that time travel isn't possible, but then again science used to insist that Earth was flat. Who knows what tomorrow holds?

The Eternal Now

How do we experience time? We've already discussed some of the ways. We have our psychological, subjective, internal experience of it, and then there's the external or objective manifestation of time. Einstein's theories deal with the second kind. Yet our lives are full of objective time. Many of us live our lives based on the demands that objective time puts on us. It's as real to us as any material object is. In many ways, time has become a commodity in our culture. You know the old saying, "Time is money." That's probably more true now than it's ever been before.

One problem we have with time is that we have no freedom of movement within it. We can move in space—that's how we operate on a regular basis. But when it comes to time, we have only one option, and that's to move forward. Time is linear. It moves like a straight line in one direction. Einstein's theory shows that time begins to slow down as an object approaches the speed of light, but no one has ever traveled that fast, except in *Star Trek* and other science fiction stories. Time travel has captured the imagination of many writers, and much of our fascination with time travel stems from the fact that we have no control over time.

Can you see the difficulty with the concept of time stopping when an object travels at the speed of light? We have no idea what it would be like not to experience time going by. Even when we're not aware of time going by, that's not the same as time stopping. Or is it? When you sleep at night and dream, how often are you aware of the passage of time? In a dream state, you often experience events unfolding, but you don't have the same conscious awareness of time going that you have when you are awake. When you wake up, do you know how much time has gone by? Usually not. You look at the clock; you assume it's the next morning, and maybe you even sometimes think it's the wrong day. But regardless of whether we're aware of it, external time keeps ticking away.

Can you conceive what it would be like to experience the stoppage of external time? Where is your attention focused most of the time? Do you spend your time thinking about all the things you need to do: projects at work, kids to pick up from school, things to plan? In this case, you're thinking about the future. If you spend your time thinking about all the things you didn't get done, remembering pleasant experiences, or wishing you were anywhere but where you are now, you're focusing on the past. We spend almost all of our time thinking about the past or future, with little thought spent on the present. When we do manage to bring our awareness into the present, it's usually because we're concentrating on doing something that requires our attention. Ironically, that's also when we aren't aware of time going by. We seldom ever just sit around doing nothing while contemplating the present moment. Why? Probably because we don't have the time!

The idea of time stopping has a broader interpretation if we consider that rather than stopping, it

Nuclear Meltdown

We live in a world where poverty is a product of not having enough wealth, but a new type of poverty has arisen in our midst: the poverty of time. There never seems to be enough of it. One of the promises of our technology was to allow us to have more free time to pursue things that nurture us, rather than having to deal with things that stress us out. But we always manage to fill those empty spaces with more things that are supposed to give us more time. Somewhere along the line, we changed from human beings into human doings. Is this what life is all about?

may become infinite or eternal. When there is no time, then all we have is the present moment, and that moment lasts indefinitely. We experience time, even in the moment, as the movement of one moment to the next: *now* and *now* and *now*. You are in *this* present, now *this* present, now *this* present. But as you approach the speed of light, the intervals between moments become slower and slower, until all of a sudden you're in the present moment forever. Can you see why this idea was so hard for people to swallow? Very few people were able to conceive of time stopping. It drove them crazy.

Boy, That's Short

The other concept that's hard to conceive of has to do with the length of an object becoming zero. This concept is just as weird as trying to understand time stopping, but you have one under your belt, so let's go for the other one.

Time intervals aren't the only thing affected by the speed of light. Length, which is a relative measurement, is also influenced by great speeds. Einstein's relativity factor is used to account for changes in length as well as changes in time. Any material object is defined by relative measurements, and a frame of reference is needed to define and give meaning to these quantities of measurement. When any of these quantities deal with the speed of light, changes in how they're measured need to be adjusted using Einstein's factor.

Mind Expansions

As humans, we have the unique ability to project our consciousness into the future or the past. This ability is what allows us to think about and plan the future and to reflect upon the past. As far as we know, most other animals don't have this ability, because it requires a sense of self or self-identity. Now most pet owners would argue that animals do indeed have personalities, and they most likely do, but that's different from a sense of self-identity. We are self-referential systems that comprehend the world as it refers to our self-identities.

Matter is composed of atoms. The structure of the atom is made up of particles that have an electrical charge attached to them. (The modern concept depicts the shape of an atom as determined by waves of vibration, not individual particles, but we'll get to that in few chapters. For now, think of atoms as particles.) Each atom has a certain distance between the cloud of electrons buzzing around it and the nucleus. If the distance between the nucleus and the electrons changes, then the overall length of the object changes too.

Now lengths don't change for the person approaching the speed of light. For them, objects appear to be the same length as always. The change is perceived by an observer who is watching an object approach the speed of light. Remember that the special theory deals with two frames of reference: the person approaching the speed of light and the observer at rest.

Let's use another analogy to help show what happens to the length of an object approaching the speed of light as perceived by an observer at rest. Suppose you were to go outside and stand in an open space. The sun is just rising on the horizon, and you see that you're casting a long shadow on the ground. As the sun rises higher in the sky, your shadow becomes shorter and shorter. The condition when the sun first rose is the equivalent of traveling at slow speeds. The condition when the sun reaches the midpoint in the sky is the equivalent of traveling at the speed of light. All this time, you've been watching your shadow get shorter and shorter, until it disappears when the sun is directly overhead. That's what happens to the length of an object as seen from the observer's point of view as it approaches the speed of light.

Many physicists say that relativity involves a physical reality, not a perceived mental association. It also applies regardless of senses or minds. Physics seeks to define the principles under which the universe operates. The only trouble with these statements is that none of this has any meaning if there is no one in either frame of reference to perceive what is happening. It's a great triumph for humanity to come closer and closer to unraveling the secrets of the universe, but *we* do the unraveling; it has meaning to *us*. Whether these principles operate has no meaning if there are no people to perceive and understand what they mean. As we begin to explore the subatomic universe, you'll see how human perception and consciousness plays a significant role in how we understand the underlying principles according to which the universe operates. The next chapter continues the discussion of relativity, and you'll come face to face with $E = mc^2$.

The Least You Need to Know

➤ All forms of measurement are relative in nature. The meaning of any unit or quantity can be understood only when compared to a standard of measurement.

➤ The relativity of time and the constant speed of light lie at the core of Einstein's theory of relativity.

➤ Two of the most important implications of the relativity theory are that time stops at the speed of light and that the length of an object becomes zero to the observer when it reaches the speed of light.

➤ Principles in physics help to define our understanding of the physical universe. They stand alone without the need of human interaction. However, in and of themselves they have no meaning unless they are understood to somehow enrich our interaction with the world in which we live.

It's Still Relative

Who would have thought that an unknown 26-year-old patent clerk would turn the world of physics upside down? Einstein's third paper, which was covered in the last chapter, was published without reference to any previous sources. This was unheard of, because all proper scientific papers always referred to previous contributions and theories by other scientists in the field. But no one had ever written anything on the ideas that Einstein put forth in this paper. It was unique and stood on its own. There was only one small reference to Michele Besso, thanking him for listening.

Few physicists paid much attention to this paper at first. Not only was it a radically new idea, but also few people could understand the concepts he had developed. The paper contained very little math, which again is not typical, and it contained the thought experiments that Einstein used as the basis for explaining how his theories worked. Who did this guy think he was? In many ways, part of the problem for his contemporaries was that the concept was too simple. There had to be more to it. A discovery as important as this seemed as though it would require a lot of mathematics and be at least two or three times longer than it was.

This paper was the forerunner of Einstein's fourth and final paper of 1905. The theory of special relativity led directly to the development of Einstein's most famous equation: $E = mc^2$. This chapter will explain exactly what that equation means, as well as explain Einstein's other significant theory, the theory of general relativity.

Who Is This Guy?

Before diving into the other theories that Einstein developed, let's take a closer look at how his papers were received. As I mentioned, no one had ever heard of Einstein before, outside of his small circle of friends. He hadn't received his Ph.D. yet, was not working at a university, and was very young—only 26 years old. He didn't exactly fit the profile of a respectable up-and-coming genius.

Einstein was very disappointed by the reaction to his paper on relativity. He was certain that it would induce a strong reaction, but quite a few years went by before it achieved the recognition he felt it deserved. Any paper by a recognized physicist would be reviewed by a panel of peers, who would check the necessary references to works by previous scientists. One of the reasons Einstein's work went unnoticed was that nobody cared to read such an unprofessional paper.

The only reason Einstein's paper was published in the first place was because he had previously published his other papers in the same German journal. Its policy was that if an author had already published a paper in their journal, he or she could have future papers published automatically, without review by a panel of peers to examine the contents for references. This is why Einstein was able to write statements such as:

> *The introduction of a luminiferous ether will prove to be superfluous because the view here to be developed will not introduce an absolutely fixed space that has special properties…*

Albert Says

In this quote, Einstein expressed his knack for asking profound but simple questions:

All I have tried to do in my life is ask a few questions. Could God have created the universe any other way, or had he no choice? And how would I have made the universe if I had the chance?

Is this arrogance or a deep desire to know the mind of God? Whatever you think, he seems to have peered into the universe pretty accurately a few times.

Imagine that you were a physicist who had devoted your life to the study of ether. Wouldn't you want to see references to other sources that backed up what Einstein was proposing? Here's a person from nowhere who calls the point of your entire professional life superfluous. Where were the other people whose views supported this idea? It sounds a bit arrogant, doesn't it?

Einstein was sometimes perceived as arrogant, but his work came from a position of confidence and self-assurance rather than from arrogance. He felt that his theories were so harmonious and could be explained so simply that they had to be right. This attitude would come into play throughout his life and would encourage others to pursue their work in the same way, but at the same time, this attitude alienated other people. One of Einstein's assistants related the story of Einstein's reaction to the news that some observations had confirmed Einstein's General Relativity Theory:

> When I was giving expression to my joy that the results of the observations coincided with his calculations, Einstein said, quite unmoved, "But I knew that the theory was correct;" and when I asked what if there had been no confirmation of his prediction, he countered, "Then I would have felt sorry for our dear Lord—the theory is correct."

For the rest of his life in physics, Einstein's ideas about what worked and what didn't work always proved to be extremely accurate. He had an incredible sense of recognizing which experiments were correct and which ones weren't. In the years to come, some experiments that originally appeared to disagree with his theories were proven to have been inaccurately formulated. Einstein had known that intuitively when he first saw them.

Max Planck, who discovered the quantum, was one of the first to realize the importance of Einstein's theory. He became so excited by it that in 1906 he asked a student of his, Max von Laue, to prepare a lecture series on it for the University of Berlin physics department, where Planck was teaching. Von Laue decided that he wanted to meet the man who had formulated such a wonderful theory, so he made the trip to Switzerland to meet Einstein. Von Laue was taken aback when he met Einstein; he was expecting to meet a much older physicist. Einstein was the same age as he was: 26. Einstein was just as excited to meet von Laue. He remarked later that this was the first physicist he had met in person. Planck began teaching relativity theory soon after the lecture series and ended up teaching the first student in history to be awarded a Ph.D. for a dissertation in this specific area of study.

If Einstein was destined to win only one Nobel prize, you would think that the theory of relativity would be the one that did it for him. As you know, he received the prize for his paper on the quantum nature of light. This paper was more of a traditional scientific paper. It had lots of mathematics and lots of references. The paper on the special theory of relativity was mostly theoretical. The Nobel committee had set certain criteria that a paper had to meet just to be nominated, much less to win the prize. The paper on relativity just didn't meet the requirements, but the paper on light quanta did.

Mind Expansions

You would think that Einstein would have received numerous Nobel prizes for physics. But as it turns out, he was awarded only one. The committees at the time were very wary of awarding prizes for speculative theories. Most Nobel prizes were given for experimental rather than theoretical physics. Even as late as 1922, which was when Einstein received his prize for the discovery of light quanta, the special theory of relativity was considered too speculative. All in all, Einstein would be nominated for Nobel prizes five times, but he never made a particularly concerted effort to win them. Most of the prize money that he received for the one prize he won went to supporting his first wife Mileva after his divorce.

The Most Famous Equation in the World

You now have some understanding of the reaction—or more accurately, the nonreaction—to Einstein's theory, as well as some insight into how he felt about his ideas. It's time to look at his final paper of 1905, *Does an Object's Inertia Depend on Its Energy Content?* Contained within the three short pages of this paper was the equation that would become his most well-known: $E = mc^2$. To understand what this equation means, you need to know about a few more concepts that deal with relativity theory first.

Chapter 13, "The Special Theory of Relativity," noted that time and length are relative. *Mass*, one of the more interesting qualities of an object, is also relative. Normally, mass is thought of as the quantity of matter in an object. To understand Einstein's equation, you need to understand what mass is from a physicist's point of view.

Two Types of Mass

There are two ways that you can measure an object's mass:

➤ You can measure it on a balance.

➤ You can determine how much force (or push) is required to accelerate it.

Relatively Speaking

Mass is the amount of stuff there is in an object. You can define mass in two ways: either in terms of how it interacts with gravity, known as its **gravitational mass,** or by measuring its resistance to being pushed around, which is called its **inertial mass.** The two are basically the same. Two other names are given to mass, but both are used only in reference to relativity theory. **Relativistic mass** is the mass of an object measured by an observer who is moving relative to the object. **Rest mass** is measured by an observer who is not moving relative to the body. Mass can be a mess.

The first method relies on gravity. You have probably used a balance if you took chemistry in high school. Remember there are two trays. The two trays hang from a beam. The beam is supported at the center. There is a pointer attached to the middle of the beam. The pointer points at 0 when the masses on each tray are equal. You put the thing of unknown mass on one tray. On the other tray, you put a bunch of calibrated or known masses until the pointer again points to 0. The balance, or scale, works by equalizing the force of gravity on each mass.

The second method works like this. If you compare a billiard ball with a bowling ball, which do you think will require more force to push it? That's right: the bowling ball. It's more massive than the smaller billiard ball. Remember Newton's equation $F = ma$, or force equals mass times acceleration? Another way this equation can be expressed is $F/a = m$, which means that the ratio of force to acceleration is equal to an object's mass. That's how you can figure out the mass of any object. This mass is called the object's *inertial mass*. This measurement is different from an object's *gravitational mass*, which comes into play in Einstein's general theory of relativity.

Measurements to determine inertial mass involve time and distance. Why? Because the force that you exert on the ball requires it to increase its speed, which is distance per unit of time, such as miles per hour. Acceleration always contains two units of time, as in feet per second per second, because the speed continues to increase in each second.

The special relativity theory stated that time and distance vary according to the relative speed of the observer. Inertial mass varies, too. Imagine you're on board a spaceship (physicists love to use spaceships in these analogies because the only place that all this makes sense is in outer space), and you have taken along your pet rhinoceros, Herman. Relative to you, regardless of the speed at which you're traveling, Herman will always have the same mass. This mass is called *rest mass*. However, if you remain on Earth and send Herman into space by himself (poor fellow), Herman's *relativistic mass* will vary from your perspective. In other words, relativistic mass varies, and rest mass is always the same, and both are measurements of inertial mass.

Mass, length, and time are the three measurements that come into play in the special theory of relativity. As you approach the speed of light, time slows down, and length becomes shorter, but mass increases. Let's see what happens when we deal with observers and objects under observation that are moving *relative to each other* at velocities close to the speed of light. For example, if the relative speed of two spaceships is close to 161,363 miles per second (or about 90% the speed of light), the passengers on either ship will observe the other ship's clock running half as fast, its hours twice as long, the ship would appear half as long and twice its original mass. Yet inside their own ship, everything would appear completely normal.

In the example above, and in this one, the important point to remember is that both spaceships are moving frames of reference. The two reference frames are moving at a constant velocity relative to each other, in a straight line, either towards or away from each other. At the speed of light, an observer in one ship would think that the other ship had slowed to a full stop. Furthermore, the other ship would appear to have zero length and an infinite mass. Why an infinite mass? Because as the ship goes faster and faster, its relativistic mass keeps increasing. The more the mass increases, the more force is required to keep it accelerating. The speed of light can never be reached, because an infinite amount of force would be required to move an infinite amount of mass. Of course, you and Herman would not be aware of any changes inside the ship, other than the fact that the

Mind Expansions

Einstein's understanding and theory of mass increasing as an object approaches the speed of light is beautifully proved using present-day particle accelerators. The faster such particles move, the greater the force that is needed to accelerate them. That's why physicists keep building larger machines for accelerating particles. They need stronger fields to boost the increasing mass closer to the speed of light. Electrons can now approach speeds equal to 0.9999999999 the speed of light. These electrons have a relativistic mass almost 40,000 times their rest mass.

universe was hurtling backward at the speed of light, time would stand still, and all the stars would be flattened disks with infinite mass. What does all this have to do with Einstein's famous equation? You'll see in a moment.

The Energy in Mass

Einstein's fourth paper was presented very much like his other three. It contained hardly any math and presented a few good thought experiments. The thought experiment that dealt with his famous equation went like this: Imagine an atom that is decaying or breaking down due to radioactivity. Radioactivity is a natural process that occurs in some atoms, such as radium. (Chapter 11 touched on this.) It occurs when an unstable nucleus or particle spits out one or more particles, whereby it transforms itself into a stable nucleus or particle.

Einstein considered the emission of the radioactive particles and the mass of the products. In the case of radium, an alpha particle (or helium nucleus) is emitted and the element polonium is left over. In fact, the mass of the alpha particle and the polonium did not add up to the total rest mass of the radium. Mass had been lost! Knowing the total energy released, E, and the missing mass, m, Einstein came up with a simple formula.

The formula looked like this:

$$m = E/c^2$$

Nuclear Meltdown

Contrary to popular belief, Einstein's equation $E = mc^2$ is not the formula for the atomic bomb. Just try to build one using it! It only gives the description of the relationship between mass and energy. It did, however, lead to the physics that made the development of the bomb possible. Einstein has been called the father of the atomic bomb, but he hated his association with it. Although he was a catalyst in convincing Roosevelt to build the bomb, he deplored its use at the end of World War II.

In English, this formula states that mass (m) equals energy (E) divided by the speed of light (c) squared (^{2}). The more familiar form of this equation is the following:

$$E = mc^2$$

That's all there is to it! But what does the formula mean? Essentially, what the formula says is that all objects with mass have an equivalent amount of energy of mass, which equals the object's mass times the square of the speed of light, which is an extremely large number. In other words, even the smallest amount of matter contains an incredible amount of energy. If your mass is around 150 pounds, you contain enough energy of mass to power a small city for a week, if only you could convert it.

In almost all cases, this energy is unavailable for use, which is one of the reasons why no one discovered it before Einstein did. It couldn't be converted into other forms of energy such as heat, electricity, or light. However, almost 40 years after $E=mc^2$ was published, the first nuclear bomb converted the energy of the mass of a small amount of uranium into kinetic energy, demonstrating that it was in fact possible to convert matter into energy.

This brings us back to the relationship between the relative nature of mass and Einstein's famous equation. The mass of a body reflects its energy content. They're basically two sides of the same coin and are sometimes collectively referred to as mass-energy equivalence. The speed of light squared ties energy and mass together. On the one hand, it explains that there is a huge amount of energy in a very small amount of mass, and on the other hand, it explains why mass changes as you approach the speed of light. If mass has energy, then energy also has mass. Four and a half pounds of sunlight hit Earth every second!

The Fourth Dimension

Before covering Einstein's last significant theory, we need to look at how space and time combine in the relativity theory. Every event that takes place in space also takes place in time. As you sit reading this book, you aren't moving through space relative to Earth (unless you're in a car or train), but you are moving through time. To find a fixed point in space, you can measure its relative distance from another object. To determine when this object occupied that point in space, you can measure its relative time with a clock. The unusual aspect of all this is that space changes with time.

The space that you're sitting in now could have been an open field 200 years ago. Two million years ago, it could have been underwater, and who knows what will be in this space in the future. Space and time are intimately connected. This connection is often called the *space-time continuum*. This is a favorite subject in science fiction because of its association with time travel.

Another name for the space-time continuum is the *fourth-dimensional continuum*, because there are four coordinates that can locate any event. Space is defined by length, width, and height; combining these three with time gives you the fourth dimension. We live in this four-dimensional universe, but we're usually not conscious of it. Whenever we agree to meet someone at a certain location and a specific time, we are making a four-dimensional statement. But we have a tendency to separate time from space. There are things we do in space and things we do in time, but we do end up unifying them unconsciously. Otherwise, we would never get anywhere on time.

Nuclear Meltdown

There was a young lady named Bright
Whose speed was far faster than light;
She set out one day,
In a relative way,
And returned home the previous night.

This limerick defines a common mistake concerning the space–time continuum. Due to the relative nature of space and time, some people assume that time travel is possible. Most physicists would argue that it's not; others theorize that it is. At the moment, time travel is mostly in the realm of science fiction, but stranger things have come true.

In 1908, one of Einstein's former mathematics teachers, Hermann (no, he's not a rhinoceros) Minkowski, took Einstein's special theory of relativity and developed a very formalized mathematical expression of it. Most advanced physics courses stress this formulation of the theory. Minkowski didn't have many fond memories of Einstein, because Einstein very seldom attended his math classes. He thought that Einstein was basically lazy. Minkowski was quite surprised at Einstein's relativity theory, but Einstein didn't think much of his former teacher's mathematical elaboration of his theory; he felt that it cluttered the idea up too much.

It wasn't until Einstein developed his general theory of relativity that he came to appreciate the mathematics that Minkowski had derived. The mathematics clearly showed that both space and time transform relative to moving frames of reference. This idea was stated in Einstein's original paper, but Minkowski showed the formal mathematics that demonstrated why this was true. In addition, Minkowski's lectures and presentations helped Einstein's relativity theory become accepted by mathematicians.

Max von Laue, whom you met earlier in the chapter, published the first detailed mathematical textbook on the special theory of relativity in 1914. Einstein's lighthearted initial reaction to it is expressed in the following quote:

> *I myself can hardly understand Laue's book.*

This comment reflects another statement made by the leading mathematician of the time, David Hilbert:

> *Every boy in the streets of our mathematical Gottingen [where he taught] understands more about four-dimensional geometry than Einstein. Yet, despite that, Einstein did the work and not the mathematicians. Do you know why Einstein said the most original and profound things about space and time that have been said in our generation? Because he had learned nothing about all the mathematics and philosophy of space and time.*

The union of space and time reflects other similar unions in this book. Electricity and magnetism were brought together by Maxwell and named electromagnetism. Einstein's famous equation brought together mass and energy, and his theory of special relativity resulted in the union of space and time. Things were coming together nicely. Only one force still needed to be fit into the picture: gravity.

The General Theory of Relativity

The idea of general relativity began in 1907. The theory wouldn't be worked out for another eight years. Einstein's theory of special relativity took only six or eight weeks to put together, but the mathematics of general relativity presented a lot of difficulties. This book focuses on the ideas rather than the math.

This theory of general relativity is essentially an expanded version of Einstein's special relativity theory. Einstein explained the relationship between the two theories this way:

> *The theory of relativity resembles a building consisting of two separate stories, the special theory and the general theory. The special theory, on which the general theory rests, applies to all physical phenomena with the exception of gravitation; the general theory provides the law of gravitation and its relation to the other forces in nature.*

Newtonian physics gives only a partial explanation of gravity. Its principles explain how to calculate the attractive force that is caused by the mass of a body. It doesn't explain *why* matter causes gravity.

Einstein's continued search for unifying principles and ideas in physics led him to the general relativity theory, as well as his special theory and his later pursuit of a unified field theory. The main difference between his two relativity theories can be summarized this way:

➤ The special theory deals with frames of reference that are moving in a straight line, toward or away from each other, and are not accelerating, but instead are moving at a constant or uniform speed.

➤ The general theory defines the relationships between objects throughout all of space, moving in any direction, with or without acceleration.

Mind Expansions

Einstein's belief in a unifying principle was the glue that held his inner and outer life together. It was the driving force in his life. It makes you wonder what the source of that belief was. It went beyond merely being an idea; it became the motivation of his life. Maybe he was inspired by the very thing he searched for.

What the general relativity theory does is unify and simplify principles regarding the constant speed of light when it encounters a strong gravitational field. This theory also explains why all bodies, regardless of their mass, fall freely in a gravitational field with the same acceleration. Let's spend some time exploring what all this means.

"The Happiest Thought of My Life"

The special theory of relativity applied only in special cases, in systems with uniform motion between two frames of reference. But how often does uniform motion occur? We're most familiar with nonuniform motion, which is acceleration and deceleration. Accelerated motion didn't appear to be subject to relativity, and this bothered Einstein. For example, remember the thought experiment about sitting in a car with super shock absorbers and traveling at a uniform speed? With the windows tinted black, you can't tell if you're moving or not. What happens if your driver steps on the gas? You're immediately pushed back into your seat. Even without looking outside, you can now tell that you're moving. If you can recognize this fact without a frame of reference, this means that nonuniform motion seems to be absolute motion. It's not relative to anything. Because Einstein had just shown that absolute motion wasn't possible, this inconsistency perturbed him.

In 1907, while still working at the patent office and two years after the publication of his four famous papers, Einstein claimed he had the happiest thought of his life. He had been working on trying to understand how gravity fit in with relativity. Max Planck had warned him not to tackle the problem, because it was too difficult. Besides, he said, no one would believe him even if he did figure it out. It was at this point that the thought came to him. What would he experience if he fell off of a roof? The image led him to the realization that while falling, he wouldn't be aware of gravity. He would be weightless (at least until he hit the ground).

From a height, everything falls at the same rate. Your pet rhinoceros falls right next to you, along with the coins that have come out of your pocket. Remember that Galileo discovered that all objects fall at the same rate, regardless of their mass. Freefall cancels the effects of gravitation. I'm sure you've seen pictures of astronauts in space. Their experience of weightlessness is exactly the same as that of freefall. But they don't flail their arms and scream, because they don't see the rest of the world rushing past them and the ground flying into their faces.

Mind Expansions

The training that familiarizes astronauts with the weightlessness they will experience in outer space is very similar to a ride in a falling elevator. A jet plane with a large cargo area flies in a great parabola, on the back end of which it dives for a few minutes to simulate freefall. The astronauts get a first-hand experience of what it's like to float in mid-air and also get to see their breakfasts freefalling next to them if the weightlessness doesn't agree with them.

Let's take this idea further. Get ready; you really have to let your imagination go with this one. Suppose that you get into an elevator, and suddenly the cable snaps. You scream and begin to freefall, floating up off the floor of the elevator and wondering when your life will come to an end. Unknown to you, a friendly alien magically transports the elevator with you in it into outer space. Your elevator has been converted into a spaceship that is on its way to rendezvous with the alien's spaceship. (Don't worry; this analogy won't end with alien probing.)

You sink to the floor and find yourself feeling pretty normal as your spaceship begins to accelerate. Thinking that somehow your elevator was saved from crashing into the basement floor, you think that you're still on Earth because everything feels the same. You have no idea that you are traveling through space towards an alien space-craft. That's the point of this story. Acceleration and gravitation are equivalent. Not knowing that you're in space, you assume that you're still on Earth waiting for the elevator doors to open. You can't tell the difference between gravitation and acceleration. You don't know that what you're experiencing is acceleration and not gravity.

That was Einstein's happiest thought: You can't tell the difference between gravity and acceleration without a frame of reference. So nonuniform motion is relative, too. This idea is known as the principle of equivalence. With this realization, Einstein knew that to fully incorporate acceleration into relativity he would have to create a theory of gravity. This theory is what the world knows as his general theory of relativity, which is Einstein's theory of gravitation.

Headlines Read: Gravity Bends Light!

With this new realization, Einstein was able to predict a number of events, including the following:

➤ Light and all forms of electromagnetic radiation are deflected or bent by gravitational force.

➤ A clock on the surface of a massive object will run slower than a clock in open space.

➤ Gravitational waves radiate out at the speed of light from large masses that are accelerating.

All three of these predictions have been proven to be true by scientific experiments and observations. However, even though it's been shown that some stars emit gravity waves, detectors here on Earth have never seen them.

Understanding the first prediction is enough to grasp the significance of the theory, so let's get back inside the accelerating elevator that's been transformed into a spaceship. A window has appeared in one of the walls, so now you can look outside. (As you can see, these aliens are rather considerate.) To your surprise and possible dismay, you see nothing but stars and realize that you're no longer in Kansas. A bright light comes beaming in through your window, but something odd has happened to it. Because you're accelerating through space, in the time it takes the light beam to pass from the window to the opposite wall, that wall has traveled a little bit. So the light beam strikes the wall lower than the height of the window. It looks as though the light beam is bent.

Because the *principle of equivalence* states that gravity and acceleration are identical, this must mean that the beam of light that is bent within the confines of the accelerating elevator can also be bent by gravity. Einstein's prediction was verified during a solar eclipse in 1914. The Sun is the closest source of a strong gravitational field, and there are billions of stars whose light can be

Relatively Speaking

The **principle of equivalence** is one of the two founding ideas that make up Einstein's general relativity theory. It basically states that acceleration is equivalent to gravity. Without a frame of reference, it's impossible to tell the difference between the force of acceleration and the force of gravity.

measured against this field. If you mark the position of a specific star at night and then its position during the day, which is why this experiment needs to be done during an eclipse, you can measure the degree to which the light from the star is displaced from its position when the sun moves closer to it. Pretty clever, don't you think?

This Is So Warped

The last, and what is considered by many to be the most brilliant, part of the general theory of relativity is the cause of gravity itself. How does gravity manifest across space? Einstein came to the following conclusion: It doesn't. Gravity is the result of a distortion of space. The distortion is caused by the presence of mass. In other words, gravitation is the direct result of a mass distorting space. This idea and the principle of equivalence are the two concepts underlying the general theory of relativity.

You've gotten so good at doing thought experiments, let's use one more to explain how gravity warps space. What you need for this experiment is a swimming pool filled with clear Jell-O, a bowling ball, and a good-sized marble. You're lying on the end of the diving board, overlooking the pool filled with Jell-O. You roll your big marble across the top of the Jell-O. It rolls across the surface in a straight line. Next, you gently place your bowling ball on the surface, and notice that although the Jell-O is support- ing it, the bowling ball has created a depression in the surface. Take your large marble and gently roll it towards the bowling ball. You will see it start circling around the bowling ball and continue to circle until it eventually spirals in to the center, as if it were in a funnel. If the Jell-O were perfectly clear like space is, you wouldn't be able to see the impression and would assume that something was attracting the marble to the bowling ball.

The distortions in space work just like your Jell-O. Planetary orbits aren't caused by any form of attraction between bodies, but are a direct result of the Sun and the planets themselves distorting space. It gets pretty mathematically complex, but it all works out. His theory was proven to be correct when it was used to define Mercury's unusual orbit around the Sun. There is a small glitch in its orbit that Newtonian theory could never account for, but it was explained using Einstein's gravitational mathematics.

So there you have it. You've delved into the secrets of Einstein's universe and come away with a swimming pool full of Jell-O. But seriously, you now have a working understanding of some of the most significant theories ever developed by the mind of any one individual. Welcome to the heart of Einstein's mind. These theories go far in explaining much of the unusual phenomena in our universe, such as black holes and quarks, which later chapters will cover. But for now, in the next chapter, we'll go behind the scenes and find out a little more about Einstein's personal life.

The Least You Need to Know

➤ Einstein's self-assurance and confidence was often mistaken for arrogance, especially by his detractors.

➤ The famous formula $E = mc^2$ is not the formula for the atom bomb; it just explains the dynamic relationship between energy and matter.

➤ Einstein's "happiest thought" was that in freefall, you aren't aware of gravity. This simple insight led to the development of his theory of general relativity eight years later. This idea is known as the principle of equivalence.

➤ The two cornerstones of the general theory of relativity are the principle of equivalence and the concept that gravitation is caused by the distortion of space.

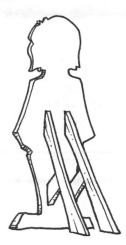

Chapter 15

Behind the Scenes

In This Chapter

➤ The conflict between Einstein's interest in science and his human relationships

➤ Einstein's appointment as a full professor

➤ The Solvay conferences

➤ Einstein's relationship with Elsa

➤ The outbreak of World War I

Congratulations! You now have a pretty good understanding of Einstein's most significant contributions to physics and our world. But these contributions are only part of the picture. A brilliant mind doesn't necessarily accompany an equally developed emotional nature. Excessive focus on mental work can mean that personality issues must be put on hold. It all depends on what your priorities are.

Einstein is often remembered as the absent-minded, kindly old professor of his later years. He was indeed by then a much wiser and more insightful man, who had learned important truths about human nature. That comes with age, hopefully. The young Albert was a less perceptive man who was not as aware of the needs of others, and this lack of awareness was a deliberate choice he made. This chapter takes a closer look at Einstein's life during the years prior to World War II. By the end, you'll have a deeper understanding of Einstein as a person and see how events in his earlier years influenced the man he eventually became.

From Patent Office to Professor

When you last visited Einstein's personal life in Chapter 10, he had gotten married to Mileva in January 1903. Their marriage appeared to be a happy one, for Albert had found an intellectual partner with whom he could discuss his theories. It's even believed by some scholars of Einstein's life that Mileva worked with Albert on his theory of relativity. Here's an excerpt from a letter that he wrote to her in March 1901:

> *You are and remain a sanctuary for me into which no man can penetrate; I also know that of everyone you love me the most deeply and understand me the best. I also assure you that no one dares or wants to say anything bad about you. How happy and proud I'll be when we have carried our work on relative motion together triumphantly to the end! When I see other people, it becomes so clear to me what I think of you.*

Here Albert states that he and Mileva worked on the theory of relative motion together, but this is the only such record indicating that they collaborated. Einstein didn't publish his paper on special relativity until four years after he wrote this letter. There is only one other reference to their collaboration, and that comes from their son, who said that Mileva read Albert's paper on relativity while Albert was sick, checked all of the equations, and enthusiastically proclaimed it was a fine piece of work. She then mailed it off to the editor at the German journal where it was published.

In the spring of 1904, their first son, Hans Albert, was born. Also during that year, Albert finally landed a permanent position at the patent office. Things were looking pretty good. He had a full-time job, a wife, and a son, and he was able to pursue physics in his spare time.

Mind Expansions

Einstein was never one to pass up learning as much as he could about anything he was involved in. This was even true at the patent office, where he learned the correct procedures for applying for patents. In the late 1920s, Einstein took out several patents, one of which was for a noiseless refrigerator. The idea would have worked, but someone else came up with a better design.

During the seven years that Einstein and his family lived in Bern, Switzerland, they moved seven times. No one knows why they moved so often, but it seems that each move either brought them closer to Albert's friends, who were so important as sounding boards for his ideas, or landed them in a completely furnished, more inexpensive place to live. Albert was also supporting his mother at the time, and he and Mileva both loved to take vacations, so it seemed prudent to find inexpensive apartments whenever they could so they could save their money.

It was while all this moving was going on that Albert wrote and published all his most famous work, except his paper on general relativity. In a letter that Mileva wrote to a friend of hers, she said:

> *My husband no longer spends all his free time playing with the boy. To his credit, I have to say that this is by no means his only occupation outside his official work; the treatises written by him are piling up quite frighteningly.*

Time for Something Else

After the publication of Einstein's papers, the world of physics slowly began to recognize his achievements. Two of the most influential figures to appreciate Einstein were Wilhelm Wien, the editor of the German journal that published his papers, and Max Planck. Each of these gentlemen helped Einstein in his quest to pursue more important, fulfilling work.

In April 1906, Einstein was promoted to technical expert, second class, with a 30 percent raise in salary. Of course he didn't receive the promotion for his scientific work, but for diligently performing his patent office duties, even while engaged in his own pursuits. But Einstein knew that a whole new world of possibilities lay before him.

One of the best ways to meet influential professors and important individuals in the academic job market was to attend formal meetings called *congresses* in one's area of specialty. This remains true today. Einstein avoided most of these, and in so doing missed many an opportunity, but he had his reasons. Even though good connections were important, most of the jobs that were offered were as underpaid teaching assistants. Einstein made much more money at the patent office than he could in entry-level academia. Besides, he figured that one day soon the universities would come looking for him, and therefore he needn't play their academic games.

Around June 1907, Einstein applied for a position as a *privatdozent*, or private teacher, at the University of Bern. One of the requirements to be considered for the job was the submission of a formal thesis for review by the faculty members. Einstein included with his application the 17 papers he had written to date, but he was rejected for not turning in a thesis specifically written for the department to which he was applying.

It was also around this time that Einstein was approached to write a comprehensive article on his theory of relativity. He devoted a lot of time to writing it, clarifying ideas and improving the mathematics. This opportunity was a stroke of good luck for Einstein, because the publication of this article presented his theories to a wider audience. It also got him the privatdozent position he had been rejected from earlier. The administrators at the University of Bern decided they could afford to bend the rules a little bit to allow someone of Einstein's caliber to teach for them. Einstein had three students in his class in his first semester. This amount was barely enough to pay for his meals, let alone for a family, so he kept his position at the patent office and held classes in the evenings. The following year he had only one student, so he canceled the class altogether.

Relatively Speaking

The **privatdozent,** which means private teacher, is a position that dates back to the Middle Ages. This strictly European idea has never been adopted in American universities. The privatdozent's fees are paid by the students who attend his or her classes, not by the university.

On to Zurich

The University of Zurich created a new post in its physics department, and some think it was made just for Einstein. It was now 1909, four years after the publication of his famous papers, and news was spreading of Einstein's theories. So in July of that year, Einstein resigned from his post at the patent office and moved his family to Zurich. He didn't make any more money at the University than he had at the patent office, but he achieved his academic dream of teaching full time. He was also awarded his first honorary degree, from the University of Geneva. Over the years, Einstein would receive over two dozen honorary degrees from some of the most famous and prestigious universities in the world. To mention just a few, he was awarded degrees from Oxford, the Sorbonne, Princeton, Cambridge, Harvard, Buenos Aires, and Brussels.

His stay in Zurich lasted only three semesters. He knew that he wanted to obtain a regular departmental chair, rather than just a teaching position. He was never fully accepted by the faculty because he held only an associate professorship, which in those days was seen as a secondary position. Albert and Mileva were very fond of Zurich, but he kept his eyes open for a more substantial move up the academic ladder.

Problems in Prague

The departmental chair of theoretical physics had been vacated at the University of Prague in Czechoslovakia, and Einstein headed the list of possible candidates. To lend weight to his application, which ultimately would be accepted or rejected not by the university but by the imperial and royal authorities, the department asked Max Planck to write a letter of recommendation for Einstein. As you know, Planck was very taken with Einstein's relativity theory, so he wrote this letter to the Emperor Franz Joseph:

Albert Says

Emperor Franz Joseph was the monarch of one of the largest empires in Europe: Austria–Hungary. Fanatical nationalism had been on the rise in Europe for over a century, and even the best scientific minds in Europe had a hard time not succumbing to its atmosphere. One of the main reasons Einstein chose to become a Swiss citizen was to avoid the social pressure of having to publicly voice his political positions.

> *In boldness, it probably surpasses anything so far achieved*
> *in speculative natural science and indeed in philosophical cognition theory; non-Euclidean geometry is child's play in comparison. And yet the relativity principle, in contrast to non-Euclidean geometry, which so far has been seriously considered only for pure mathematics, has every right to claim real physical meaning. This principle has brought about a revolution in our physical picture of the world, which, in extent and depth, can only be compared to that produced by the introduction of the Copernican world system.*

Imagine having your theory compared to Copernicus's by the father of quantum mechanics! That's quite an achievement.

Age-Old Problems

Einstein received an invitation to discuss appointment terms. The fact that he was a Swiss citizen caused no problems, but his nondenominational religious views were unacceptable to the Franz Joseph empire. The old monarchy was suspicious of anyone who had no allegiance to a religious group. How could they be loyal and swear an oath of allegiance to the emperor, if they belonged to no denomination? Einstein relented and stated that he was a Jew on the questionnaire.

This early run-in with the demands of the empire marked the beginning of Einstein's personal and social problems. His scientific career was about to take off, but he would have to pay a price for his fame and love of science.

The move to Prague was difficult for Mileva, who by now had two sons to take care of as she set up a new place to live. Living in Prague began to undermine the Einsteins' relationship. Einstein began spending more and more time away from his wife, engrossed in his pursuit of general relativity theory. This period of time brought Albert's choice to focus on his theories instead his personal relationship with Mileva to a head.

The environment of the University of Prague didn't help at all. The University of Prague was the oldest in central Europe and had become a place of strong hostility between Germans and Czechs. In 1888, the animosity between these two cultures caused a split within the university system into German-speaking and Czech-speaking sections. There was also a significant population of minorities—Jews, Slovaks, Poles, Croats, Hungarians, and Serbs—which complicated the cultural issues.

Nuclear Meltdown

Almost 100 years have passed since the first events that eventually led to World War I occurred at the turn of the century. Yet even today the same sad, dangerous ideologies are held by many of the same ethnic groups that were involved in World War I. Hatred and dogmatic nationalistic and religious beliefs continue to drive cultural groups, who see their way as the only correct way in an increasingly diverse world. Will tolerance and mutual understanding ever take the place of "my way or the highway"?

Albert never cared for the dictates of proper social conduct, especially those that stressed appropriate social behavior and the need to belong to a group. Albert, by design or accident, insulted many of his German colleagues by failing to make social calls, which was expected of a person in his position. Mileva was not accepted by the community either, due to her Serbian heritage. She felt like she was in exile and became more and more depressed as time went on. Her dreams of becoming a new type of woman in a world of men were dashed when she failed to graduate from the Polytechnic in Zurich. In a letter she wrote to a friend, she stated:

I am starved for love, and I almost believe wicked science is guilty.

Her feeling of being a social outcast deepened as more and more of the German community came to reproach her not only for having a questionable cultural heritage (according to the Germans anyway), but also for being married to an unusual man who had renounced his German citizenship. Her husband was a Swiss Jew who chose to be considered Jewish instead of German. (Albert never believed in the old adage, "When in Rome do as the Romans do.") Following are some insightful quotes that show Albert's emotional nature around this time:

> *I have learned to isolate myself from the unpredictability of human relations. Life tends to get clogged up, especially marriage.*

> *I'm not much good with people, and I'm not a family man. I want my peace. I feel the insignificance of the individual, and it makes me happy.*

He would not compromise his love of physics for his love of Mileva. The search for the meaning of gravity became his sole obsession. For his entire life, science would always come first. On one occasion, he said:

> *When I think seriously day and night, I cannot engage in loving chatter. In the same way, one cannot play the violin if he has just been working with a large hammer.*

Mind Expansions

Franz Kafka (1883–1924) lived in Prague for most of his life. He's considered one of the most important existentialist philosophers. He died an early death from tuberculosis and requested that his many stories be burned upon his death, rather than published. However, one of his friends felt that they were important and had them published after Kafka's death. Many of these stories dealt with rising European dictatorships and foresaw the impending economic crises that America and Europe would soon endure.

The Religious Nature of Science

It's a common trait of human nature that what we seek or desire in our lives is that which gives us pleasure, meaning, or fulfillment. For some people, it's material possessions; for others, it comes in less tangible forms. Spiritual or religious traditions mean more to some people than to others. Albert drew an interesting correlation between religion and science. What follows is one of his shorter essays on the topic:

> *You will hardly find one among the profounder sort of scientific minds without a peculiar religious feeling of his own. But it is different from the religion of the naive man. For the latter, God is a being from whose care one hopes to benefit and whose punishment one fears; a sublimation of a feeling similar to that of a child for its father, a being to whom one stands to some extent in a personal relation, however deeply it may be tinged with awe. But the scientist is possessed by the sense of universal causation. The future, to him, is every whit as necessary and determined as the past. There is nothing divine about morality; it is a purely human affair. His religious feeling takes the form of a rapturous*

amazement at the harmony of natural law, which reveals an intelligence of such superiority that compared with it, all the systematic thinking and acting of human beings is an utterly insignificant reflection. This feeling is the guiding principle of his life and work, insofar as he succeeds in keeping himself from the shackles of selfish desire. It is beyond question closely akin to that which has possessed the religious geniuses of all ages.

Although he wrote this in 1934, he later realized that, unfortunate as it was to his relationship with Mileva, his religious feelings born of his scientific inquiry were the motivating force in his life.

The Solvay Conference

Einstein was now a full professor at the University of Prague and a department head. His arrival was announced with headlines in the newspaper. His fame was beginning to spread. Any time he gave a public lecture, the audience was standing room only. At one of these lectures, he met Max Brod, a friend of Franz Kafka. Brod had been working on a novel entitled *The Redemption of Tycho Brahe*. He was looking for someone to model the character of Johannes Kepler after and, upon seeing Einstein, realized he had found his model. Of Einstein, he wrote:

Time and again he filled me with amazement, and indeed enthusiasm, as I watched the ease with which he would, in discussion, experimentally change his point of view, at times tentatively adopting the opposite view and viewing the whole problem from a new and totally changed angle. He seemed to take a downright delight in exploring all possibilities of the scientific treatment of a subject with unflagging boldness; he would never tie himself down; with virtuosity and jocularity, he never avoided any multiplicity and yet always retained a sure and creative grip.

This description of Einstein represents a fine insight into his creative genius. The ability to look at the same issue from a number of different views, and not being attached to any one particular interpretation, reflects Einstein's personal desire to understand, rather than to invest his ego in any one perspective.

A wealthy industrialist by the name of Ernest Solvay and a chemist friend of his, Walter Nernst, had the idea of putting together a summit meeting of the greatest minds in science. It was by invitation only. Travel expenses, hotel accommodations, and food were all paid for. In April 1911, the meeting was held, and of course Einstein was invited to participate.

Albert Says

There isn't enough room in this book to introduce you to all the important people who contributed significant theories to physics. One that I haven't mentioned so far is Hendrik Lorentz, mainly because his theories are almost purely mathematical. But his work had a tremendous influence on Einstein's ability to mathematically investigate and provide evidence for his work on relativity.

Einstein met his old friend Max Planck at this conference, and quite a few other well-known people, including Marie Curie, Ernest Rutherford, Louis de Broglie, James Jeans, and Hendrik Lorentz. In the years to come, Einstein attended over a dozen Solvay conferences, where he continued to see old colleagues and meet new ones.

During this conference, Einstein became good friends with Marie Curie, the only woman in attendance. She, however, got involved in a scandal at the conference. The French newspapers got hold of some love letters she had written to a married physicist, Paul Langevin. Marie's husband Pierre had been killed five years earlier, in 1906, when he was run over by a horse-drawn carriage.

Einstein stood in defense of her actions, because Langevin was planning to divorce his wife anyway. He denounced the publicity, because he felt that what Marie Curie did with her life was nobody's business but hers. This stance would soon be relevant to his own life, as his relationship with Mileva continued to worsen and his involvement with another woman was just around the corner.

While in attendance at the conference, another topic occupied Einstein's attention. His alma mater, the Polytechnic in Zurich, offered him a full professorship in theoretical physics. This position was a definite move up the academic ladder. Although the position was not that different from his previous two, the Polytechnic position was much more prestigious because it was an international institute, instead of a national one.

Einstein decided to return to Zurich, for a number of reasons:

➤ His best friend Marcel Grossman was a professor of mathematics there.

➤ On some level, returning in triumph to the school that wouldn't even consider giving him an assistant position six years earlier was irresistible.

➤ Mileva pressured him to return to a more socially hospitable climate.

Mind Expansions

Einstein's popularity had grown considerably in a very short period of time. His invitation to the Solvay conference was just the beginning of widespread recognition. People other than scientists wanted to meet the boy wonder as well. Einstein met Sigmund Freud and Carl Jung, the founding fathers of two very popular forms of psychology. Neither Einstein nor Jung were able to follow what the other was talking about, so their discussions never lasted very long. Does that mean that psychology and physics don't mix?

Albert and Elsa

Returning to Zurich was like a homecoming for Albert and Mileva. Both relaxed into a more comfortable lifestyle, but Albert was already beginning to think of where he wanted to go next. Less than two years after returning to Zurich, he left the Polytechnic and his 10-year offer to teach. Max Planck and Walther Nernst had been planning as far back as 1910 to offer Albert the directorship of the newly created Institute for Physics in Berlin. Berlin was the place to be in Europe prior to the First World War if you wanted to meet and teach some of the greatest minds in physics.

Planck and Nernst visited Einstein in the summer of 1913 to discuss the position. Their offer was not only for the directorship, but also involved a doubling of Einstein's salary and membership in the Prussian Academy of Sciences. Einstein made up his mind to accept the offer, but he kept his decision to himself. He said he needed time to think about it and recommended that they meet later. If he decided to take them up on their offer, he would wear a red rose, and if not, he would wear a white one. When they next met, Planck and Nernst were very happy to see Einstein's red rose.

But Albert had other motives for moving to Berlin. While he was still in Prague, in April 1912, he had visited Berlin with some friends. During this visit, he had stayed with his cousin Elsa, who was divorced. The only thing she had in common with Mileva was that they were both three years older than Albert. In all other ways, they were quite different. Elsa was not the independent spirit that Mileva once was. She had no interest in science and had strong maternal qualities.

Elsa shared many attributes with Albert:

➤ They had the same cultural heritage and shared many fond memories of their childhood.

➤ Neither had the slightest interest in the latest fashions or social fads.

➤ They both loved sausages. If you're German, you can understand that.

➤ What Albert sought in marriage was domestic comfort, not an intellectual partner. Elsa felt the same way.

Albert and Elsa.

During his 1912 visit, Albert told Elsa of his marital problems with Mileva. He couldn't resist her understanding ways and quickly fell in love. It wasn't until 1993, when their letters were made public, that the nature of their early relationship was revealed. Albert decided by 1914 to end his marriage with Mileva, but he and Elsa had fallen in love two years before.

The period of time between 1908 and 1914 was probably the most intense time of Albert's life, at least in terms of emotional upheaval and creativity. His obsession with the general theory of relativity kept him away from Mileva, and this drove them farther apart. Yet at the same time, he still sought emotional refuge with a female presence, and his first cousin Elsa filled that need.

Mind Expansions

Hans Albert, the oldest son of Mileva and Albert, had a good education in Switzerland. He came to the United States, where he taught engineering at the University of California at Berkeley and died in 1973. Einstein's niece still lives in Richmond, California today. Of all the places that Albert lived, the Golden State was one of his favorites.

Mileva still felt that she could salvage their marriage, but it was already too late. With the outbreak of World War I, Mileva stayed in neutral Switzerland while Albert continued on in Berlin. When Albert informed her in 1916 that he wanted a divorce, Mileva underwent a physical and mental collapse. It took her several months to recover. Their divorce was finalized in 1918, and Elsa and Albert married in 1919. Their marriage proved a happy union for both of them and lasted for the next 20 years, until Elsa's death.

Albert pledged the money he received for his Nobel Prize to Mileva for her financial support and the support of his two sons. Mileva lived in Zurich, the city where she and Albert first met as students back in 1896, for the next 34 years, until her death in 1948.

The Stupidity of War

Shortly after Einstein moved from Zurich to Berlin, World War I broke out. Europe changed from a relatively peaceful world into a troubled, embittered battlefield of nationalism, which continues to be played out in some areas even today. Modern warfare was thought to be capable of such sweeping destruction that it was thought the war would end in just a few months. The speed at which the German armies swept through Belgium seemed to lend credence to this concept. But when the Battle of the Marne halted the advance of German troops at the gates of Paris, no one could have known that the war would last for another four years.

Einstein initially spent this period in earnest protest against the war, but when he saw that his resistance had no impact, he returned to his research and creativity. It was during this time that he put the finishing touches on his general theory of relativity.

As you know, from his earliest childhood Einstein disliked Germany's tendency toward a militaristic attitude. He had deeply rooted pacifist beliefs that would be a major political focus all his life. So when Germany put forth its Manifesto to the Civilized World, Einstein refused to have anything to do with it. This decree was put forward by the government to raise public support for its military efforts. It was circulated by prominent German intellectuals to defend against accusations that Germany had committed atrocities in Belgium and also to justify their invasion of Russia. Ninety-three well-known Germans signed this document, including Max Planck. Einstein refused. In a letter he wrote to a friend, he said:

> *Europe in her insanity has started something unbelievable. In such times, one realizes to what a sad species of animal one belongs. I quietly pursue my peaceful studies and contemplations and feel only pity and disgust.*

In reaction to the first manifesto, Einstein helped draft a response called the Manifesto of Europeans, which focused on the commonality of European countries and asked for the formation of a European League to counter the war. But his efforts were to no avail.

The advent of the First World War ends the coverage of this portion of Einstein's life. Upcoming chapters further explore his life during World War II and his life in the United States. Let's return to physics by jumping into the world of quantum mechanics

The Least You Need to Know

➤ Albert's singular passion for physics was more important to him than his relationship with his wife Mileva.

➤ It would appear that Einstein's ambition to seek more and more prominent positions in the academic world was the result of a need to be recognized. Not true. What he sought were jobs in which he could pursue his research and teaching and, of course, receive a little more money!

➤ What Einstein wanted from marriage was domestic comfort, not an intellectual partner.

➤ Elsa fulfilled Albert's need for a life companion and at the same time gave him the freedom to pursue his love of physics.

➤ Einstein was very outspoken against Germany's involvement in the First World War.

Part 4

Anybody Know a Good Quantum Mechanic?

No field in the study of physics is more bizarre than quantum mechanics. If you put 10 physicists in a room and asked them each what quantum mechanics was about, you would get at least 8 different answers. Why? Because quantum theory can't be explained. We can examine how it was developed and show how it is applied, but beyond that, it's like trying to make sense of Alice in Wonderland. *That's an unusual statement to make about science, which prides itself on clear-cut methodology. But you'll understand why quantum mechanics is so weird in a little while.*

The mathematics that define the principles under which quantum theory operates are based on probability rather than certainty. The reason physicists continue to use quantum theory, even though they can't explain why it works, is because it yields very accurate results. Remember that scientific systems are used because they work, not necessarily because they make sense. Many systems of thought explain how *something works, but not* why.

In this part of the book we're going to spend some time in Wonderland and see how the world of quantum mechanics becomes curiouser and curiouser. Things will not always appear as they seem, and just when you think you have it figured out, they will disappear and come out the other end of the rabbit hole. Let's begin our chase after the white rabbit by conducting a brief review of the subatomic world and then follow with the development of newer theories to bring us up to date.

Chapter 16

What Does an Atom Look Like?

In This Chapter

➤ Ernest Rutherford's model of the atom

➤ The discovery of subatomic particles

➤ Defining the quantum leap

➤ The theory of matter waves

Although Einstein's most significant works were already completed, he continued to actively participate in investigating the tiny world of the atom. Many advances in the theories of quantum mechanics occurred because of friends of his, and he vigorously debated their ideas and theories with them.

This chapter takes a look at how different models of the structure of the atom developed. Each of these models became outdated as more and more information about the nature of the particles that make up the subatomic universe was discovered.

Henry Ford? No, Rutherford

Remember J. J. Thomson and his discovery of the electron in Chapter 11, "A Warm-Up to Relativity," He used a CRT (cathode ray tube) in conjunction with a magnetic field to focus and steer a beam of electrons, and in the process he discovered their electrical charge and mass. Do you remember Thomson's model of the atom? He pictured it as something like a tiny bit of raisin pudding. Embedded within this pudding were tiny electron raisins. The number of electrons depended upon what the substance was. For example, Thomson thought that a hydrogen atom, which was the simplest atom

Mind Expansions

Models are powerful tools to help solve physics problems. Many physicists believe that a problem is not fully understood until there is an accurate visualization and a simple set of formulas to describe the visual model. Stephen Hawking formulates theories almost entirely by geometrical visualization before he dives into the math. Richard Feynman developed a methodolgy for solving problems in quantum theory based on diagrams, called Feynman diagrams, which are still used and taught in classes on advanced theoretical physics. Understanding, like symbolism, is a multilevel interpretation of underlying meaning. This simply means that there are a lot of ways of looking at any concept.

Relatively Speaking

The **nucleus** is the central part of the atom. It contains two fundamental particles called the **proton** and the **neutron**. The nucleus carries a positive electrical charge equal to the total negative electrical charge of the orbiting electrons. Almost the entire mass of the atom is contained in the nucleus.

known, had one electron raisin, which had a negative electric charge. This charge was somehow balanced out by a positive electric charge.

An atom in its normal state is electrically neutral. So in addition to the negative electric charge of the electrons, there must also be an equal amount of positive charge somewhere in the atom. Thomson said that the additional positive charge in the atom was distributed in the 'pudding part' of the atom. No one had discovered any positively charged components within an atom though. The mass of the electrons was found to be negligibly small in comparison to the total mass of the atom. So it followed that the positively charged part must represent almost all of the atom's mass, and thus should be most of its bulk as well.

Thomson's idea of raisin pudding was the first model of the atom. This model was difficult for physicists of that time to accept, because they didn't believe that there could be anything smaller than an atom. There was no experimental evidence to support Thomson's theory, at least not until Ernest Rutherford came along.

Ernest Rutherford (1871–1937) was a New Zealander who came to England in 1895 to work at the famous Cavendish laboratory. Three years later, he left for Montreal where he worked for the next nine years. In 1907, he returned to England, which is where he made most of his significant discoveries.

While in Montreal, Rutherford performed a number of experiments on radioactivity, that strange and unusual phenomenon discovered by Marie Curie and her husband Pierre. Rutherford named the two types of radiation given off by radium *alpha* and *beta*. He soon realized that the beta radiation consisted of the electrons that Thomson had recently proposed. But he had no idea what alpha radiation was. After a series of experiments, he found that alpha radiation was composed of positively charged particles. These particles could penetrate through thin layers of solid material, such as the glass walls of a tube. With this information, he decided to investigate the internal structure of atoms by shooting these tiny, positively charged particles through them.

In 1907, with the help of his assistant Hans Geiger (name sound familiar?), he developed an apparatus to study the effects of alpha particles on various substances. They bombarded a thin piece of gold foil with alpha particles emitted by a piece of radium. Behind the gold foil, they placed a glass plate coated with a chemical that would emit a flash of light every time it was hit with a particle. Geiger later refined this apparatus into the first successful detector of alpha particles. You know this instrument as a Geiger counter.

Almost all of the particles passed through the gold foil, but every once in a while, a particle veered off to the side or even bounced straight back toward the source. After many long hours in a darkened room observing the deflections, Rutherford and Geiger estimated that 1 in 8,000 particles rebounded from the sheet of gold foil. But what was that 1 hitting that the other 7,999 particles missed?

After accumulating an abundance of data, Rutherford realized that the entire atom wasn't positively charged, as in the pudding model. Instead, the positive charge appeared to be concentrated in a very tiny central part of the atom called the nucleus. This nucleus was much smaller than the entire atom, but astonishingly, it accounted for almost all of the atom's mass. As a matter of fact, the nucleus is only one ten thousandth the size if the whole atom. Rutherford named the positively charged particles in the nucleus *protons*. Protons, along with electrons, were the second building block of the atom.

For about 15 years, it seemed that all atoms were composed of these two particles, and therefore that the entire structure of the universe was constructed of electrons and protons. This was by far the simplest concept of the structure of matter ever devised by humanity. It was much simpler than thinking that the universe was composed of the 90 elements of the periodic table.

The model put forth by Rutherford was known as the planetary model, because the structure of the atom was described as being similar to that of our solar system. The nucleus is like the Sun, and the electrons orbit the nucleus like the planets. However, the atom is almost entirely empty space. If you were to take an atom and blow up its nucleus to the size of a golf ball, the atom would be the size of Earth, with an electron the size of a grain of rice in orbit. Matter is mostly just empty space. It appears solid only because the group of electrons orbiting each atom resists intrusion by the electrons of adjacent atoms, so each atom has its own space. The speed at which the electrons

Mind Expansions

Rutherford discovered an interesting correlation between the number of protons in the nucleus of an atom and its atomic number. Mendeleev had arranged the periodic table of elements based on the atomic weight of each atom. An atomic number of each element was assigned based on its sequence in the table—nothing more significant than that. However, Rutherford found that the atomic number was exactly the same as the number of protons in the nucleus and the number of electrons in orbit. For example, gold, which has an atomic number of 79 because it was number 79 in the periodic table, also happens to have 79 positive protons in its nucleus and 79 negative electrons in orbit. Rutherford found this to be true for the atomic number of every element on the table. Believe it or not!

orbit the nucleus is also what gives matter its solidity. This, at least, is what was thought at this time.

A couple of problems couldn't be accounted for in this particular model. The main problem was stability. If the electron was negative and the nucleus positive, why didn't the nucleus draw the electrons toward the center? Remember, opposite forces attract. Although Rutherford's theory explained some things about the structure of atoms, it didn't look like the final answer.

Nuclear Meltdown

The planetary model put forth by Rutherford and Bohr is the most commonly held concept of the atom. It's still taught that way in most high schools and introductory college courses. Most people don't know that it was dropped as a true representation of the atom about five years after Niels Bohr received the Nobel prize for his theory in 1922. It is at heart a classical mechanical model of the atom, and quantum mechanics doesn't deal with classical mechanic explanations, but with probability, uncertainty, and discontinuity. The relativistic quantum mechanical explanations are too difficult for many people to grasp clearly, so the older model still has some use. However, the fortunate readers of this book will be among the select few who know how the quantum world of the atom operates.

The other problem was with the way atoms released energy. If electrons were just whirling around in a cloud, something like the whirling of a fan blade, and constantly changing direction and speed, the atom would have to devote itself to this orbiting and would not be able to emit any energy. At least it couldn't emit energy continuously, because that would spell disaster for Rutherford's model. Any planet that would give up energy continuously during its orbit would, according to mathematical prediction, spiral into the Sun. That would mean that every time the electron emitted its light energy, it would crash into its nucleus, and all matter would collapse. This obviously doesn't happen, so again the model didn't appear to be complete.

If the electron couldn't emit energy continuously, then how was it able to radiate light? For light or any other form of electromagnetic radiation to be emitted, the electron would have to radiate energy at some time.

He's Such a Bohr

Enter Niels Bohr (1885–1962). Bohr was a native of Denmark. He came to England on a scholarship to work with J. J. Thomson. He and Thomson didn't get along very well, because Bohr told Thomson that his theoretical model of raisin pudding was incorrect. Thomson had a hard time with anyone who criticized his theory. So Bohr left and went to work as an assistant for Rutherford. He learned quite a lot from Rutherford, and the two decided that they needed to develop a better picture of the dynamics of atomic theory for Rutherford's model to be fully accepted.

Bohr came up with a theory based on the notion of quanta put forth by Max Planck and Albert Einstein. Remember that light is emitted in chunks of energy called quanta. Bohr theorized that the emission of quanta is a basic property of all types of energy. If this theory were correct, it would explain why atoms could emit light and yet not collapse. Bohr came up with two great explanations for the stability of the atom and for the emission of energy:

> ➤ Electrons travel in fixed orbits around the nucleus, with each orbit at a specific distance from the center.

> ➤ A quantum of energy is released or increased when an electron jumps from one orbit to another. This jumping is where the familiar term *quantum leap* comes from.

The Old Shell Game

Bohr spent a lot of time studying the periodic table of elements, to see whether he could recognize a correlation between the structure of atoms and the periodicity of the properties of the elements. He found that electrons arranged themselves in orbits, or shells, one outside of the other. As each shell becomes full, the next electron occupies the next furthest shell. Bohr developed the following sequence based on the maximum number of electrons in each shell:

Shell number:	1	2	3	4	5
Number of electrons:	2	8	18	32	50

The shell game involves something called *angular momentum*. Angular momentum can be thought of as momentum moving in a circle. To understand what this means, imagine that you have a ball attached to a string. When you hold the loose end of the string and swing the ball around like a cowboy ready to rope a calf, you are using angular momentum to keep the ball swinging in a circle. Bohr's electrons were traveling in an orbit around the nucleus, exhibiting angular momentum, just like the swinging ball. The invisible rope that keeps the electron in its orbit is the electrical attraction between the opposite charges of the electron and the nucleus, which counteracts the centrifugal force that would tend to send the electron flying away from the nucleus.

But how much energy was in the electron's angular momentum? Bohr thought that Planck's constant might make a good unit of measurement for the electron's energy. To illustrate what this energy is, go back to the swinging ball for a minute. If you swing the ball faster, you'll feel a greater force pulling on your hand. The faster you swing the ball, the greater its angular momentum is.

Relatively Speaking

Angular momentum is a property of rotating objects, just as plain momentum is a property of objects moving in a straight line. The angular momentum of a spinning object depends on its mass, size, and the speed at which it is spinning. In classical physics, it defines the movement of any object on a curved path. In quantum physics, angular momentum is quantized, which means it can change only in whole chunks.

Now picture the electrons in orbit about the nucleus. Mathematically, the speed of the electron can be determined if the force holding it in its circular path and a given fixed amount of angular momentum is known. The radius of the orbit can also be determined. In other words, if you compare this to the analogy with the swinging ball, if you knew how much force you were exerting to swing the ball, along with the mass of the ball, you would know the angular momentum. From this you could calculate the length of the rope, or radius of the circle, and also the speed at which the ball was swinging in its circle around you.

Using this information, Bohr was able to calculate the diameter of each electron orbit, along with the maximum amount of electrons that each orbit or shell could hold. The electrons in each successively larger shell had greater angular momentum. The model worked. Bohr had explained why the structure of the atom remained stable.

The only other concept that needed an explanation was how the electron was able to radiate energy. This discovery goes back to the nature of the quantum. Remember that quantized energy is emitted only in whole chunks, not just pieces of a chunk. This form of energy is not continuous, but discontinuous.

Bohr had to come up with a principle to describe how the electron radiated light. The theory that he developed was similar in nature to his explanation of how the stability of the structure worked. He simply made it up! No one knew how any of this really worked, so many theories were just guessed at. There was no physical evidence to explain why the electron stayed stable, but Bohr's theory worked. So he tried out another idea to explain how an electron emitted light, and in the end, that one worked well, too. The electron emitting the light was not doing so by oscillating or orbiting around the nucleus. So Bohr theorized that the electron jumped from one shell to another in orbit around the nucleus. The energy that it emitted while performing this jump was released in quanta, or whole chunks. This quantum leap occurred as the electron jumped from one place to another and this release of energy was the radiation that was emitted in the form of light.

Bohr's theories accounted for the stability of the atom, as well as how atoms emitted light. But the model of the atom was far from being complete. There were still a number of unanswered questions:

➤ What causes the electrons to make the jump from orbit to orbit?

➤ Is the electron a wave or a particle?

➤ What forces hold the nucleus together?

➤ Is there more than one type of particle in the nucleus?

Rutherford had predicted back in 1920 that another particle would be found inside the nucleus. Sure enough,

Relatively Speaking

The **neutron** is one of the three elementary particles found in the atom. It resides in the nucleus along with the proton. It has a neutral electrical charge and almost the same mass as the proton.

in 1932, the *neutron* was discovered. The neutron had no charge; it was electrically neutral. But it did help account for the mass of the nucleus.

The Subatomic World

First of all, what does *subatomic* refer to, other than some form of evil death ray invented by a mad doctor of science fiction? The word *atomic* refers to things composed of atoms. If you explore what the atomic world is made up of, then you're dealing with pieces of matter smaller than atoms. These smaller pieces are therefore known as *subatomic*. If you were to go even smaller and examine what the subatomic world is made up of, then you would be in the sub-subatomic universe. (Yes, in a little while you'll be going there, too.) Everything just keeps getting smaller and smaller. In a way, the atomic universe was found to be something like the little bottle on the table in *Alice's Adventures in Wonderland* that said, "Drink me." When physicists did, things became smaller and smaller, just like Alice.

What have you learned about atoms so far? The atom is made up of three different kinds of particles:

➤ The electron, which is the negatively charged particle that orbits the nucleus

➤ The proton, which is the positively charged particle in the nucleus of the atom

➤ The neutron, which is the neutral particle in the nucleus of the atom with the proton

Do you notice anything odd about the arrangement of these particles? In the nucleus, positively charged particles are being held together. But like charges are supposed to repel each other, so why don't the protons repel each other? What is holding them together? The answer to that would have to wait until a binding force, which was neither electrical nor gravitational in nature, would be discovered.

How are these particles related? And what are their relative sizes?

➤ The proton has a mass 1,836 times greater that that of the electron.

➤ The neutron has a mass 1,839 times greater than that of the electron.

➤ The diameter of the nucleus is about one hundred thousandth of the diameter of the whole atom.

➤ The volume of the nucleus, the space it takes up, is one trillionth of the whole atom.

As you can see, even though the nucleus is very small compared to the whole atom, it still accounts for almost all its mass. But as you can also see, the atom is virtually empty space. Its total size is only a description of the cloud-like space through which its electrons vibrate and move with tremendous speed.

To put these sizes in a better perspective, imagine that you could enlarge the diameter of Earth to an incredible size. Say the diameter of Earth was now equal to the distance from the Earth to the Sun, which is 93 million miles. If you were to increase the size of all of the atoms in proportion to this expansion, the nucleus of the largest known atom would *still* be microscopic. You might be able to see it with a magnifying glass, but it would still be tiny.

Do you think that any of this is a little weird? The most fundamental building blocks of matter are mostly empty space. They are almost surreal, with an illusory quality.

Let's keep looking at how the understanding of the structure of the atom continued to change.

Waves of Matter

Mind Expansions

Some of Niels Bohr's theories were confirmed through a series of experiments that proved that the high-energy light given off when electrons move from a high-energy state to a low-energy state was in the form of x-rays. A mathematical correlation exists between the x-ray frequency given off by atoms of each element and the element's atomic number. If the frequency of the x-rays is known, the amount of electrical charge in the nucleus can be determined. This measure of x-ray energy is specific for each of the elements and is commonly used to identify the elements that comprise an unknown substance. It's analogous to a fingerprint. This measure is unique for each element.

Chapter 12, "Albert's World of Friends, Photons, and Molecular Motion," discussed the dual nature of light. Remember Thomas Young's experiment in which the interference patterns caused by light beams passing through slits in a board indicated that light behaved like a wave? Then Einstein came along and said that light is made up of particles that he called photons. Some physicists remained unconvinced that light was made up of particles, so they performed some experiments to see what the heck light was.

By using a number of photographic plates exposed at different rates, physicists could see the particle nature of light revealed. If you take a photographic negative and expose it for the proper length of time, you get a fully developed positive picture. However, if you expose the negative to a very small amount of light, the print will show an unrecognizable pattern of a few dots. If you continue to expose the print to successively greater amounts of light, the dots begin to statistically group themselves into to bare outlines of images. Eventually, with increased exposure to light, the dots become overwhelming, and the final print is obtained. Each dot represents the absorption of a photon at a specific point, confirming that light is made of particles.

The dual nature of light has been one of the central paradoxes of twentieth-century physics. As you will see, this dual nature applies to other things as well. Niels Bohr even developed a *principle of complementarity* that included the entire subatomic world. Complementarity just states that you can't define the nature of one thing without reference to its complementary opposite. Good and bad, up and down, left and right—all of these complementary concepts can be understood only in terms of their relationship to each other. What does up mean without down, or good without bad? You can't have one without the other; otherwise, you have no frame of reference to compare a concept to.

Within 10 years of the proposal of the Rutherford-Bohr theory of the structure of the atom, a French physicist named Louis de Broglie came up with some reasonable explanations for Bohr's theories. de Broglie was convinced that Einstein's relativity theory was correct, and therefore he believed that mass and energy were equivalent. Because both mass and light are understood as forms of energy, he felt it might be possible to describe them in the same terms.

Einstein had shown in his description of the photoelectric effect that light exhibited both wave-like and particle-like properties. de Broglie reasoned that matter must exhibit both wave and particle properties, too. He then took this a step further and argued that because Einstein had shown that there was an intimate relationship between space and time (remember four-dimensional space?), there must also be a deep connection between energy and momentum. By doing some simple algebra with Einstein's equations, he was able to show that the momentum of the photon was equal to Planck's constant divided by the photon's wavelength. What this meant was that as the wavelength of light decreases, its momentum increases.

Because de Broglie was able to define the relationship between the energy and momentum of light, and light and mass are both forms of energy, he reasoned that this relationship must work for matter as well. He thought this was especially true of the electrons in an atom. De Broglie believed that electrons get from one place to another by acting like a wave, just like the wave nature of light determines how photons get from one place to another. So by doing just a little math, de Broglie inverted the formula for the relationship between wavelength and momentum and used the velocity of the electron (because momentum is mass times velocity) to calculate the electron's wavelength.

Relatively Speaking

The **principle of complementarity** was developed by Niels Bohr to explain the dual nature of some quantum phenomena. It basically states that waves and particles represent complementary aspects of the same physical phenomenon. You also can apply this principle to the world in which we live. The idea is that the world is almost entirely composed of dualities: male/female, positive/negative, convex/concave, light/dark, and so on. This concept lies at the heart of Chinese philosophical understanding of the universe and manifests as yin and yang. You've probably seen the symbol of a circle with light and dark swirling halves, each having a dot of the opposite color that looks something like an eye. That's complementarity. They are not opposites, but each is a codependent half of the whole.

According to Bohr's theory of the atom, a certain number of electrons can exist in each orbit or shell. De Broglie turned his attention to how his theory of the wave nature of electrons could be related to Bohr's concept of each orbit being able to accommodate only so many electrons before it was full. Remember the discussion of blackbody radiation that used the analogy of a sitar player in a room? During that discussion, you were given an explanation of what a standing wave was. Standing waves are complete waves that move at a frequency that makes it appear as though they're standing still.

De Broglie pointed out that if a whole number of standing waves could fit exactly around the circumference of an orbit, as the electron circled around the orbit under the influence of its wave nature, the wave would sustain itself in this pattern. If, however, a whole number of standing waves couldn't fit the orbit, as the waves moved around in their orbit they would cancel each other out, and the orbit would collapse. Basically, what this means is that the reason only certain orbits are allowed in an atom is because the wave nature of the electrons can only establish a stable wave pattern for certain orbits. For all other possible orbits, the wave pattern can't be stabilized. If you're having problems visualizing what this might look like, take a look at the picture of de Broglie's wave pattern.

De Broglie had developed this theory for his Ph.D. dissertation, and when it was completed, he had it published. The faculty committee that reviewed his thesis certainly thought it was original—perhaps a bit too original. The study of the atomic and subatomic world was a relatively new branch in physics, and there was no experimental data to back up his unusual ideas. Bohr's ideas were pretty far out to begin with, and to have a Ph.D. candidate come along and explain why some of Bohr's theories worked was just a little too much for the conservative faculty members, so they called in Einstein to evaluate de Broglie's paper. Einstein welcomed de Broglie's theory:

> *I believe de Broglie's hypothesis is the first feeble ray of light on this worst of our physical enigmas. It may look crazy, but it is really sound!*

Thanks to Einstein's comments, de Broglie received his Ph.D.

After a long while, experimental evidence of Bohr's theories began to pop up in England, Germany, and the United States. Waves, as you know, have a property known as diffraction. Through a number of interesting experiments it was discovered that electrons could be diffracted in the same way as x-rays. While working with crystals, which are good at diffracting all types of wavelengths, scientists found that it was possible to measure the wavelengths not only of x-rays, but also of electrons. The wavelength of a beam of electrons could be calculated from the velocity of the beam, the mass of the electron, and de Broglie's wave theory equation.

Albert Says

As discussed in previous chapters, advances in one area of science often have unpredictable effects on other areas of science and human endeavor. A new invention can be responsible for the discovery of a principle that wouldn't have been found if the technology had not been developed. Or a theory may be proven to be correct because of a new invention. The experiments done to show the existence of waves of matter led directly to the invention of the electron microscope, which is designed according to principles advanced by Broglie's wave theory. These microscopes in turn showed physicists the world of molecules, and for the first time, people were able to see the things they previously had only theories for.

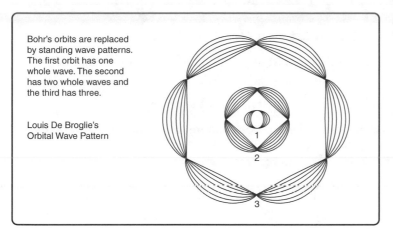

Bohr's orbits are replaced by standing wave patterns. The first orbit has one whole wave. The second has two whole waves and the third has three.

Louis De Broglie's Orbital Wave Pattern

Louis de Broglie's orbital wave patterns.

That's just about it for de Broglie's wave theory of matter. To summarize, matter and light are both forms of energy and can both be described in terms of the propagation of waves as well as in terms of particles moving under the influence of various forces. In the same way that Einstein viewed light waves as the means to get photons from one place to another, matter waves were the means to get particles, in de Broglie's case electrons, from one place to another.

Matter waves became very widely accepted, so much so that physicists began to question whether particles even existed. They thought that maybe if they could get two waves to interfere with each other, they could produce a particle. A lot of questions still needed to be answered. Although de Broglie's theory helped explain how Bohr's orbits were formed, it didn't address other questions, such as:

➤ What is continuously going on inside of the atom?

➤ How does an electron change from one orbit to the next?

➤ What allows the electron to radiate away its excess energy as light?

Newton's laws of motion and any principles based on his laws wouldn't work to calculate the motion of particles or the wave nature of matter. New principles and equations were needed. The picture of the atomic universe was still incomplete and needed help from a few more innovative minds. The individuals who contributed to elucidating the nature of the quantum world and took us closer to understanding the unusual characteristics of this world will be the focus of the next chapter. Our journey through Wonderland is about to become a little more weird.

The Least You Need to Know

➤ The model of the atom that is most widely taught is the planetary model, developed by Rutherford and adopted by Niels Bohr.

➤ The nucleus of the atom contains nearly all of the atom's mass.

➤ Positively charged particles called protons and neutral particles called neutrons form the nucleus of the atom.

➤ An atom is made up mostly of space.

➤ Because mass and light are both forms of energy, and light can be understood as a particle or a wave, Louis de Broglie theorized that matter could be seen as a particle or wave also.

Chapter 17

Can We Really Know Anything for Sure?

In This Chapter

➤ Erwin Schroedinger's wave function

➤ The development of probability theory

➤ How paradoxes operate

➤ The central mystery of quantum mechanics

➤ Is what you perceive really there?

So far in this book, you've been introduced to a few unusual ideas and theories, but nothing too bizarre. But this chapter goes right to the core of the inherent uncertainty of the quantum world. It goes deeper into the wave-particle duality and discusses how the observer in an experiment changes the outcome just by the act of observing what happens.

Our journey through Wonderland, which began in the last chapter, will now take us inside a universe that is spooky, goes against common sense, and is filled with paradoxes. Be prepared to leave your everyday experiences behind and enter the world of Alice and the hookah-smoking caterpillar.

Schroedinger's Wading Pool

At the beginning of the twentieth century, it was firmly believed that light was an electromagnetic wave and electrons were particles. But within the next 30 years, experimental evidence showed that light was made of particles called *photons* and electrons moved in waves. So which is it? You'll be surprised to know the answer.

You're almost there, but you and Alice still have a few more tea parties to attend first. Let's meet Erwin Schroedinger and his three-dimensional matter waves.

Schroedinger was an Austrian mathematician who published a paper in 1926 on the propagation of matter waves. Chapter 16, "What Does an Atom Look Like?" introduced Louis de Broglie's theory of matter waves, which offered a glimpse inside the atom. However, more of an explanation was needed. Some model or theory was still required to account for the shifting patterns of waves that produced light by changing their energy. Neither Niels Bohr's quantum-leaping electrons nor de Broglie's wave patterns explained how light came from these changing wave patterns.

de Broglie's wave theory was basically one-dimensional: It described waves traveling just around the circumference of an orbit. They didn't move up and down or left and right, only back and forth. That's what a violin or guitar string looks like when it vibrates. Dropping a pebble in a pond produces two-dimensional waves. They not only ripple out in one dimension, but they also move up and down, which adds a second dimension. Schroedinger developed a mathematical equation that described wave motion as being three-dimensional.

The best way to picture a three-dimensional wave is if you drop a number of pebbles in a pond and watch the waves moving toward each other. When the waves run into each other, they begin moving in different directions, giving a turbulent, choppy look to the surface. You now have waves going up, down, left, right, and back and forth. This is something like what Shroedinger's three-dimensional waves look like, but not exactly. It's very difficult to accurately describe three-dimensional waves, but that didn't matter, because his wave equation was just an abstract mathematical function. In other words, Shroedinger's equation couldn't describe a real model that anyone could imagine—it was just a mathematical *function* that described how mathematical waves moved.

So what's a mathematical function? Let's use another concrete example. If you've ever been in a shallow wading pool with a bunch of little kids, you may have noticed that some parts of the pool are warmer than others, because kids pee in the pool. As you move around in the pool, you notice that the temperature is different between the cooler spots and the warmer spots. If you were to express this mathematically, temperature would be a function of location in the pool. Over time, the temperature will change at any given point in the pool, so temperature is also a function of time. So temperature in this example is a function of space and time.

Schroedinger's wave was also a mathematical function of space and time, just like the temperature in the wading

Relatively Speaking

A **function** in mathematics is a quantity that depends on another quantity or quantities. For example, a quantity like temperature may depend on another quantity, like time. Suppose you're heating some water. The temperature of the water when you first turn on the gas is 75 degrees. After three minutes, the temperature becomes 90 degrees. The temperature of the water depends on the amount of time that has gone by, so in this case, temperature is a function of time.

pool. It described the possible location of an electron within a three-dimensional wave. These waves took the place of Bohr's orbits or shells, just like de Broglie's wave did. Schroedinger's waves, however, accounted for three dimensions inside the atom, rather than just one like de Broglie's waves did. An important concept here is that these shells are not completely distinct from one another; they penetrate and overlap one another. The electrons are in something like clouds around the nucleus, rather than in distinct shells.

There was one main problem with Schroedinger's wave function. No one knew how to look for the warm spots and cold spots in the wading pool or, in other words, the crests and the troughs of the waves.

Analyzing big systems (more than two interacting objects) has always been a problem. Conceiving of the atomic orbitals of plutonium in the context of the solar system model would have been just as difficult as any complicated problem in quantum mechanics. In fact, the so-called 'three-body problem' in classical mechanics, which is three objects orbiting around each other held together only by the force of gravity, has never been solved! People have worked on this problem for over 100 years and still can't solve it exactly even with classical mechanics. If physicists at the time objected to the idea of an entire atom being represented by a single wave function—well, that's all you can really say about it. Superconductors and superfluids are described by a wave function that contains 10^{26} particles.

An objection (or mystery) to Schroedinger's wave equation is that you need to square it to get the probability for a particle's location (or momentum). No one really knows why this is the case even after 80 years of scratching our heads about it. But no one objects to the fact that it works!

Even though no one could visualize what it looked like, Schroedinger's wave function worked like a charm. It explained many types of physical phenomena at the quantum level. His theory showed that an electron was a wave, or at least that it appeared to behave like one. It also explained why atoms are stable.

Think of the change in the electron wave as the change in sound when you pluck one guitar string and then another. The two notes together create a third sound, which is a harmony between the two individual strings vibrating at the same time. Schroedinger's theory stated that light was produced or energy was emitted as a result of a harmony between the upper and lower vibrations of the electron waves. Gradually the upper

Mind Expansions

Erwin Schroedinger was a very smart physicist and a notorious womanizer. He was often inspired in his physics work by his most recent love interest. In 1925, during the Christmas holidays, he developed the most significant theory of his career during a passionate involvement at his favorite romantic hotel in Austria. He had been thinking about waves. (Which makes one wonder, did this hotel have waterbeds?) So sex is intimately related to science. It requires chemistry, of course it's physical, and it inspired one of the great theories of physics.

frequency matter wave tone becomes quiet, leaving only the lower one still vibrating. Light stops being radiated, and the atom just continues to vibrate its electron wave at the lower frequency.

Waves of Probability

These ideas are all leading up to some very curious concepts. Like Alice, you're still just exploring this strange new world, so just because the mushroom you've eaten hasn't kicked in yet, be patient. The Cheshire cat is about to swallow himself up!

Schroedinger's wave function described many types of atomic phenomena. It answered important questions about the movements of electrons and the spectrum of light emitted from a hydrogen atom. However, there were some types of experiments that it couldn't explain.

Max Born had been performing collision experiments at his institute in Germany. These experiments were similar to those that first led J.J. Thomson to the discovery of the electron. With much more sophisticated equipment, Born was able to conduct detailed studies of collisions between atoms and electrons. Even though it had been shown that electrons move in waves, Born's collision experiments gave convincing evidence that electrons were particles.

Schroedinger's wave function equation worked for electrons confined to orbits around the nucleus of an atom. In Born's experiments, electrons were able to move freely in space, controlled by magnetic fields that focused a beam of electrons at a target. Born realized that no one could locate a single electron within a beam of electrons. He tried applying Schroedinger's equation, but it didn't work.

Maybe it was time to develop a new concept of what a wave was. Born came to the conclusion that a wave was not a real particle. A wave was just a probability in space for finding an electron. It was a probability function.

Probability functions are used to describe the distribution of likely occurrences. For example, when you toss a coin in the air, the probability function for it to land tails-up is .50. Once it lands, the probability changes. If it lands heads-up, the probability function becomes 0. If it lands tails-up, the probability function becomes 1.

Probability is used a lot in our everyday lives. Insurance companies use it to describe the distribution of car accidents. In metropolitan cities, the probability of any one car colliding with another is much higher than in the country, where there is less traffic and no bike messengers. So the probability density, or distribution,

Relatively Speaking

A **probability function** is a specific type of mathematical function. **Probability** is the number of times something is likely to occur over a range of possible events. Time is a common function of probability. It allows one to predict how many times a certain event may occur over a period of time. Max Born used space as a function to predict the probable location of a single electron.

of car accidents is lower in the country. If you could view an entire state from high up in the air, you could watch all the cars driving about and predict where the collisions are most likely to occur.

Born pictured the flow of electrons in his experiments in much the same way that we imagined to view of the cars when looking down on an entire state. Wherever there was a greater concentration of electrons, Born predicted a higher probability of a collision between an atom and an electron.

But what happens when you want to predict the probable location of just one electron? What determines where the electron will be found? Is this just all a game of probability? The answers to these questions will plunge you into the mysteries of the quantum world. The piece of mushroom that you ate in Wonderland is about to kick in.

Albert Says

Visualizing a paradox is a good way to understand the contradictory nature that lies at the heart of a conceptual paradox. No one was better at doing this than M.C. Escher. His ability to incorporate contradiction into his art gave his work the feeling of illusion. He tricked the viewer into seeing his world as a topsy-turvy, surreal realm where anything was possible. If you've never seen any of his posters or drawings, you should check them out!

The Paradox of the Box

One of the most important and most paradoxical concepts of quantum physics regards the role of the observer in experiments dealing with the subatomic world. Let's examine this role a little more closely.

We often encounter paradoxes as ideas or concepts. Jokes are a perfect example of the use of paradox. What is it about a joke that makes you laugh? Normally a joke is a story that relates a series of events. At the end comes the punch line. If it's a good punch line, delivered with good timing, you laugh. You laugh because you're given an idea or concept that you don't expect. The series of events that unfolds doesn't seem like it should lead to that conclusion. It's inconsistent with what your mind has led you to anticipate, given the nature of the previous chain of events. The more absurd the conclusion, the funnier the joke is. That's one form of a paradox.

Here's another paradox: Take out a piece of paper and write the following message on one side in nice big letters: *The statement on the other side of this paper is true.* Now, turn the paper over and write this message in the same nice big letters: *The statement on the other side of this paper is false.* Read either statement to yourself, and then read the other side of the paper. Try to reason it out for yourself. Keep going back and forth until your brain turns to mush or until you realize that you're caught in a paradox.

Jokes and the paper trick are examples of mental paradoxes. Is it possible to observe a paradox out in the world? Not usually, unless of course you're in Wonderland with Alice. Why not? Because when you observe something, you see either one fact or its opposite, but not both at once. To illustrate how this works, look at the cube in the following figure. Which side is facing you? You may see the left upper square in front,

219

as if you were looking up at the cube. Or if you change your perception, you can see the lower right square as being in front, as though you were looking down on the cube. Because you're the one doing the observing, you control which way you see the cube. It's either one way or another; you can't see it both ways at the same time. You choose how you want to resolve the paradox.

The paradox of the box.

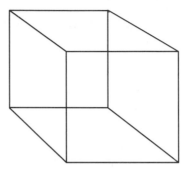

You can look at this box in yet another way. The way the illustration is drawn makes it appear to be a three-dimensional cube. But in reality it isn't. It's really just 8 points connected by 12 lines on a flat surface. When your mind perceives it as a cube, you're forced to make a choice about which face is in front and which is in the rear. However, when you don't see it as a cube, but just a pattern of connecting lines, you have no choice to make regarding how you perceive it.

In the illustration's flat, abstract form, both front surfaces are visible at the same time. When you view the illustration as though it were a cube, you mentally change this two-dimensional drawing into a three-dimensional cube. The very act of your observation creates the concept in your mind that this is a cube. So you have two basic ways of seeing this drawing: either as a flat, abstract picture of connecting lines or as a three-dimensional cube with a front and rear face. Only when you shift your perception to see the drawing as a cube does the image become paradoxical. The mind is conditioned to see things as being solid, three-dimensional objects, and when you apply this perception to the box, you end up creating the paradox.

This idea of creating a paradox comes directly into play when physicists begin examining quantum events. In the next couple of pages, you'll begin to grasp exactly how weird things can become and how important the role of the observer is. Next stop: uncertainty!

The Double-Slit Experiment

Richard Feynman was a brilliant theoretical physicist who was never content with what he knew or what others knew. He ceaselessly questioned scientific truths and became a legendary figure among his fellow physicists. He worked at Los Alamos on the first atomic bomb and was awarded the Nobel prize for work that gave physicists a new way to describe and calculate the actions of subatomic particles. Feynman felt there was a mystery lying at the core of quantum physics. Feynman said of the double-slit experiment:

> ... *it encapsulates the central mystery of quantum mechanics. It is a phenomenon which is impossible, absolutely impossible, to explain in any classical way and which has in it the heart of quantum mechanics. In reality, it contains the only mystery ... the basic peculiarities of all quantum mechanics.*

So what exactly is this mystery? In order for you to understand it, you need to have a brief review of Thomas Young's double-slit experiment.

Young's Experiment Revisited

In the double-slit experiment, light is shone onto a surface with a small hole or narrow slit cut in it. The light passes through this hole, and shines on a second surface that has two holes or slits cut in it. The light spreading out from these two holes finally lands on a third flat surface, making a pattern of light and dark. This pattern of light and dark is called an interference pattern and is the result of light spreading out from the two holes (which is called diffraction) in a series of overlapping waves.

These patterns of light and dark are a result of the nature of how waves interact. If you throw two rocks in a pond and watch what happens to the waves, some waves run into each other and cancel each other out, and others will build up together to form an even larger wave. The light pattern works that way, too. Where light waves cancel each other out, there is darkness, and where the waves combine to increase in size, there is a patch of light.

An interesting phenomenon occurs with these patches. The brightest spot of light is not behind either of the two holes, but exactly halfway between them, behind a spot where there is no hole at all. Mathematically speaking, this spot is precisely four times brighter than the light passing through either hole alone. As you move away from this bright spot, the intensity of each successive patch of light decreases. This pattern of decreasing intensity from one bright spot to progressively less bright spots is a classical wave pattern.

Mind Expansions

Richard Feynman was a philosopher extraordinaire, so it might be best to let his own words give you an idea of who he was:

> You see, one thing is, I can live with doubt and uncertainty and not knowing. I think it's much more interesting to live not knowing than to have answers which might be wrong. I have approximate answers and possible beliefs and different degrees of certainty about different things, but I'm not absolutely sure of anything and there are many things I don't know anything about, such as whether it means anything to ask why we're here ... I don't have to know the answer. I don't feel frightened by not knowing things, by being lost in a mysterious universe without any purpose, which is the way it really is as far as I can tell. It doesn't frighten me.

Now this isn't the way you would expect a stream of particles to behave. If you were to stand in front of a wall with two holes in it and threw stones through the holes, you'd end up with a pile of stones behind each hole, right? There would be no interference. So if light were made up of particles, you would expect to find two bright patches of light behind each of the holes, with no interference pattern.

But very curiously, if you carry out the exact same experiment using a beam of electron particles instead of a beam of light, you get the same results. The electrons fired in the double-slit experiment produce an interference pattern. No, this isn't the big mystery—that's still coming up.

If electrons fired through the holes make an interference pattern, they must be traveling as waves. But as each electron arrives at a particular spot on the surface, it makes a single point of light. Remember in the previous chapter that photographic plates showed how light behaved as a particle when they were exposed slowly? This is the same type of surface that was used in the double-slit experiment with electrons. When the electrons landed on the final screen, they arrived as particles. This experiment shows that quantum particles travel like waves, but arrive as particles. These two experiments suggest that that light can be both a wave and a particle at the same time.

Nuclear Meltdown

From the scientific point of view, facing a paradox that has no solution is always difficult. Many scientists thrive on the unanswerable questions that they often face; others refuse to even consider them because they go against what is assumed to be true. Even when innovative solutions are offered, these solutions might go too far outside the box of accepted beliefs about how the world operates to gain widespread acceptance. The arrogance of science is that it thinks it has a monopoly on objective truth. But fortunately, there will always be free thinkers who question the certainty of assumptions. This constant questioning is the wisdom of science. The difficulty lies in being able to question what you accept to be true.

Those Smart Electrons

We now come to the central mystery of quantum mechanics. The previous experiment may have caused a little worry for physicists, but they could explain it as a statistical effect due to the large number of particles in the beams. However, when single particles are fired through the slits one at a time, the pattern they create is allowed to build up slowly.

A single particle travels through the experiment, making its own single spot on the final screen. It would be reasonable to assume that each particle must pass through only one of the two holes. But as more and more spots build up on the screen, the pattern that emerges is the classical interference pattern for waves passing through both holes at once. Thus, electrons not only seemed to be able to pass through both holes at the same time, because they created the pattern by interfering with themselves, but they also seemed to have an awareness of past and future. They made their contribution to the overall pattern in just the right places to build up the pattern. It was as if each particle knew how the pattern looked before it arrived and which spot it should land in to help complete the overall pattern.

If you think that's a little weird, it gets even stranger. If you set up a detector at each hole to find out exactly which hole each particle goes through, all of the weird behavior discussed in the previous paragraph disappears. Your detector registers just one particle going through each hole, one at a time. The pattern created on the screen is just two blobs of light with no interference pattern, just like when you throw stones through two holes in a wall. These particles seem to know when they're being watched! They adjust their behavior according to how the experiment is set up and whether you're watching them.

If you set it up again with no detector, sure enough, the mysterious wave pattern appears. Then if you decide to set up the detector again, and watch as it counts each particle as it goes by, no interference pattern forms. You just get two blobs of light. How do they know how to behave? And why are they being so secretive? These are just single particles. What's going on? Nobody knows, that's why this is the great mystery of quantum physics. This unexplainable behavior applies to all subatomic particles, not just photons and electrons.

Particle Predictions

An experiment called the delayed-choice experiment originally began as a thought experiment, but it later was actually performed. It's basically the same as the double-slit experiment, except instead of placing the detector by the hole to count each particle as it goes through, the detector is set up halfway between the two holes and the last screen. This way you can see which route a particular particle took after it has already passed through the hole, but before it reaches the screen.

Based on quantum theory, if you turn off the detector and don't look at the particles, you should end up with a typical interference pattern. But if you look at the particles to see which hole they pass through, even after they've already passed through one hole or the other, you end up with blobs of light and no interference pattern. This means that your choice whether to look at a photon after it has already passed through the hole, it seems to affect how the particle previously behaved at the time it was passing through the hole, a fraction of a second in the past.

Back in the 1980s, this experiment was performed independently at the University of Munich and at the University of Maryland. The experiment was set up, and the decision whether to use the detector was made while the particles were already in flight. The particles in these experiments exhibited similar behavior. When the detector was on, there was no interference pattern, but when it was not used, the interference pattern appeared. This implied that the particles knew whether a detector would be used before they passed through a hole. Particles with precognition?

In these experiments, the distances are so short that the time scales involved are only a few billionths of a second. But imagine that you could do this experiment on a cosmic scale by using the light from a distant object outside of our galaxy, which has reached you by two different routes. This occurs as a result of something called *gravitational lensing*.

223

Relatively Speaking

Gravitational lensing is a term that applies to the effects of gravity on light waves. Einstein theorized in his paper on general relativity that a strong gravitational field could bend light waves passing close to it. His theory was proven correct, and the term gravitational lensing was developed as a direct result. It occurs when light from a distant astronomical object is bent around a galaxy on its way towards Earth, so that you see two or more images of it.

Gravitational lensing occurs when light is bent around a massive galaxy that it must pass in order to reach you. A number of different paths of light are available, and you can choose any two to create an interference pattern. This pattern would show that the light on both paths reached you as a wave. You could also monitor each photon, which would then give you blobs of light and no interference pattern. These distant objects outside of our galaxy are at least 10 billion light years away, which means the light has taken 10 billion years to get here from its source.

According to the lesson you learned from the double-slit experiment, you would expect that your choice to watch or not watch the photons to affect how the photons appear on your screen before they get here, which means that they would have to have known whether you would detect them five billion years before our solar system was even formed. If this particular delayed-choice experiment is ever carried out, it would show, incredible as it may seem, that the past of the quantum world is influenced by factors associated with how we choose to conduct our observation now. (The Red Queen just shouted, "Off with their heads!")

What You Can Know

Quantum physics teaches that not only do you observe reality, but you are also a participant in it. Observation is not a passive experience. You probably never thought that by the act of just watching something, you can alter its behavior. But are you really altering it or merely perceiving that you are? This tricky question needs closer examination.

Our upbringing has conditioned us to perceive the world as something that happens outside of ourselves, regardless of whether we're aware of it. We usually think that we interact with the world only through our bodies, our five senses. Perception is assumed to be an experience of what is real. Yet on the quantum level, which is the very essence of everything that appears to be real, things don't appear to be as real as the objects in our everyday world. Maybe our everyday perception of reality is only one possible way of experiencing what we think of as the real world.

Remember the paradox of the box? The moment you chose how to look at it and the three-dimensional box appeared, what happened to your awareness of choosing it? At the instant that you noticed the box, you projected its form in your mind. You suddenly realized that it was a box and experienced yourself as being separate from it. The picture of the box in your mind suddenly became the box "out there." This all happens so quickly that you're not even aware of it. It's almost automatic. It's how our minds are conditioned to operate from a very early age.

Imagine that you could never see the box as an abstract, two-dimensional image of 8 dots and 12 lines. Suppose you were preconditioned to see it only as a three-dimensional box. You either see it as though you were looking up at it or as though you were looking down at it. There's no intermediary way of looking at it, other than these two perceptions. The cube seems to jump from one appearance to another without you doing anything but observing it. Yet the cube changes because of the act of observation. After a while, you try to figure out why it keeps jumping back and forth, looking for some kind of hidden reason.

It could very well be that what goes on at the quantum level is similar to what happens with the cube. We're preconditioned to observe the jumping back and forth of the nature of quantum particles because we don't know of any other way to perceive them. We've lost touch with the fact that perception occurs inside of us first, before it occurs "out there." Why? Because we choose to spend our time out there, not within.

That's enough for now. I still have some more bizarre things to tell you about in the next chapter. And of course Einstein has a thing or two to say about all this. For now, go back to your reality, take a deep breath, and watch out for the Red Queen.

Albert Says

So many of our perceptions happen so automatically that we're not even aware of how we perceive what we do. A great example is when you read comics. (Yes, those graphic stories that your parents said would rot your mind.) You're presented with a series of pictures that tell a story. But it's impossible to include every act and piece of information from scene to scene, so your mind fills in the needed information to provide a continuous story. You've put something in that actually isn't there, but needs to be there so you can understand what's going on.

The Least You Need to Know

➤ Not all quantum theories can be visualized as models. Some can be expressed only as mathematical equations.

➤ It's not possible to know the exact location of an electron, only its probable location.

➤ Paradoxes can help you understand unusual behavior in the quantum world.

➤ Quantum particles seem to have the ability to choose how they want to appear to us.

➤ How we perceive the world is conditioned within us. There may be other ways of experiencing the world that we're not aware of.

Uncertainty Is Certain

You would think that a question like, "What is reality?" belongs in a freshman philosophy class. However, it's also a fundamental question that is asked about the quantum world. Can everyone agree on what reality is? Or is it like beauty, found in the eye of the beholder?

This book has covered some pretty unusual ideas about the nature of subatomic particles. Electrons can't be located at any particular point in a nuclear orbit. The electron exists as a standing wave around the nucleus, but you never know exactly where it is—you know only the probability of where it can be found. The determined world of classical physics has given way to chance at the quantum level. Particles also seem to know when they're being observed, and the perception of this behavior may be a result of how our minds are conditioned to comprehend the world.

Now that you've had some exposure to a few of these strange concepts, the journey through Wonderland ends with coverage of a couple more of the most significant theories in quantum universe today. Einstein comes back into the picture, and you'll hear his reactions to these mysterious quantum phenomena. Let's get weird again.

The Principle of Uncertainty

Many of our beliefs about the present, such as that it is a product of the past, are based on the philosophy of *determinism*. In other words, who you are today is determined by your past history. It seems that we live in a world of causality, in which all events are connected by a cause-and-effect relationship. Are there any actions that you can think of that aren't the result of previous actions? Yet as we delve deeper and deeper into the microcosmic world, the very particles that make up trees, rocks, planets, stars, and our bodies don't seem to behave in the same causal way that the rest of the world does.

This discrepancy between the quantum world and the world in which we eat, sleep, work, and play brings up some important questions about the nature of reality:

Relatively Speaking

Matrix mechanics is a mathematical system developed by Werner Heisenberg. It was the first complete, self-consistent description of quantum mechanics. Matrix mechanics described states of quantum entities by using a matrix of numbers. The easiest analogy is to picture a checkerboard. By assigning numbers to each square and also to each checker, you can mathematically define the state of the game at any given time according to the correlations between the numbers on the squares and the numbers on the pieces. The same thing can be done to locate the probable position and other properties of subatomic particles or waves.

➤ If things operate so strangely in the subatomic world, why doesn't this have a greater impact on the larger world?

➤ What is the relationship between these two worlds?

➤ Is the fact that the quantum world doesn't appear to follow the same principles as the larger world merely due to a shortcoming in how we perceive the quantum world?

➤ Is our perception of the larger world incorrect? Does the larger world in fact operate under the same principles as the quantum world?

These questions have yet to be answered, because knowledge of the quantum world is incomplete. The mathematical principles used in quantum mechanics very accurately describe how subatomic particles interact, but they don't tell us why. Until we know why and can explain the strange behavior of the quantum world, these questions will remain difficult to answer.

Before I ask too many more questions without answers, let me add to your increasing uncertainty by highlighting the contributions of another physicist, Werner Heisenberg, to the understanding of the workings of the quantum world.

Matrix Mechanics

In May 1925, Heisenberg came down with a killer case of hay fever. He went to an island in the North Sea to recuperate. While he was there, without hay fever or distractions, he formulated the first complete, self-consistent description of quantum mechanics and called it *matrix mechanics*. This theory is too complicated to discuss completely in this book, but it is important within the chain of ideas in quantum mechanics.

Around this time, the understanding of the quantum universe was in a very confused state. It was a hodgepodge of ideas, hypotheses, theories, and principles. Nobody knew how to put it all together. The main problem was that every problem in quantum particles could be solved only by first working out a similar problem in classical physics, and then adjusting the answer by using formulas that represented quantum behavior. This process had to be massaged constantly until the solution matched experimental results. Heisenberg's matrix mechanics introduced a whole new way of working with the behavior of quantum particles that didn't require any manipulation or the use of classical mechanics.

Although this step was important in itself, the reason why Heisenberg developed matrix mechanics in the first place is just as important. Heisenberg felt that any physical theory should concern itself only with things that can be observed experimentally. We can't see atomic particles, their orbits, or waves. We can observe the transition from one state to another, but not the states themselves. Energy is released during the transition, and this energy can be measured and observed.

In Heisenberg's view, worrying about what happens during the transitions and thinking about orbits are just remnants left from classical images of planetary models. Heisenberg was a hard-liner. He didn't care how the particles got from place to place. All he cared about was the mathematical relationship between the two states of quantum particles. His focus on these two states led to his *uncertainty principle*, a cornerstone theory of quantum mechanics.

Mind Expansions

Heisenberg and Bohr were very strongly influenced by a philosophy called **positivism.** This philosophy states that sense perception is the only admissible basis for human knowledge. What we know is simply what we sense. Both Heisenberg and Bohr reflect this philosophy in their interpretation of the quantum world. Since their theories are widely taught and accepted, does that mean that positivism is correct? Science may think so—after all, what makes science scientific is its ability to explain what we sense. But is that all it does? What do you think?

Position and Momentum

Heisenberg realized that the simultaneous measurement of certain pairs of subatomic particles had a fundamental limitation. In essence, he discovered:

➤ There's no way to accurately pinpoint the position of a subatomic particle, unless you are willing to be very uncertain about the particle's momentum.

➤ There's no way to accurately determine the particle's momentum, unless you're willing to be very uncertain about its position.

➤ There is a fundamental limit in nature to how accurately both position and momentum can be measured.

It's not that you can't measure both the position and the momentum of something like an electron at the same time because of deficient measuring instruments. The electron simply doesn't have both a precise position and a precise momentum at the same time. At any given moment, the electron itself can't know both where it is and where it is going.

To make this idea a little clearer, let's take another look at the double-slit experiment. When an electron is passing through one (or both) of the slits (depending on whether you try to detect it), there is a substantial amount of uncertainty about its position, but less uncertainty about the direction it's going in. After the electron has been detected on the final screen, there's very little uncertainty about its position, but now there's considerable uncertainty about the path it took to get there. Which slit did it come through? Did it come through both slits?

Are you uncertain about this uncertainty? A simpler way of stating the principle of uncertainty is: If you observe it, you disturb it. Remember the principle of complementarity from Chapter 16, "What Does an Atom Look Like?" This principle sheds some more light on uncertainty.

Complementarity refers to the dual nature of quantum particles. It means that they can be described either as a particle or as a wave. Position is a property of a particle. When you see the spot of light an electron makes on a screen, you know exactly where it is, but not how it got there. A wave lacks a clearly defined position, but it has a well-defined direction of motion. From this perspective, you know that when an electron is traveling, its position is uncertain even though it knows where it is going.

That's why you can never know both the position and momentum of a particle at the same time. Position is a property of a particle, but momentum is a property of a wave. You can't observe a quantum entity like a particle as a wave and as a particle at the same time. It's like trying to see both sides of a coin at the same time or the cube and its inside-out version at the same time. You can't. That's what the principle of complementarity is. Both waves and particles define what an electron is, in the same way that both sides of a coin define what a coin is. Each can be viewed or experienced independently of the other, and both are required to define the whole thing, but you can't experience or observe them both at the same time.

The Copenhagen Interpretation

Are you ready for even more weirdness? Good! Uncertainty, complementarity, probability, and the collapse of a wave function combine into something called the *Copenhagen interpretation.*

You're entering the 1920s, a very heady time in physics. A new group of enthusiastic physicists was coming onto the scene to question the older guard. Max Planck was past

60, Albert Einstein was already in his 40s, and Niels Bohr was around 35. Bohr was heading up a brand new institute of physics in Denmark called the Copenhagen School. Werner Heisenberg was among many bright young men who would influence the development of physics over the next 60 years.

In 1927, the fifth Solvay conference brought together the great minds of physics for a showdown the likes of which the world had never seen. The attendees comprised a comprehensive who's who in the world of physics. Niels Bohr and Albert Einstein led the list of major figures, but the attendees also included:

➤ Louis de Broglie

➤ Werner Heisenberg

➤ Max Planck

➤ Erwin Schroedinger

➤ Max Born

➤ Wolfgang Pauli

➤ Enrico Fermi

The debate between Einstein and Bohr is coming soon. Before I go into that, let's get back to the Copenhagen interpretation. Niels Bohr is given most of the credit for advancing this idea, because as the name implies, it came from his school in Copenhagen. Bohr combined a number of different principles and ideas into a workable package that really didn't have any one underlying theory to explain how the whole thing worked. This combination has never been set down in any coherent fashion, although it's always presented as an introduction to quantum theory and weirdness. It was, however, a consistent theory that helped physicists come up with answers to some very difficult questions about the quantum universe. For a long time, no one cared about how strange it was because it worked.

Relatively Speaking

The **Copenhagen interpretation** is the standard explanation of what goes on in the quantum world. It is still taught in most textbooks and university courses, but it is by no means the only way to interpret quantum mechanics. Up until the 1980s, it was the main theory, but since then other more interesting theories have been given equal treatment. However, it is important to understand the Copenhagen interpretation because it is one of the cornerstone theories of quantum mechanics. Very simply put, it states that there is no meaning to the objective existence of a quantum particle unless it is observed. Until you see it, it does not exist. This rather unusual theory led Einstein to ask, "Does the Moon disappear once we stop looking at it?"

One of the key aspects of the Copenhagen interpretation is that it stresses the role of experimentation to understand the quantum world. The only thing we can really know, Bohr insisted, is what we measure with our instruments. Consequently, the answers we get depend on the questions we ask. The questions we ask are influenced by our everyday experiences. When we set up an experiment to measure the momentum of an electron, we obtain an answer that we interpret as representing momentum. We can set up another experiment to measure something else, and we'll get an answer

Albert Says

Physicists fall into two main categories: experimenters and theorists. Of the two, you'll usually find that theorists ask more questions about the weirdness of interpretations of quantum behavior. Many experimenters who are involved in the workaday world of physics could care less about why these theories are so strange. For them, the only thing that counts is that the theories work, because they help to derive very accurate results from their experiments, and for this breed of physicist, that's the whole point.

Mind Expansions

Einstein hated the whole idea of action at a distance, which refers to the Copenhagen interpretation explanation of how two waves light years apart could know which box was being opened so that the correct box's wave could collapse into a particle. This concept suggests that this knowledge would have to travel faster than the speed of light, and there's nothing that moves faster than that—at least not in this reality.

that we again interpret to be what we're looking for. In other words, all we can really know is that if we probe a system a certain way, we'll get certain answers.

The answers that we get will be reflected in our meters, dials, detectors, and any other instruments we use for measurement. But such measurements belong to the world of classical physics, and classical physics can't accurately describe the quantum world. So according to Bohr, the only way we can measure anything is by probing the quantum world with our instruments of classical physics, thereby disturbing it. This leads to a basic idea of the Copenhagen interpretation: It is meaningless to ask what atoms and particles are doing when we aren't looking at them.

Let's go back once again to the double-slit experiment. If we apply the Copenhagen interpretation to it, we would conclude that there is no meaning to the concept of the objective existence of an electron at a point in space, for example at one or both of the slits in the experiment, independent of observation. The electron seems to spring into existence as a real object only when we observe it. Therefore, reality is, in part, created by the observer.

Now consider an empty box. Common sense tells you that an electron in the box would have a specific location, even if you don't know exactly what it is. The Copenhagen interpretation would say that the electron exists as a wave filling the box and could be anywhere inside. The moment you look in the box for the electron, the wave function collapses into a particle at a specific location. This is similar to what happens to an electron in the double-slit experiment. When the electron is observed, it stops behaving like a wave and becomes a particle, according to the Copenhagen interpretation.

If you put a partition into the box without looking, common sense tells you that the electron must be in one half of the box or the other. The Copenhagen interpretation would say that as long as you don't look, the electron wave occupies both halves of the box. When you look, the wave will collapse into a particle in one side of the box or the other. As long as you don't look, the wave fills both halves of the box, even if the two halves are separated and moved far apart. You could move the two halves of the box light years apart, and still only when we look at the electron will the wave instantaneously collapse and the electron decide which half of the box it's in.

The idea is that the electron is in both halves of the box at the same time, but appears only in the half that you look at. This spooky thought was called action at a distance. Doesn't it all sound a bit strange? Some physicists thought it was downright absurd.

Schroedinger's Cat

Erwin Schroedinger was very upset about the Copenhagen interpretation. He felt that his wave equations had been misused and wished that he had never developed them. To show the absurdity of the Copenhagen interpretation, he came up with the following thought experiment. This experiment is a little bizarre, but then again, you're still in Wonderland and everything here is a bit on the weird side. So let's take a look at Schroedinger's famous cat. (Note: No actual cats were harmed during the creation of this experiment.)

Imagine that a live cat is put in a box with a radioactive source, a Geiger counter, a hammer, and a sealed glass flask containing a deadly poison. When radioactive decay takes place, the Geiger counter triggers a device that releases the hammer, which falls and breaks the flask. The poison fumes will then kill the cat.

Now suppose that quantum theory predicts a 50 percent probability of one decay particle each hour for the radioactive source. After an hour has passed, there is an equal probability that either the cat is alive, or that the cat is dead. The Copenhagen interpretation would predict that exactly one hour after the experiment began, the box contains a cat that is both alive and dead—a mixture of the two states. This is the equivalent to what is called the *superposition* of the two wave functions, if you were to apply this analogy to the quantum world rather than the feline world.

As soon as you lift up the lid of the box to find out the outcome, your act of observation collapses the superposition of the two wave functions to a single one, making the cat entirely alive or entirely dead. Schroedinger felt that this experiment proved his point. How could the cat be alive and dead at the same time? Regardless of how strange it seems, the Copenhagen interpretation is still taught as a way to understand quantum theory and the superposition of states.

Albert Says

One of the problems of quantum physics deals with defining exactly what is meant by **reality.** To Niels Bohr, reality was nothing more than a word. It had no bearing on what reality was. To Bohr, reality meant something totally different from what it meant to Einstein. It's so ambiguous that talking about it is as hard as getting everyone to agree on what it means. Like a symbol, it has many levels of interpretation. On its most fundamental level, it simply pertains to what is real. But this definition leads people into a very lengthy discussion about how to define what is real. Not everyone can agree on that, either. So you're left with a conundrum that begs to be resolved, yet there seems to be no solution. Is reality totally relative?

233

Some physicists thought that human consciousness played a central role in determining the nature of reality in the Copenhagen interpretation. Other physicists could have cared less. Why should a theory that worked be questioned? It gave accurate, practical answers to many theoretical questions. But others weren't so easily convinced. Albert Einstein was one of the skeptics. Let's take a look at what he thought about all this.

The Einstein-Bohr Debates

Einstein and Bohr met on a number of occasions at the fifth Solvay conference. During these meetings and after listening to many presentations on quantum theory, Einstein remained unconvinced of the validity of quantum mechanics. He understood the theories, but he felt that they provided an incomplete picture. Bohr was shocked and disappointed with Einstein's reaction. He had been confident that Einstein would accept his interpretation, because it was tied to experimentation. Einstein had used this method to defend his theory of special relativity, which challenged common sense just as much as Bohr's theories.

Einstein said:

> *I do not like the probability theory and believe the path followed by Born, Heisenberg, and yourself is only temporary, of heuristic value, so to speak.*

In a letter to Schroedinger, Einstein wrote:

> *The Heisenberg-Bohr tranquilizing philosophy, or religion, is so delicately contrived that, for the time being, it provides a gentle pillow for the true believer from which he cannot very easily be aroused. So let him lie there.*

God's Dice

One of Einstein's most famous quotes, which he was heard to repeat over and over, was, "God does not play dice with the world." He just couldn't believe that the most fundamental properties of the universe were based on probability and uncertainty. He set out to disprove Bohr and the Copenhagen interpretation by identifying apparent paradoxes in the theory.

Einstein was a down-to-earth thinker, so he phrased his arguments in terms of hypothetical experiments designed to point out holes in Bohr's interpretation. For example, Heisenberg's uncertainty principle had defined certain limits about what could be known about a quantum object. Einstein agreed that nature may put practical limits on what we can know and measure, but felt that these limits didn't imply that a deeper, objective reality didn't exist.

Einstein's thought experiments were presented to Bohr, who after a few days of thought could always find a flaw in the argument. No matter how ingenious Einstein's proposal, Bohr always identified a flaw. Bohr responded to each of Einstein's thought experiments, one after another, even after Einstein's death. Bohr's friends pointed out that Bohr began each day reviewing all of his arguments in real or imaginary dialogues with Einstein to make sure he had left nothing out.

In 1948, Bohr wrote a summary of the discussions he had with Einstein over the years. At the end of the work, he wrote:

> *Whether our actual meetings have been of short or long duration, they have always left a deep and lasting impression on my mind, and when writing this report I have, so to say, been arguing with Einstein all the time even when entering on topics apparently far removed from the special problems under debate at our meetings.*

The thought experiments and arguments that Einstein presented are subtle and carefully thought out. Bohr recounted in his summary one of their most celebrated discussions, which took place at the Solvay conference in 1930. Einstein described a box full of light and suggested that both the energy of a single photon and the time it was emitted could be determined precisely. The box had a number of clocks and scales set up inside it. Einstein said that first the box needed to be weighed. After that, a single photon could be released at a particular instant through a shutter operated by a clockwork mechanism inside the box.

Then the box could be weighed again. Knowing the change in mass, he could calculate the energy of the photon by using his famous equation, $E = mc^2$. The change in energy would then be known, as would the precise time when the photon was emitted. Einstein felt that this example was the end of the uncertainty principle.

After a sleepless night, Bohr discovered that when the photon was released, there would be a recoil, causing uncertainty about the position of the clock in Earth's gravitational field. This uncertainty would then produce a corresponding uncertainty in the time recorded, due to—and here's the catch—the fact that clocks run slower in a gravitational field, as Einstein had shown in his general theory of relativity. Einstein had forgotten his own theory in trying to catch Bohr in one of his thought experiments.

These debates lasted for over 30 years, and neither the physicists nor their theories would budge. In 1948, Max Born summarized the attitude of many physicists when he wrote of Einstein:

> *He has seen more clearly than anyone before him the statistical background of the laws of physics, and he was a pioneer in the struggle of conquering the wilderness of quantum phenomena. Yet later, when out of his own work a synthesis of statistical and quantum principles emerged which seemed acceptable to almost all physicists*

Nuclear Meltdown

Bohr and Einstein could never resolve their differences regarding the theories of quantum mechanics. Once they even attended a party thrown by mutual friends, but Einstein and his friends stood on one side of the room, and Bohr and his followers stood on the other side of the room. The entire evening went by with neither group talking to the other, something like a junior high school dance. Paul Ehrenfest knew both men very well and witnessed many of their discussions. His torment over his own conflicting feelings about who was right was one of the main factors that led to his suicide in 1933.

he kept himself aloof and skeptical. Many of us regard this as a tragedy—for him, as he gropes his way in loneliness, and for us, who miss our leader and standard bearer.

Just before Born wrote this, he received a letter from Einstein which stated:

In our scientific expectation we have grown antipodes [exactly opposite]. You believe in a dice-playing God and I in perfect laws in the world of things existing as real objects, which I try to grasp in a wildly speculative way.

The EPR Paradox

There's one more thought experiment worth mentioning in relation to Bohr and the Copenhagen group. It was formulated by Einstein, Boris Podolsky, and Nathan Rosen and came to be called the EPR paradox (after their names). Let's examine this paradox by means of another analogy. Imagine quantum twin brothers. Both are sons of an important king and have been raised in secret inside a castle. The two princes look alike, talk alike, and dress alike. One day they set out on a great adventure. Prince Charles heads south, and Prince Edward rides north. After a few days' ride, Prince Charles gets captured by an evil warlord and is thrown into a dungeon. The evil warlord tortures the prince to find out what he knows. After a short while, he finds out that Prince Charles has a twin and realizes that the more he tortures Prince Charles, the more he can find out about the twin. Because these guys are identical twins and have been raised in the same way, the evil warlord can find out about Prince Edward by questioning Prince Charles.

Prince Edward had no idea what's happening to his brother. As Charles is tortured and slowly begins to die, the evil warlord obtains more and more detailed information about Prince Edward. In the end, the warlord has gained full knowledge of Prince Edward's reality without ever disturbing his existence. The evil warlord now sets out to capture Prince Edward, because he knows how to find him.

The EPR paradox works basically the same way. Because measurements disturb a quantum system, the idea was to perform measurements on one quantum system and use the information obtained to learn about a twin system some distance away. In other words, two similar quantum systems, let's call them C (Prince Charles) and E (Prince Edward), are allowed to interact until their internal states become correlated. Now put the two systems on opposite sides of the room. Einstein, Podolsky, and Rosen argued that if you performed experiments on C, you could learn information about E without disturbing E.

Mind Expansions

The EPR thought experiment was eventually carried out in the 1980s. It established that nature does seem to behave in a non-commonsensical way. It also proved that Einstein was wrong in his belief that action couldn't happen at a distance. Measurements made on one particle appeared to instantaneously affect the counterpart particle, no matter how far apart they were. So something does appear to move faster than the speed of light in the quantum universe. But there is no way to send useful information faster than light using these quantum connections—at least not yet.

The main point of the EPR concept was to leave system E alone, to avoid interacting with it in any way. Using measurements of system C, one could make deductions about E's independent reality. Einstein, Podolsky, and Rosen believed that they had elucidated the position and velocity of E, which was something that Bohr and the Copenhagen interpretation said couldn't be done without somehow affecting the measurements.

The EPR thought experiment hit Bohr and the Copenhagen interpretation like a bombshell. The Einstein camp thought it had the game wrapped up this time. But Bohr came back and refuted it. It was a pretty subtle rebuttal. Bohr basically felt that the EPR experiment just wasn't a strong enough argument and consequently couldn't overturn the Copenhagen interpretation view. The EPR experiment made certain assumptions about the nature of reality that Bohr felt quantum mechanics had proven were not valid. Bohr thought that quantum mechanics demanded a radical reinterpretation of what was meant by physical reality. In quantum theory, reality no longer had a precise meaning. The only thing that was important was the conditions under which a quantum experiment was made. If one chose one set of conditions, a certain aspect of a quantum system was revealed. If one chose different conditions, a complementary aspect was revealed. None of these conditions or elements of reality were described in the EPR thought experiment, so it wasn't relevant to refuting the quantum theories of Bohr or his Copenhagen interpretation.

So where does all of this leave us? Maybe Wonderland isn't all it was cracked up to be. The quantum world is definitely a very strange place. Einstein never accepted its theories because of his belief in an underlying order not based on probable and uncertain principles. Meanwhile, quantum mechanics continues to predict and work very accurately for the systems to which it is applied. So is that the end of story? Far from it. The next chapter delves into the depths of human consciousness, and you'll see how relevant it is to the study of the quantum universe. Who is it who knows what the knower knows?

The Least You Need to Know

➤ In the quantum universe, it is impossible to know two properties, such as position and momentum, at precisely the same time. You can know one or the other, but not both.

➤ Heisenberg felt that the only thing that was important in physical theory was what was observable via experimentation.

➤ Bohr and the concepts that came out of his school defined interactions in the quantum world.

➤ Einstein never accepted quantum mechanics as a sufficiently complete system to describe what happens at the most fundamental level of the physical universe.

How Conscious Are We?

In This Chapter

➤ The split between mind and body

➤ Immanuel Kant's categories

➤ Sigmund Freud's contributions to psychology

➤ The relationship between the conscious and unconscious mind

Human consciousness can be defined in a number of different ways. Science seeks to explain consciousness as a function of the human brain. Religion often considers it part of the soul or spirit. Philosophy continues to delve deeper into trying to comprehend it. Psychology endeavors to do all these things and adds its own unique perception of how consciousness operates.

As you can see, there are quite a few approaches to the study of human consciousness, so it's not easy to define exactly what it is. Like reality, it's open to many levels of interpretation, and you lose sight of it if you limit your interpretation to just one perspective. Like a polished jewel, its beauty can be truly appreciated only when all of its facets glimmer in the light.

The quantum universe has given us some interesting concepts to deal with. Many of its founding fathers, whom you have spent a little time getting to know, wrote volumes of letters to one another, wrestling with the deep issues and paradoxes that the quantum world seemed to reveal. Central to many of these questions was the role of the observer, as well as a quest for a better understanding of human consciousness. You've already examined a few of these concepts, but you need to pursue a few more important ideas of what consciousness might be. Wonderland has been left behind. You're about to enter a world only you can make sense of.

I Think, Therefore I Am

One of the themes in this book has been the development of a larger perspective from which you can see the unfolding of ideas in the evolution of physics. Albert Einstein has been of key importance regarding that evolution in the twentieth century. But he, like many others, is just one piece in the puzzle. To understand Einstein, you needed to see what came before him, so you could recognize the relevance of his discoveries. You're not finished with him yet, because your understanding of Einstein is not complete. But you are at a point in your study where you need to address some of the questions that resulted directly or indirectly from his theories. I'll come back to Einstein later. For now, I'm going to focus on what it means to be conscious.

The history of science can be understood as an interpretation of historical events that unfolded over time. That concept is inherent in the definition of the history of science. You can also understand it as an evolution of connecting ideas, which is one way that the philosophy of science can be studied. Another way you can understand it is as a combination of these two, by examining the history and philosophy of science as results of the evolution of human consciousness.

The history and philosophy of humanity can be seen as a backdrop in the study of the development of consciousness. We've already looked at how ancient humanity's awareness changed from a myth-based interpretation of the world to one in which reason became the primary tool to seek explanations of events in the natural world. Consciousness shifted to a more active, participatory mental focus from a more emotional, *allegorical*, and somewhat religious focus.

Relatively Speaking

An **allegory** is a story in which people, events, and things have hidden or symbolic meanings. They're often used to teach or explain ideas and moral principles. You find them a lot in mythology and religion, as well as in philosophy and psychology.

Epistemology is one of the main branches of study in philosophy. It seeks to discover the origin, nature, methods, and limits of human knowledge.

This is not to say that one worldview replaced the other. What changed was the concept of how much an individual could influence and comprehend the world in which he or she lived, from his or her own unique perspective, rather than from a collective interpretation. There was a liberation of how one could think about oneself in relation to the outside world.

One of the most fundamental properties of human consciousness is self-awareness. The expansion or development of consciousness is always inherently related to an expansion of self-awareness. More than any other quality, self-awareness governs how you see yourself, as well as how you see the world around you.

Self-awareness is a subjective quality that is intimately linked to your beliefs about yourself, about others, and about the world. From it, you obtain your sense of personal identity, which in turn provides you with the tools that you need to interact, however capably you can, with the world in which you live. In short, you perceive yourself and the rest of the world to the degree that you are consciously aware of them, no more and no less. I'll come back to this idea a little later. For now, let's take a look at what some important people in this area have thought about what you can know.

The Search for Knowledge

In the study of philosophy, there is a branch of inquiry called *epistemology*. This area of study asks two fundamental questions:

➤ What is it that we know?

➤ How do we know that we know these things?

Sounds like a typically loaded philosophy question, right? So it must be unanswerable in 5,000 words or less. Actually, it's not that bad. Traditionally, philosophers agree that there are six different ways that you can know something:

➤ *Empiricism* is based on sense perception. It forms the core of the scientific method. If you can't measure something, then you can't experience it through one of the five senses, and science doesn't want anything to do with it.

➤ *Rationalism* is the application of reason to the object that you want to learn about. It lies at the core of the mind's ability to think about something.

➤ *A priori* is also known as the doctrine of innate ideas. It basically means having knowledge of something before having an experience of it. This is a tricky one to examine in detail, but it usually refers to things like mathematical truths, where you don't necessarily have to have experience to know that 2 + 2 = 4. It also sometimes refers to instinctual biological behavior, such as knowing that one needs food.

➤ *Authority* is based upon knowledge gained by being told that something is true without questioning it or confirming it for yourself through some other means. For example, people entrust religion, teachers, and parents to provide them with knowledge about things.

➤ *Faith* often falls into the realm of religion, but it can also pertain to other belief systems, institutions, or people.

➤ *Intuition* is on the opposite end of the scale from empiricism. This pertains to knowledge that is obtained by feeling, but it goes beyond having a gut feeling or woman's intuition. It concerns a deeper level of knowing that can be developed by a continual awakening process of trusting your feelings about things. This concept is difficult to put into words, because it is based on pure experience, or what is sometimes called direct knowing or cognition.

Of these six ways of knowing, five are self-referential, meaning that they involve your ability to arrive at knowledge of something independently, but the other one, authority, is knowledge given to you.

241

Dualism vs. Materialism

The birth of modern philosophy began with a French philosopher and mathematician by the name of René Descartes (1596–1650). You met him briefly back in Chapter 4, "Newton's Clockwork Universe," during the discussion about developments in mathematics. Descartes formulated the first truly original concept of what it meant to be a conscious being. He is the person responsible for the famous quote that is the title of this section. *Cogito ergo sum,* or I think, therefore I am. Descartes began by throwing away everything that he knew could be reasonably doubted and arrived at the one idea that he felt he had knowledge of beyond refutation. The fact that he was aware, or conscious that he was thinking, meant that he had to exist. He formalized the key concept of self-awareness.

Albert Says

One of the basic premises of this book is that philosophy and physics are very closely linked in terms of the types of questions they ask. Each of these areas of study also can affect the development of the other. Descartes, for example, expressed the spirit of his time with his declaration that he existed because he thought. With this intoxicating power to think, Newton turned his immense intellect outward and saw a world ripe for thought and examination. His work in turn influenced other philosophers to probe the workings of the mind in similar ways. Physics and philosophy continue to feed off each other.

Descartes was also responsible for setting up what's known as the mind/body problem. He argued that our minds were entities distinct from our bodies. This argument set up a raging debate between philosophy and science. Science claims that there is no evidence to support the idea that the mind is separate from the body. The mind is nothing but the direct result of the electrochemical processes in the physical brain. Mind and body are made of the same stuff. When the body dies, the mind dies, too. This argument is the philosophy of *materialism* at its best.

Meanwhile, many philosophers have argued that the mind is distinct from the body. Although they have no physical evidence, they cite claims of people who have out-of-body experiences or near-death experiences, in which the physical body has died but the person remained in some form of conscious state. The scientific method is not able to address these experiences of reality. The belief in the separation of mind and body is known as the philosophy of *dualism.*

The Mind of Immanuel Kant

Descartes lived at the end of the high Renaissance, so it makes sense that he would consider what it meant to be conscious of oneself. With a few exceptions, humanity had fallen asleep during the Dark Ages, was groggily waking up in the Middle Ages, and completely reawakened during the Renaissance. If you remember back to Chapter 2, "Great Minds of the Renaissance," a major paradigm shift was occurring during that time in relation to how people saw themselves and their place in the world. Humans were becoming much more self-aware not only of themselves, but also of the world in which they lived. This time was as much a reawakening of self-awareness as it was a rebirth of developments in the arts and sciences.

With the advent of self-awareness came a more detailed inquiry into the mind itself. While many philosophers became either empiricists or rationalists, one individual sought to unify these two schools of thought by developing a coherent synthesis. Immanuel Kant would do for philosophy what Copernicus did for astronomy. His insight has even been referred to as Kant's Copernican revolution.

The empiricists argued that knowledge could only be acquired through the experience of the senses. The rationalists thought that knowledge was obtained by the mind's ability to think about things, without necessarily having an experience of them. Kant brought these ideas together by stating that humans inherently organize reality. He argued that physical reality is experienced only through inner structures in the human mind that organize the world into sensory perceptions. These structures, which he called *inherent categories*, mediate between the physical world and the inner world of our thoughts. The fact that human beings are able to structure physical reality sufficiently to process sensory perceptions at all strongly suggests that inborn, identical psychic structures exist in all of us.

Of course, like all highly original ideas, Kant's categories present a more straightforward picture than is found in reality. Current research shows that our sense organs themselves organize reality. Further organization goes on as sensory data is presented to the mind. But the full chain of operations is far from being understood at this point.

If Kant was right, then an understanding of the nature of the mind and consciousness was critical for determining the nature of reality. The study of the human mind, *psychology*, needed to become a field of its own and not just an adjunct to other disciplines. However, the time was not yet ripe, and modern psychology wasn't to begin until the second half of the nineteenth century, when the twin poles of experimental and clinical psychology came into existence. Kant had the most profound effect of any philosopher since Plato, but the world of the late eighteenth century was not yet ready to accept the full impact of his teachings.

Mind Expansions

The empiricists and rationalists waged one of the longest philosophical battles in philosophy. The empirical position, which was mostly British, was voiced by philosophers such as David Hume, John Locke, and Bishop Berkeley. Simply put, they stated that all ideas that we have about the world come directly from experience. The rationalist group, made up of philosophers from continental Europe such as Descartes, Gottfried von Leibniz, and Baruch Spinoza, argued that ideas come from the use of reason applied to the logical relationships found in the world. Thinking supplies knowledge, not experience. Immanuel Kant would resolve the dispute by arguing for his categories, which represented a synthesis of both positions.

Different Forms of Consciousness

The understanding of human consciousness is essentially a function of the expansion of consciousness itself. In other words, the more consciousness expands, both individually and collectively, the more self-awareness increases, and the understanding of the fundamental properties that define how consciousness operates becomes deeper.

Albert Says

The study of consciousness is difficult to pin down. Because we don't really know what consciousness is, we need to draw upon many different areas of study to help us understand it. This is known as an interdisciplinary approach. The more we learn about consciousness, which is a fundamental trait of humanity as well as of other species of living organisms, the more it will provide us with insights that we can use in other areas of study. A reciprocal dynamic process forms the corner-stone of mutual understanding.

This change can be summarized in the following theory about human consciousness: When you grasp the truth about an event, concept, idea, or theory and apply that understanding to an immediate problem or challenge within your environment, your consciousness expands through the use of that understanding of truth.

Have you ever worked on a problem, or tried to understand a theory or concept (for example, Einstein's theories of relativity)? Maybe you've even tried to resolve a dispute in a relationship. What you're doing is grasping for some type of solution or understanding of an issue. Then out of the blue, you get the *ah-ha!*—a flash or leap of insight that becomes the answer or understanding you were looking for. This insight is what the theory means by an expansion of consciousness. You suddenly see more and understand more that you did before.

Many philosophers and psychologists have contributed to the understanding of human consciousness, but this book only has room to mention a few. The following sections examine Sigmund Freud and Carl Jung, two psychologists who made major discoveries about the structure of human consciousness.

Sigmund Freud

Early ideas in psychology represent an important link in the chain of connected ideas that comprise the expanding knowledge about human consciousness. Volumes of material have been written about the psychoanalytical theories developed by Sigmund Freud (1856–1939). His main contribution was the discovery of the unconscious mind.

Through his work with neurotic patients, Freud realized that an unconscious dynamic was working beneath their conscious actions. He tried to use hypnosis to tap into this unconscious well, but found it unsatisfactory. Freud developed a tool that he called free association, which allowed him to get beyond the conscious mind. Dreams were another rich source of unconscious material. He decided that dreams represented a kind of wish fulfillment and that their complexity was due to an inner censor that made a final attempt to prevent forbidden thought from coming into consciousness.

This unconscious material appeared to be overwhelmingly sexual to Freud. He found that his patients were invariably sexually disturbed and that they all exhibited some sort of repressed sexual perversion in their desires. According to Freud, children pass through three transitional stages, usually by age five: the oral, the anal, and the phallic. He theorized that only by going through proper sexual development before age five could proper adult sexuality develop. Freud felt that all human problems and even all human achievements could be traced back to disturbances in childhood sexual development.

Freud stood at the end of a long line of thought that can be traced back to the Renaissance. It begins with humanity glorying in newly discovered ways to observe the world. With Descartes, humanity became synonymous with intellect, and the great split between mind and body was clearly stated. Humanity developed an ever greater ability to stand separate from the world around it, dissecting the world into smaller and smaller parts. Science grew from this ability, and this growth brought humanity new power to dominate the world. Humanity increasingly turned this power onto itself, separating its rational mind from its body and emotions. The body's needs and wisdom were increasingly hidden from consciousness, which saw itself as a disembodied intellect. As this separation between the rational mind and the emotions and our bodies grew over hundreds of years, a counterbalance also began developing. This counterbalance was a pull toward harmony, relationship, and connectedness rather than separation. It started slowly with things like hypnotism, which allowed a glimpse into the relationship between our minds, emotions and bodies, and would grow further as psychology developed into its own discipline.

Nuclear Meltdown

A sad but true aspect of western culture is that many of the worst crimes against people are motivated by repressed sexual tendencies. The western cultural belief system has given sexuality a negative connotation and associated it with guilt, fear, and punishment. It's no wonder that one of the most basic, natural, and instinctual desires has become so difficult to deal with and understand. Clarity, caring, and understanding are much better tools to alleviate our social woes than hatred, anger, and condemnation. Society needs to apply a bigger picture here, don't you think?

Freud was confronted with a dilemma. On the one hand, he saw that our conscious lives are largely controlled by unconscious motivations. Yet this unconscious motivation was a primitive, instinctive, unreasoning force. He decided that the best humanity could do was to understand the split inside itself and hold on to rationality when faced with the urgings of the unconscious.

The way out of this dilemma was found in a synthesis of the conscious and unconscious, of the mind and the body. It depended upon relationship, rather than separation. This concept wasn't ever proposed in the nineteenth century—humanity wasn't capable of understanding it yet. This breakthrough would have to wait until the twentieth century, with the psychological concepts of Carl Jung (1875–1961).

Realities of the Unconscious

Freud and Jung had been comrades in the early years of psychology, but Jung soon questioned many of the theories put forth by Freud. Jung stated very clearly in one of his lectures that:

> Freud derives the unconscious from the conscious … I would put it the reverse way. I would say the thing that comes first is obviously the unconscious … in early childhood we are unconscious, the most important functions of an instinctive nature are unconscious, and consciousness is rather the product of the unconscious.

Jung felt that the science of psychology was too young to develop a theoretical structure, so he concerned himself with describing what he encountered. As a clinical psychologist, he was concerned with the practical issue of how to resolve the problems he encountered in his patients. What he discovered was previously unknown to western science. Because he confined himself to description, Jung was able to produce a detailed and accurate picture of the deepest levels of the human psyche. Via his encyclopedic knowledge, he was able to relate what he encountered in the dreams and fantasies of his patients to the art, literature, philosophy, mythology, and religion of the world.

Jung realized that the unconscious is a world that can only be viewed indirectly, through its byproducts and relationships. This is very similar to the world of quantum mechanics, where physicists study subatomic particles that can only be viewed indirectly through their probable and uncertain relationships with other particles.

One tool that modern physicists use is known as a *bubble chamber*. This is a container in which the movement of particles can be traced when they collide with one another. To a nonphysicist, the bubble chamber just shows a bewildering complexity of lines. The nonlogical, seemingly incoherent picture presented by dreams is also bewildering. But both the bubble chamber and dreams can provide a lot of information to those who observe them with openness and understanding.

In the nineteenth century, it was still assumed that nature could be observed directly. The twentieth century showed otherwise. Jung's theories reflected the changes taking place in how consciousness was understood; Freud's theories were a product of the earlier

Mind Expansions

The word **psyche** originally comes from Greek and Roman mythology. Psyche can be found in Book XI of Homer's *Odyssey*, and also in Apuleis, the Roman who wrote *The Golden Ass*. In the latter, Psyche is a maiden who, after undergoing many hardships due to Venus's jealousy of her beauty, is reunited with her true love Cupid and made immortal by Jupiter. Apuleis was a Platonist, who believed that our souls are intrinsically divine. We lose our divine nature during our lives on Earth and may despair of ever regaining it, but we can do so if we desire it enough to weather the trials and tribulations that come our way, never losing sight of our ultimate goal. Today the word **psyche** means the whole self, all the emotional, intellectual, and spiritual aspects that make up a human being.

nineteenth-century paradigm. The difference between Jung and Freud reflects another shift in understanding and awareness.

Jung found a very strange world in the unconscious mind. It was just as strange as the world of quantum physics. Both seemed to be free of the restrictions of time and space that define waking consciousness and the everyday world. Jung made this stament in a lecture:

> We cannot directly explore the unconscious psyche because the unconscious is just unconscious, and we have therefore no relation to it. We can only deal with the conscious products which we suppose have originated in the field called the unconscious … We do not know how far the unconscious rules because we simply know nothing of it. You cannot say anything about a thing of which you know nothing. When we say the unconscious we often mean to convey something by the term, but as a matter of fact we simply convey that we do not know what the unconscious is.

Relatively Speaking

The **bubble chamber,** which is similar to the cloud chamber, was invented by Donald Glaser at the University of Michigan in the early 1950s. The bubble chamber is a device that records the tracks of high-energy particles moving through a fluid by tracing the trail of bubbles along the line of flight of the particles. It played a crucial role in the investigation of the particle world, and led to the development of the standard model, which you'll learn more about later on.

This statement sounds a bit like some of the comments made by physicists about the quantum world when it was first being explored. The behavior of subatomic particles is so strange that it defies common sense, and we can only estimate some of the principles under which the quantum universe operates. We still have an incomplete picture of its underlying operating mechanism. Surprisingly, because the role of the observer plays an important part in the study of the quantum world, there may be a close relationship between the dynamic structure of the quantum universe and the nature of the unconscious.

Probing the Depths of Consciousness

Models are often used in physics to help explain an idea or concept. Models can also be used in psychology to help describe complex theories. Jung wanted to avoid using one specific type of model, because he preferred working with acquired knowledge rather than theory and often felt that models were incomplete. He therefore proposed many different kinds of models throughout his work. In one of the models, Jung proposed that the totality of the human psyche can be broken down into three categories:

➤ Personal consciousness

➤ Personal unconscious

➤ Collective unconscious

Personal Consciousness

Our personal consciousness, of which we are so proud, is a very transitory affair, consisting of whatever occupies our conscious awareness at a given moment in time. There's nothing permanent about it, including the sense of identity or ego, which just sort of comes and goes. Consciousness is merely a sliding frame that moves along, sometimes lit by awareness, sometimes not. (Of course, a lot of this depends on how big of a couch potato you are.)

Jung had this to say about consciousness:

> *Consciousness ... is an intermittent phenomenon. One fifth, or one third, or perhaps even one half of our human life is spent in an unconscious condition. Our early child-hood is unconscious. Every night we sink into the unconscious, and only in the phases between waking and sleeping have we a more or less clear consciousness. To a certain extent, it is even questionable how clear that consciousness is.*

Albert Says

Much of the information supplied in this chapter concerning Freud and especially Jung is very simplified and doesn't come close to providing you with the fullness of the insights that either of these individuals gave to the world. If this material interests you, I recommend that you read Freud's *Theory of Sexuality* or Jung's *Memories, Dreams, and Reflections.*

Everything passes into consciousness by way of the unconscious. Even sense perceptions, which many regard as primary, are processed somewhere inside of us in a way of which we are largely unaware and then pass into consciousness. Objects and events become conscious momentarily, and then pass out of consciousness again as our awareness either shifts to something else or turns itself off for awhile. Things that pass from the conscious are either recorded in some fashion or simply lost.

Jung thought that:

> *... the sum total of unconscious contents fall into three categories: first, temporarily subliminal contents that can be reproduced voluntarily (memory); second, unconscious contents that cannot be reproduced voluntarily; third, contents that are not capable of becoming conscious at all ...*

Personal Unconscious

Jung was deliberately simplifying the very complex relationship between conscious and unconscious material in order to stress that consciousness is not the whole of the psyche. As examples of material in our memory, we can all recite the alphabet, the multiplication tables, and Einstein's famous equation. Consider, however, the fact that a great deal of unconscious material can be accessed without ever passing into conscious awareness. For example, you don't have to be conscious of tying your shoes in order to correctly tie them—however, you can make yourself aware of how you tie them if necessary. You can even do so in your mind without touching a shoe or shoelace. Do you see how complex the relationship between the conscious and unconscious is?

Consider the huge gray area between that which is in our memory and that which is part of our consciousness but cannot be recovered as memory. We can all recall into consciousness events of emotional significance in our lives. The amount of conscious recall of such events largely depends on the significance of the event.

But many memories are on the borderline of conscious recall. These memories might have been too insignificant to pay close attention to. You could try tricks of mental association to try to recall these hazy memories. If the tricks are clever enough, such as deep hypnosis, you can usually recall every detail of such events.

Other memories are harder to recover not because they have too little emotional significance, but because they have too much. For example, if an event was too painful for you to accept, you may have recorded the memory fully, but erected a psychic barrier to prevent you from re-experiencing the pain. This pain could be physical, such as a broken bone or childbirth, or an emotional pain, such as an incident when you were deeply humiliated. Freud called such memories *repressed memories*.

Another example could be an event that was too threatening to handle. If, while crossing the street, you suddenly see a car about to hit you, you might block out the perception even before it causes you pain. Your conscious memory would end with the sight of the car coming at you and would not include the car coming closer, the impact, or the pain. You might also have no conscious memory of an event that threatened your view of reality. For example, if a person had constructed a rigidly rationalistic view of life, that person might not see a ghost, if such a thing existed, even if one appeared to him or her. The sight would be too threatening, in that the person's whole view of reality would be jeopardized. So the person wouldn't consciously see the ghost, nor would the person be able to bring that memory into consciousness. However, the memory might be accessible using special techniques such as hypnosis.

All of these memories are part of what Jung calls the personal unconscious. They constitute the totality of Freud's unconscious. But Jung contended that there was an entity much bigger than the personal unconscious, which he termed the collective unconscious.

The Collective Unconscious

Jung defined the collective unconscious in these terms:

> *While the personal unconscious is made up essentially of contents which have disappeared from the consciousness through having been forgotten or repressed, the contents*

of the collective unconscious have never been in consciousness and therefore have never been individually acquired, but owe their existence exclusively to heredity … there exists a second psychic system of a collective impersonal nature which is identical in all individuals. This collective unconscious does not develop individually, but is inherited. It consists of pre-existent forms, the archetypes.

Jung held that everything emerges into consciousness from the unconscious. When you call up memories of how to drive a car, it is not your consciousness that organizes those memories and makes them available—clearly, something in the unconscious is able to organize the memories that are necessary and make them available to your conscious mind. The remembered behaviors may never even reemerge into consciousness. You may drive the car with no conscious awareness of doing so.

Relatively Speaking

Archetype literally means the original pattern or model from which all things of the same kind are made. In Jung's psychology, it is similar in meaning but more far-reaching. It pertains to the patterns or symbols that reside in the pool of the collective unconscious that everyone draws from when they dream. It also refers to objects and events that have had meaning for a large number of people for a long period of time and leave a record in the human psyche.

This whole process is a mystery to which we have grown so accustomed that we have come to see it as commonplace. The part of your psyche that organizes and presents you with all the memories needed to drive the car, psychic and muscular, has to be on a higher level than the memories themselves. That is, there has to be something that organizes these memories. Such an organizer is inherently on a level of organization different form the memories it organizes.

Let's look at an analogy. The operating system of a computer operates on a higher level than any of the other programs. The operating system is like a foreman in a factory—it keeps things running smoothly. It knows which programs are running in the computer, which are waiting to run, and which have already run. The only thing it doesn't do is make the decision regarding which programs to run—that's up to the computer operator.

An archetype is like a computer operating system, but it is much more. By studying the dreams of his patients, thousands of them, Jung found that certain images could be found in everyone's dreams. It didn't matter if the dreamer was Chinese, Italian, Jewish, or South African; people's dreams revealed similar images from art, literature, mythology, and religion, just about any expression of human endeavor, regardless of the time period the images came from.

This pool of imagery is also thought to be the source of human creativity. From the depths of this collective reservoir come leaps of insight, innovative ideas, and inspiration. It is also the repository of all the darker aspects of humanity.

Carl Jung's contributions to unlocking the secrets of human consciousness are immense. I have only barely scratched the surface of the tremendous amount of information he contributed and that others elaborated on. The purpose of this chapter has been to provide you with an overview of the continuing unfolding of human consciousness and the ideas that have provided us with insight in this area. Besides offering you a framework to see the bigger picture, hopefully it has also raised questions about the very nature of consciousness that we are seeking to understand.

There's more to observing the quantum world than meets the eye. Perception is a very complex process, and the seemingly paradoxical nature of quantum events may be directly linked to how consciousness operates on a very fundamental level that we are still seeking to understand. With that idea, the part on quantum mechanics comes to a close.

The next chapter resumes the biographical information by taking a look at the remainder of Einstein's life. The last part of the book concludes the coverage of Einstein by examining how his theories have provided us with insights into the cosmos.

Mind Expansions

During the discussion of chemistry, I noted a dream that a man named Freidrich Kekule had. He had been working for months in an attempt to discover the structure of the last remaining organic compound, benzene. One afternoon, while sitting in a study, he had a dream about a snake that had seized its own tail, forming a whirling circle. When he awoke, he made the connection between the image in the dream and the structure of the benzene molecule. Such ring patterns had never previously been discovered in organic chemistry. His unconscious solved the problem for him, and then presented the solution to his conscious mind in the form of a symbol that he could understand: a snake seizing its own tail.

The Least You Need to Know

➤ Epistemology is the study of human knowledge. It seeks to understand what we know and how we know.

➤ Descartes defined thinking as synonymous with self-awareness.

➤ Immanuel Kant first proposed that the mind has an underlying structure.

➤ Sigmund Freud realized that consciousness had two aspects: the conscious and the unconscious.

➤ Carl Jung stated that consciousness emerges out of the unconscious. The collective unconscious lies at the core of our beings.

Part 5
Einstein, Man of the World

Einstein's career is often divided into four distinct phases. The first covers his life from 1900 to 1906 and shows him as a young, brilliant, but struggling researcher, whose publication of four papers in 1905 upset the world of physics and began to thrust him into the scientific spotlight.

The second phase lasted a little longer than the first. It covers a 12-year period from 1907 to 1918. During this period, Einstein spent a lot of time teaching and lecturing at many universities across Europe and ended up in Berlin. Here he did his greatest work during World War I.

The years 1918 to 1932 mark the third period of his career. Einstein now achieved heroic status as the world's leading scientist. He was at the peak of his mental powers and was looked to for further incredible discoveries.

The last period covers the years between 1933 and his death in 1955. He spent this period in the United States, playing an influential role in physics and politics. You'll see the wise old scientist with the disheveled hair slowly fade out of the mainstream world of physics, searching for a theory that nobody believed was possible to find, except, of course, Einstein.

Chapter 20

The Post-War Years and Einstein's Trip to America

At the end of World War I, Europe attempted to recover from four long years of conflict. Einstein's stance against the military and his support of world harmony no longer seemed so strange. People were clamoring for a look at the man whose fantastic views of space and time had turned the world of physics upside down. Lecture halls were filled with physicists and laypersons who wished to hear about the relativity theory.

Einstein was 40 years old, happily married, and compiling a long list of admirers. In October 1919, a British expedition confirmed the bending of light rays by the Sun. One of the main predictions of Einstein's general theory of relativity was thus proven correct. Over the next 10 years, his fame brought him to America and involved him in the political turmoil of Judaism. This chapter's look at Einstein's life picks up in post-war Europe.

Everybody's Talking About Relativity

In the early part of 1920, Albert's mother moved in with him and Elsa. She was dying and wanted to spend her last months with her son. By then her son had become a celebrity, having received an award in London by the Royal Astronomical Society for his accuracy in predicting the bending of light rays by gravity. Reporters descended upon him in droves, seeking interviews with the man who had redefined the nature of the universe.

Einstein had a knack for dealing with the media. He very seldom came across as an absent-minded professor, although the cartoons in the newspapers often portrayed him as one. He always displayed a good sense of humor and never regarded anybody as being below him. Once a reporter wanted to impress Einstein with his knowledge of physics, and asked him the following question:

> *What is the meaning of potential; invariant; contravariant; energy tensor; scalar; relativity postulate; and inertial system? Can you explain these quickly?*

Albert Says

The media can make you or break you. Its ability to influence public opinion has put many people on public trial, in which they were presumed guilty and needed to prove their innocence. Of course, it goes the other way, too. Many people have achieved stardom because of media hype surrounding their name. A nobody from nowhere is all of a sudden the flavor of the month, to be pursued and interviewed. Within a few weeks, that person again sinks into oblivion. This was one of the things that Einstein disliked most about the press. It was as true in his time as it is now. The media defends itself by saying that it only provides what people want. That's probably true to a certain extent, for if we all thought a little more for ourselves, we wouldn't need the media to do our thinking for us.

Einstein, in his typically quick-witted style, answered:

> *Certainly. Those are all technical terms.*

Throughout his life, Einstein's charm and sense of humor in public always made him a favorite of the media. Regardless of the question or the topic that he spoke about, his clarity and seemingly wise responses attracted people seeking counsel or simply his opinion about things.

It's hard to imagine the incredible wave of publicity and admiration that overwhelmed Einstein beginning in 1920. Something about him captured the imagination of a war-weary public looking for new heroes. Relativity was on everyone's tongue. Few understood what it was all about or the implications that it held for the world in which they lived, but it seemed to have a magical quality. Newspapers and magazines poured out millions of words, each trying to explain relativity in simple terms—most of which were incorrect.

Although Einstein was cordial when he needed to be, he disliked being a world-famous figure. Everyone was treating him like a new animal at the zoo and expected him to perform. All he wanted was to be left alone so he could pursue his work, but the world wouldn't allow it. The fact that few people understood his theories, especially the man in the street, didn't stop them from treating him as a combination movie star, wizard, and heavy metal hero.

Often the publicity was just plain ridiculous. A popular song contained a lyric about the man who "attracted some attention/when he found the fourth dimension." A limerick was written in which his name was linked to Gertrude Stein, the avant-garde poet, and Jacob Epstein, an artist that sculpted distorted modern statues:

> There are three people named Stein.
>
> There's Gert, and there's Ep, and there's Ein.
>
> Gert's poems are punk,
>
> Ep's statues are junk,
>
> *And nobody understands Ein.*

Everything that Einstein said was printed in newspapers around the world. The flood of publicity inundated him for the rest of his life, although it did abate a little when he was older. Not a year passed between 1919 and his death in 1955 without the name of Albert Einstein appearing in the pages of the *New York Times*. Shortly after the announcement that his general relativity theory had been proven correct, Einstein began selling photographs of himself to journalists to help raise money for the starving children of Vienna, who were still suffering from the ravages of the war. The photographs appeared in newspapers and magazines around the world. His theories were endlessly discussed, dissected, and quoted. Albert Einstein had become a household name.

Mind Expansions

The media blitz and commercialization of Einstein's relativity theory was the 1920s equivalent of today's *Star Wars* craze. Every bookstore was filled with magazines, newspapers, and books about relativity theory. One opportunistic businessman created the Einstein Cigar, and Einstein was invited to do a season at the London Palladium. There was also a $5,000 award offered by an American magazine for the article that could best explain relativity theory. People stood in line for hours waiting to get tickets to see or hear Einstein speak. Did you dress like a Wookie and camp out for your *Star Wars* tickets?

In a book published in the 1920s, the journalist Alexander Mozkowski gave this description of the public's response to the theory of relativity:

> *Newspapers entered on a chase for contributors who could furnish them with short and long, technical or nontechnical, notices about Einstein's theory. In all nooks and corners, social evenings of instruction sprang up, and wandering universities appeared with errant professors that led people out of the three-dimensional misery of daily life into the more hospitable Elysian fields of four-dimensionality. Women lost sight of domestic worries and discussed coordinate systems, the principle of simultaneity, and negatively charged electrons. All contemporary questions had gained a fixed center from which threads could be spun to each. Relativity had become the sovereign password.*

This may sound a little exaggerated, but it probably wasn't too far from the truth. Everyone welcomed Einstein as a major distraction from the war and its effects.

Relatively Speaking

The term **Elysian fields**, or **Elysium**, comes from early Greek mythology. It can be found in Book IV of Homer's *Odyssey*. The fields were found at the end of the world, and people who were favored by the gods were sent there after death. It was a land of beauty and restfulness. It later became part of the Under-world, a section you could enter only if you led a righteous life on Earth. For centuries, writers used it as a metaphor for a place of heavenly beauty. If you've ever been to Paris and walked down the mile-long boulevard that reflects its name, the Champs Élysées, you'll know why it was named that.

Einstein's theories transported the imagination of a needy public to the realms of H.G. Wells and Jules Verne, only instead of science fiction it was proven fact. Everyone suddenly became a relativity expert. No matter where you looked or where you went, you saw, read, or heard about Einstein and his relativity theory.

Relativity Isn't for Everybody

Not everyone was as thrilled as the general public about Einstein and his theories. Opposition fell into four main categories:

➤ Nonscientific, ignorant ridicule

➤ Philosophers who didn't comprehend its meaning

➤ Physicists who were jealous and resentful

➤ Anti-Semitic political opposition

The first category was mostly comprised of harmless jabs by the news media, which thought that the whole idea was beyond relevance to the average person. Many cartoons in the daily newspapers portrayed Einstein as an eccentric figure to be poked fun at. Einstein and his theories were quickly incorporated into common slang to represent anything that was too complex to understand or appeared a little too weird. Here's an example of a typical journalistic jab:

Relativity can be lumped in with the annual sea serpent, the seven-year mutation of our bodies, the jargon of Freud, the messages from Mars … Certain troubled spirits, hearing the law of gravitation called in question, do not feel sure that the Earth may at any moment slip its Newtonian moorings and go ranging off out of gravitation into the ether—which we now hear does not exist.

A number of philosophers comprised the second category of critics. They were found all over the world, not just in the West. They tried to create their own interpretations to demonstrate that Einstein's theories were mistaken. But none of them really understood relativity in the first place. All they managed to do was inflate their own egos and, in the end, show a lot of people how much wisdom had been wasted on their philosophies.

Experimental physicists comprised the third category of naysayers. These people resented that fact that a theoretical physicist suddenly rose to such fame. They saw his ideas as mere flights of fancy, with barely any experimental data. These guys were hard-core experimenters, who never cared for any theories that couldn't be backed by hard empirical evidence in the laboratory.

All three of these groups were more or less harmless in their mostly ineffectual criticisms of Einstein's theories. It's the fourth group that began to pose a serious threat to Einstein and his ideas. These individuals were dangerous and completely fanatical about opposing relativity theory. Their mindset was a mixture of jealousy, racial hatred, and pathological egoism.

Anti-Semitism Hits Home

Einstein soon felt the sting of *anti-Semitism* in a way he had never known. Bitter attacks mounted against him from Germany, which not only hated the concept of relativity, but its discoverer as well. With the military defeat of Germany in World War I, many patriotic Germans blamed their country's humiliating defeat not on military inferiority, but on pacifists and Jews. Einstein was both. He had never made his opposition to the war a secret, and he was becoming more vocal in his support of pacifism and an international community for peace. His relativity theory was sneered at as being Jewish physics and contrary to the Aryan spirit of Teutonic man.

Most of the German anti-Semitism that flourished after the war reflected the need of some Germans to have a scapegoat to explain their country's problems. Although the growing hatred of Jews didn't reach its full momentum until the Nazi Party gained political control, there was an undercurrent of anti-Jewish sentiment already present in Europe before anti-Semitism became the official banner of Nazism.

Many large meetings were held in Berlin in the 1920s during which an organization known as the Study Group of German Natural Philosophers denounced what they felt was an attack on the German race and its superiority. Their leader was a man by the name of Paul Weyland, who had a fanatical eye for personal gain and self-publicity. He was able to win the support of a physicist by the name of Paul Lenard, who had won the Noble prize in 1905. Lenard had originally been a great admirer of Einstein and his work, but he changed his perspective after the war. He became consumed with jealousy and hatred for Einstein and stood before large audiences denouncing relativity as a sham and a Jewish theory. He had a lot of prestige and convinced a number of German scientists to speak out against Jewish physics. The meetings, however, never attacked the theory of relativity on scientific grounds. Instead, Einstein was denounced for being anti-German and a Jewish revolutionary. Here's an excerpt from a paper Lenard wrote:

Nuclear Meltdown

A seven-headed demon in Chinese mythology brings death and destruction to humanity by hypnotizing people into believing that anyone different from them should be destroyed. Although the demon is just a myth, it is a metaphor for the forces of hatred that continue to plague humanity. The emotions fueled by intolerance, self-righteousness, and certainty in the truth of one's opinions and beliefs prevent any possibility of understanding and compassion. Of all the famous quotes by Einstein, the one least understood is probably one of his most insightful: "The true value of a human being is determined primarily by the measure and the sense in which he has obtained liberation from the self."

Relatively Speaking

The word **Semite** comes from the Bible and refers to anyone descended from Shem, the eldest of Noah's three sons. It includes any cultures that speak a Semitic language. Although Hebrews, Arabs, Assyrians, and Phoenicians all speak or spoke a Semitic language (the last two cultures are long gone), in the vernacular it usually pertains only to Jews. Thus, **anti-Semitic** is another way of saying anti-Jewish.

Relatively Speaking

Zion was originally a Canaanite fortress in Jerusalem that was captured by David and was called the City of David in the Bible. Later on, it became the hill in Jerusalem where Solomon built his temple and came to be regarded as the symbol and center of Jewish national life. **Zionism** started as a movement to re-establish a state of Israel and now reflects the continuing support of this state.

The most important example of the dangerous influence of Jewish circles on the study of nature has been provided by Herr Einstein with his mathematically botched-up theories consisting of some ancient knowledge and a few arbitrary additions. This theory now gradually falls to pieces, as is the fate of all products that are estranged from nature. Even scientists who have otherwise done solid work cannot escape the reproach that they allowed the relativity theory to get a foothold in Germany, because they did not see, or did not want to see, how wrong it is, outside the field of science also, to regard this Jew as a good German.

In August 1920, Einstein attended one of the most publicized meetings of this group at the Berlin Philharmonic hall. He went despite the warnings of his friends and wife and managed to procure a box seat all for himself. Eyewitnesses stated that Einstein found the evening immensely entertaining. As the attacks on him became more and more outrageous, he appeared to be enjoying himself even more, clapping and laughing out loud at the absurdity of it all.

The next day, he wrote a letter defending his work and pointing out the group's anti-Semitism and had it published in a major Berlin newspaper. He probably would have been better off not writing anything, because he basically took their bait. His friends were disappointed that he had responded to the group's ravings, believing that he had done exactly what they wanted him to instead of giving them no satisfaction. Years later, Einstein regretted his quick response and wished he had swallowed his pride.

Return to Zion

When Einstein was young, he had no interest in his Jewish background and felt that any form of ethnic identity or religious institution was a social fabrication. But the more he bore the brunt of anti-Jewish remarks and saw the dangerous political movements against the Jews, the more responsibility he felt toward his cultural heritage.

Einstein was recruited to *Zionism*, which at that time was the movement to locate a permanent Jewish homeland. Its chairman, Chaim Weizmann, realized that if he wanted to make this dream a reality, he needed to have plenty of political and financial support.

As you know, Einstein didn't care for nationalism any more than he cared for ethnic or religious identity. In his eyes, Zionism was not all that different from the views of those who were anti-Semitic. It seemed just as fanatical. But Einstein began to develop a philosophical view that he eventually applied to many political matters. He decided to see things from the perspective of the greatest good. If a cause had some worthy aspects, and a large number of people could benefit from it, he was willing to overlook specific aspects that he didn't care for.

One of Einstein's main interests was the establishment of a Hebrew university in Palestine. He had seen many Jews turned away from universities throughout Europe due to their cultural heritage and wanted to make sure there would be a place where they could study. Regardless of his views on Zionism, Einstein felt that an education was everyone's basic right. The way things were going in Europe, the sooner he could help set up a university, the better off Jews would be.

In the early part of 1921, Einstein accepted an invitation by Weizmann to go on a fundraising tour of the United States and England. It didn't take long for the news to spread that Einstein would soon be visiting the wartime enemies of Germany. In his effort to help the Zionist movement, Einstein inadvertently incited sentiment against him not only by Germans who didn't like Jews, but by the German Jews as well. Many German Jews had died in the war, and the thought of their fellow countryman traveling to these countries caused considerable outrage. Many of them considered Zionism a totally unworthy cause. Fritz Haber, a fellow German Jewish scientist, wrote the following to Einstein regarding his planned trip:

> To the whole world you are today the most important of German Jews. If at this moment you demonstratively fraternize with the British and their friends, people in this country will see this as evidence of the disloyalty of the Jews. Such a lot of Jews went to war, have perished, or become

Mind Expansions

Just prior to Einstein's trip to America, he lectured in Prague. He stayed at the university, and a reception was given in his honor at the physics laboratory. A young man burst into the room clutching a thick manuscript and insisted on speaking with Einstein. After a lot of fuss, he finally got a chance to speak with him. He told Einstein that based on his famous equation, $E = mc^2$, it was possible to build a weapon of incredible destructive power for military purposes. According to eyewitnesses, Einstein said, "Calm yourself. You haven't lost anything if I don't discuss your work with you in detail. Its foolishness is evident at first glance. You cannot learn any more from a longer discussion." Little did he know that the application of his equation would eventually lead to the development of the first atomic bomb. Whoops.

impoverished without complaining, because they regarded it their duty. Their lives and death have not liquidated anti-Semitism, but have degraded it into something hateful and undignified in the eyes of those who represent the dignity of this country. Do you wish to wipe out the gain of so much blood and suffering of German Jews by your behavior?… You will certainly sacrifice the narrow basis upon which the existence of academic teachers and students of the Jewish faith at German universities rests.

Einstein's reply shows that a more important issue was at stake:

Despite my international beliefs, I have always felt an obligation to stand up for my persecuted and morally oppressed tribal companions as far as is within my power … Far more is involved, therefore, than an act of loyalty or one of disloyalty. Especially the establishment of a Jewish university fills me with particular joy, having recently seen countless instances of perfidious and loveless treatment of splendid young Jews, with attempts to cut off their chances of education.

So Einstein found himself caught between a rock and a hard place. The Germans and the German Jews didn't like that he was visiting their former enemies, and the growing political groups didn't like him because he was a Jew. But this was just the beginning of Einstein's troubles with the European community.

Give My Regards to Broadway

A very impressive welcoming committee was waiting for Einstein when he got off the boat in New York. The mayor, the president of the city council, tons of reporters and photographers, and even a few filmmakers were on hand. Einstein didn't know any English, so everything had to be translated for him. Of course, the first thing that anyone asked him to do was to explain the relativity theory in a few sentences. Being quite familiar with this question, he replied:

It used to be thought that if all things disappeared from the world, space and time would be left. According to relativity theory, however, space and time disappear along with the things.

This response pleased not only the reporters, but the readers as well. Einstein impressed them with his modesty, kindness, and humor. He looked more like an artist or musician than a scientist as he walked down the plank from the boat with his pipe in one hand and his violin case in the other.

He traveled in a motorcade to city hall past thousands of cheering people, where he was to be awarded an honorary citizenship. However, at the last moment, one of the city councilors had doubts as to whether the relativity theory was true or just a bunch of bologna. It took a few days to convince him that it would be in his best political interests to consent, and the celebration eventually took place.

Within a few weeks' time, Einstein had some free time to offer a series of lectures at Columbia University. He was tired of being paraded around like a prize bull and making a thousand speeches at big and small meetings. Later that week, he also lectured at the City College of New York, where his young interpreter displayed great competence not only in German, but also in the mathematics of relativity theory. This interpreter soon earned a reputation of being one of the few people who actually understood it.

He finished his academic tour at Princeton, where he lectured for four days. Then he left for an exhaustive tour of the Midwest, lecturing at the University of Chicago in between his fundraising meetings. He ended up back in New York after a two-month whirlwind tour of a good percentage of the United States and prepared for departure back to England and thence to Europe. He was a bit dismayed when he and Weizmann sat down and tallied the total amount of money they had taken in. They expected to raise between four and five million dollars, but it came out to only $750,000.

When Einstein returned to Germany, he was asked to write an article for the main newspaper in Berlin about his impressions of America. Here are a few excerpts from that article:

Albert Says

There's probably nowhere in the world with a climate more politically volatile than Washington D.C. While Einstein was there, President Harding declined to meet him because he mistakenly thought that Einstein had signed the Manifesto of Ninety-Three, the document that was passed around just prior to World War I, in which 93 German scientists indicated their support of the war. After it was realized that Einstein hadn't signed the document, he was allowed to attend a reception for Marie Curie. The hospitality shown Einstein by the government was a far cry from his reception by the American public. Either way, it didn't matter to Einstein, but it did cost Harding the support of many influential Jews in the political arena.

I must redeem my promise to say something about my impressions of this country. This is not altogether easy for me. For it is not easy to take up the attitude of impartial observer when one is received with such kindness and the undeserved respect as I have been in America. First of all let me say something on this score. The cult of individuals is always, in my view, unjustified. To be sure, nature distributes her gifts unevenly among her children. But there are plenty of well-endowed, thank God, and I am firmly convinced that most of them live quiet, unobtrusive lives. It strikes me unfair, and even in bad taste, to select a few of them for boundless admiration, attributing superhuman powers of mind and character to them. This has been my fate, and the contrast between the popular estimate of my powers and achievements and the reality is simply grotesque. The awareness of this strange state of affairs would be unbearable but for one pleasing consolation; it is a welcome symptom in an age which is commonly denounced as materialistic, that it makes heroes of men whose goals lie wholly in the intellectual and moral sphere. This proves that knowledge and justice are ranked above wealth and power by a large section of the human race. My experience teaches me that this idealistic outlook is particularly prevalent in America, which is decried as a singularly materialistic country.

What first strikes the visitor with amazement is the superiority of this country in matters of technology and organization … The second thing that strikes the visitor is the joyous, positive attitude to life … Great importance attaches to the material comforts of life, and equanimity, unconcern, security are all sacrificed to them. The American lives even more for his goals, for the future, than the European. Life for him is always becoming, never being … The prestige of the government has undoubtedly been lowered considerably by the Prohibition law. For nothing is more destructive of respect for the government and the law of the land than passing laws which cannot be enforced … the increase in crime is closely related to this. The public house is a place which gives people the opportunity to exchange views and ideas on public affairs. As far as I can see, such an opportunity is lacking in this country, the result being that the Press, which is mostly controlled by vested interests, has excessive influence on public opinions.

There's quite a bit more; these comments are only a few of the highlights. The article was reprinted by a number of American newspapers and didn't go over very well. His enthusiasm and admiration of the American people was contrasted with his criticism of Prohibition, the lack of taverns, the overvaluing of money, and political isolationism. Americans didn't like being lectured to by an Old World scholar and viewed his criticisms as being out of touch with the times. But to make matters worse, in an interview he gave to a Dutch journalist, Einstein spoke a little too recklessly and admonishingly. He not only mocked America's excitement over a scientist whose theories they didn't understand, but he also accused American men, though hardworking, of being "toy dogs of the women, who spend the money … to wrap themselves in a fog of extravagance."

It took quite some time before Einstein was forgiven by Americans, especially for his remark about the toy dogs. Einstein learned the hard way that opinions shouldn't be carelessly voiced, that opinions can alienate as well as encourage, and that when you're in the spotlight, silence often speaks louder than words.

Mixed Fortunes

Einstein's visit to England was much more of a success. Less outspoken and more congenial, he was seen as a presence that could help heal the ill feelings left over from the war. One of the London newspapers ran a series of 10 articles about him and his theories during his two-week stay.

With his return to Berlin, his life became more politically unstable. Early in 1922, he was asked to deliver a lecture in Paris to a prestigious scientific academy. It was purely scientific and had nothing to do with politics, but that fact didn't matter to many of the French, who still saw him as a representative of their hated enemies, the Germans.

264

Paul Langevin, who was a long-time friend of Einstein's, had invited him to the French academy in the hope that an exchange of ideas between the two countries would help ease the anger and resentment that remained between the nations after World War I. Langevin and Einstein shared the belief that social change could be brought about by the use of reason in place of emotion. At least in this case, they were wrong.

Einstein delivered the lecture in French, his second language, and it was very well received. But it didn't take long for the attacks from both sides to begin. The French Academy of Sciences threatened to stage a walk-out if he spoke at their next forum, and upon his return to Berlin, the Prussian Academy of Science's lecture hall was virtually empty when he appeared to give his lecture there.

By the end of 1922, Einstein was becoming concerned about the fate of Jews in Germany. Right-wing extremist groups were on the rise, and the German foreign minister, Walter Rathenau, was assassinated earlier that summer. The murder had nothing to do with his political views; he was killed because he was a Jew. Einstein's friends became concerned for his safety. To the anti-Semitic and fanatical nationalist groups, Einstein was a living symbol of everything that was wrong with Germany.

In spite of considerable verbal and emotional abuse, Einstein still hoped that a new Germany had emerged from the ashes of war. The Kaiser had fled the throne in 1918, and Germany had become a republic. There was a widespread feeling that democracy had finally come to Germany. In those days, it was possible to have a dual allegiance—in other words, one could be a citizen of two countries at the same time. Einstein wanted to show his support for the new republic, so he requested formal citizenship, which of course he was granted. So he was now both Swiss and German. Later in life, Einstein would remember this token of allegiance as "the worst folly of my life."

However, democracy was still in its infancy, and the German spirit of patriotism was much stronger. A young man was walking the streets of Munich, the same streets that Einstein had walked as a boy, calling to all true Germans to join his patriotic cause. His party would soon become a major political force, the Nazi Party, and his name, Adolph Hitler, would become synonymous with evil.

Mind Expansions

In November 1922, Einstein was scheduled to deliver a series of lectures in Japan. His arrival was treated with honor, but it didn't involve the media blitz that his trip to America inspired. This pleased him a lot. When he finished his first lecture, he realized that it had lasted over four hours, during which time the audience sat seemingly attentive throughout. Realizing that this was probably way too long for anyone to sit through anything, he decided to make his next lecture only two and a half hours long out of concern for the comfort of his audience. After his second lecture, his hosts engaged in a heated discussion in Japanese, right in front of him, which he and they both knew was rather impolite. Concerned over what could possibly be going on, Einstein asked what the problem was. He was told that the hosts were offended that their lecture was significantly shorter than his first one. Sometimes you just can't win!

For the next 10 years, Einstein remained in Europe, lecturing and supporting the pacifist cause. But in 1933, he left his homeland forever and permanently moved to the United States. This period in his life is the topic of the next chapter.

The Least You Need to Know

➤ The theory of relativity became so popular that it made Einstein a household name.

➤ Though Einstein didn't believe in nationalism, he began to relinquish his dislike of some ideologies in recognition of the overall good that could come from them.

➤ Einstein's warm welcome in the United States was later overshadowed by some reckless remarks he made in an interview, which alienated him from Americans for years to come.

➤ The political factions in Europe placed Einstein in the difficult position of being a focal point for their anger and mistrust of each other.

Chapter 21

Banned in Berlin

In This Chapter

➤ Hitler rises to power

➤ Einstein builds his pacifist stance

➤ Einstein turns 50

➤ Einstein joins the Institute of Advanced Study in America

➤ Einstein reverses his pacifist position

Fifteen years passed between World War I and World War II. That's not that much time, considering that the first one was called "the war to end all wars." During the 1920s, few people could have foreseen the disaster that was waiting on the horizon. And if they could have, they weren't paying any attention. "Let the good times roll" appeared to be the motto of America. The victors of the war were enjoying a lifestyle that reflected an insatiable appetite for living life to its fullest. But trouble was brewing at political rallies in Germany. The seeds of World War II had already been sown with the close of World War I.

The somewhat peaceful years of the roaring 20s ended climatically. In 1929, two significant events changed the lives of countless people, and a third significant event changed the life of at least one. America was hit by the Great Depression, Hitler rose to power in Germany, and Albert Einstein turned 50. This chapter examines the events that led to the Second World War and how these events affected Einstein.

Why the Treaty of Versailles Failed

In 1918, after months of negotiations, a series of agreements called the Treaty of Versailles brought World War I to an end. The agreements were highly contested, and in the next 20 years, they would totally collapse. The treaty ultimately failed for four reasons. The first had to do with how the lands of the three former empires were broken up. These Old World empires contained vast groups of minorities that had been oppressed by a dominant majority that had granted them only minimal rights. The three empires were

➤ Austria-Hungary

➤ Russia

➤ The Ottoman Empire

Minority groups within these empires tried to gain independence several times during the nineteenth century, but to no avail. The problem after World War I was how to draw up new political as well as ethnic borders. This feat was impossible, because the ethnic groups were all intermingled. Massive planned migrations was one effort to solve this dilemma. Four hundred thousand Turks moved from Macedonia to Turkey, and 1,300,000 Greeks moved from Asia Minor to Greece. These migrations did very little to ease the ill feelings of the minority groups, and in most cases the ethnic populations simply stayed where they were. Large German populations found themselves as part of Czechoslovakia, and others ended up as part of Poland. Hitler used this situation as one of the main themes of his propaganda.

A second reason the treaty failed was because the Ottoman Empire contained many German colonies and non-Turkish territories that were split up among the victors of the war. The Middle East eventually became a problem for both Britain and France as they tried to monitor and police ethnic problems.

Relatively Speaking

The Ottoman Empire existed from 1300 to 1918. It was ruled by the Turks and included within its empire were much of southeast Europe, southwest Asia, and northeast Africa. Its most famous ruler was the sultan Suleiman the Magnificent (1494–1566), who was in power during the Ottoman Empire's peak cultural achievements.

The third reason was probably the most directly responsible for the downfall of the treaty, because it involved Germany itself. Germany was perceived as being the main cause of the war in the first place and therefore was disarmed and had to pay severe economic and financial reparations. These payments for damages crippled Germany's economy and resulted in a state of near-starvation for millions due to the devastating inflation that it caused. Businesses went bankrupt, and Germany crumbled under the financial burden. German citizens considered these reparations to be extremely unfair.

The rejection of the treaty became the Nazi Party's first priority. Hitler as an individual would have had little chance of gaining such great political power if the conditions hadn't been ripe for his ascent to power. He knew exactly what his country wanted to hear and revived the notion of the superiority of the German race to compensate for the degrading position the treaty had imposed upon Germany.

The final reason why the treaty failed was the inability to enforce the territorial divisions and Germany's payments. The United States had refused to ratify the treaty, and the new League of Nations, the forerunner of the United Nations, had no powers of enforcement. In 1920, the Turks rebelled against the treaty and forced a revision.

On January 30, 1933, Adolf Hitler became chancellor of Germany. In less than a year, the Nazi Party was declared the only legal political party and soon brought most of public life under its control through intimidation and violence. By the time World War II began, Hitler had abolished the treaty, re-militarized the left bank of the Rhine River, reintroduced military service, annexed the German border region of Bohemia, and staked a claim to the Polish lands closest to Germany.

Unknown to the rest of the world, Germany's military power far exceeded that of any other country in Europe. In 1938, Hitler set out to resolve what he perceived as the problem of Greater Germany. He put the *Wehrmacht*, or war machine, into action by annexing all the lands inhabited by Germans. First came Austria, which only earned Hitler a mild rebuke from Paris and London. In an attempt to appease an increasingly aggressive Hitler, the prime minister of England, Neville Chamberlain, negotiated an agreement that essentially gave Czechoslovakia to Germany, adding this country to the growing list of the *Reich*'s annexations.

International opinion finally began to change when Hitler turned toward Poland. Neither Paris or London could back down again for fear of losing further face with the smaller European countries that Germany threatened. France joined Britain, as did the Soviets, in guaranteeing their support for Poland against Germany. Meanwhile, talks began in Moscow that led to a secret nonaggression pact between the Nazis and Soviets, which amounted to a tacit agreement to divide Poland between Germany and the Soviet Union. Hitler offered to divide the world between Germany and Britain, but Britain refused. On September 1, 1939, Hitler invaded Poland, and Europe was again at war. Three years later, Japan attacked Pearl Harbor, and World War II became a conflict of global proportions.

Mind Expansions

Reich is the German word for regal, or rule, usually combined to mean the right to rule. The First Reich was the Holy Roman Empire, regarded as the first German empire. The Second Reich refers to the German Empire from 1871 to 1919. In between the Second and Third Reich was the Weimar Republic, from 1919 to 1933. The Third Reich was the German fascist state ruled by the Nazis from 1933 to 1945.

Why Pacifism Was So Important to Einstein

The Depression hit the United States in 1929 and ultimately affected the entire world. Germany had already been ruined by its own financial crash in 1923, and while it was still trying to climb out of the basement, it was again devastated by the collapse of the American market.

Einstein survived Germany's economic catastrophe with the help of his wealthy friends, who provided him with more than enough money to live comfortably.

Nuclear Meltdown

Behind the scenes of any war, especially those of the twentieth century and most recently the Vietnam and Gulf Wars, arms manufacturers have had some of the most insidious connections to politics and national decisions. Billions of dollars of business result when a country is at war. The national debt increased a few trillion dollars during the 1980s during the arms race with the former USSR. Arms manufacturers gain large profits during wartime and weapons buildup periods. Therefore, these industries are motivated to encourage military activity. This problem is worldwide. It's difficult *not* to play the military game when other countries are playing actively and technology development has overtones of national security and espionage. In the free market system, profitability often outweighs damage to human life. One of its core beliefs is that might makes right. Our economic system works, but it's far from perfect.

Einstein had also cashed in on the success of relativity. His book *Relativity: The Special and General Theory* sold over 65,000 copies, which meant that it was a big hit. He also continued to receive salaries from his teaching position at the university and from the Prussian Academy of Sciences. Without having to worry about money to live, Einstein continued to actively participate in the political arena.

I've mentioned Einstein's pacifist beliefs several times; these beliefs were a major part of who Einstein was. From an early age, he disliked authoritarian behavior, the military mentality, and any groups, religious or political, that held themselves to be superior to others. These examples of Einstein's fundamental beliefs demonstrate the perspective from which he saw the world.

Einstein espoused pacifism so passionately because of his hope and vision for the future. His single most dominating belief was in the brotherhood, and sisterhood, of humanity. He considered international cooperation more important than loyalty to any single country, so he didn't believe in nations or nationalism. In his view, a single world government would lead to a more tolerant world, free of international bickering and national borders. With such a government in place, there would be no need for armed forces, and all the time and energy formerly spent on military actions could be spent in ensuring that everyone was provided for in every way. Consequently, there would be no wars, no starvation, and no haves and have-nots. The essence of this vision reflected the idea of unity in diversity.

Einstein was unbending with regard to pacifism, but as with some of his other views, Einstein sometimes relinquished parts of his philosophy for what he thought to be a greater good for all concerned. His stance on pacifism began to change as he recognized that the only way to overcome Germany's threat to the world was to take an offensive position against it. Pacifism would not quell the specter of Nazi fanaticism.

In a final plea for peace, Einstein wrote an article in 1934. Its conclusion still holds true:

> *The importance of securing international peace was recognized by the really great men of former generations. But the technical advances of our times have turned this ethical postulate into a matter of life and death for civilized mankind today, and made it a moral duty to take an active part in the solution of the problem of peace, a duty which no conscientious man can shirk.*

> *One has to realize that the powerful industrial groups concerned in the manufacture of arms are doing their best in all countries to prevent the peaceful settlement of international disputes, and that rulers can only achieve this great end if they are sure of the vigorous support of the majority of their people. In these days of democratic government, the fate of nations hangs on the people themselves; each individual must always bear that in mind.*

Einstein Hits the Half-Century Mark

In the late 1920s, although Einstein had his share of run-ins with anti-Semitism, he and Elsa were still in constant demand on the Berlin social scene. Einstein didn't care for these social functions. On one occasion, when asked to attend a celebrity dinner, he answered, "So you would like me to serve as a centerpiece?" In response to another invitation, he replied, "It must be feeding time at the zoo." He felt that these events embodied nothing but pretentiousness and were basically a waste of his time. Just to show what he thought of it all, he often came dressed formally, but without wearing socks. Most people thought that he was just eccentric and missed his point.

The only place he felt comfortable was at home with his family. Elsa was completely in charge of the house, and Albert fully supported this. His study was the only room in the house that no one was allowed to enter. Elsa acted as his buffer against the outside world. No one could see Albert without going through her first.

In 1928, Albert fell seriously ill. He had had a number of ailments towards the end of World War I that had left his body weak. Once, while walking home from the train station with heavy luggage, he collapsed. At the hospital, he was diagnosed with inflamed walls around his heart. This inflammation was a direct result of a lack of physical exercise; his only physical activity was a little walking. With Albert confined to his bed, Elsa insisted on taking care of everything, but it soon became apparent that she couldn't handle all of his correspondence.

Because of this illness, an important figure entered Albert's life. Helen Dukas was hired as his personal secretary. She took over many of the activities that required his attention in relation to the various peace movements he was involved with, the Zionist movement, personal letters that needed answering, and invitations to lectures and personal appearances. She accompanied the Einsteins on lecture tours and ended up moving with them to America, where she stayed with Albert until his death. Many personal insights into Einstein have been gained from interviews with Helen.

On March 14, 1929, Albert celebrated his fiftieth birthday. He received hundreds of presents, packages, and mail from around the world. He was swamped with calls of congratulations, as millions of people all around the world sent him their regards. It was a day of genuine warmth from friends and an admiring public, with a little absur-dity mixed in. Albert, of course, couldn't figure out what the big deal was. To him, it was just another birthday. But in response to all the presents and kind thoughts, he sent a copy of a short verse that he wrote along with a personal note to quite a few people. Here's the verse:

Mind Expansions

For Albert's birthday, the city of Berlin presented him with a house and a piece of land. Upon arriving at the house, Elsa was surprised to find that it came equipped with tenants who had no desire to leave. Of course the city council was shocked at the news and offered its deepest apologies. A second piece of land was offered to the Einsteins, this time with no house on it, so they decided to build one. How-ever, they found out that the zoning laws wouldn't permit the construc-tion of any new buildings there, so now they had a worthless piece of land. By the time everything was sorted out and they finally received a nice parcel of land that they were allowed to build a house on, all the goodwill behind the original gift was lost. They ended up building their own home, which used up almost all of their money.

> *Everyone shows their best face today,*
> *And from near and far have sweetly written,*
> *Showering me with all things one could wish for*
> *That still matter to an old man.*
> *Everyone approaches with nice voices*
> *In order to make a better day of it,*
> *And even the innumerable spongers have paid their tribute.*
> *And so I feel lifted up like a noble eagle.*
> *Now the day nears its close and I send you my compliments.*
> *Everything that you did was good, and the sun smiles.*

Some of his friends knew that he loved sailing, so they pitched in and bought him a sailboat. The city of Berlin purchased a piece of land for him, where the Einsteins built a new home. This home would be the last place he and Elsa lived before moving to America.

Over the next three years, Einstein spent quite a bit of his time traveling to get away from the hostility of Berlin and the growing spirit of anti-Semitism. He had been in contact with physicist Robert Millikan at Caltech in Pasadena and was invited to take up a temporary three-month teaching post. In the fall of 1930, Einstein went to America for his second visit.

Back to America

Several times, Millikan and his colleagues needed to defend their decision to offer Einstein a teaching position to nonscientists and representatives of the military. One military commander went so far as to accuse Einstein of teaching treasonous ideas to the youth of America. Although Millikan didn't agree with Einstein's political views, he hoped that while Einstein was teaching at Caltech he could be persuaded to permanently join the faculty. But many people were concerned about Einstein's extreme pacifism and his views on international politics.

While in California, Einstein had an opportunity to visit a number of America's physicists. He spent time at the Mt. Wilson observatory, which at that time housed the most powerful telescope in the world. There he met Edwin Hubble and his team of scientists, which developed the early theories of cosmology in the 1920s. (Hubble's contributions are covered in Chapter 24, "Big Bang or Bust.") Einstein also had a chance to go to Hollywood, where he spent an evening with Charlie Chaplin. Each of them had wanted to meet the other for many years, and they enjoyed their encounter immensely.

This visit was the first of two visits to teach at Caltech. Einstein returned to teach there in 1931. But his final move to America brought him to the East Coast of the United States, not the West Coast. His decision to teach in Princeton, New Jersey, at the newly formed Institute of Advanced Study was strongly influenced by its founder, Abraham Flexner. In an academic power play, Flexner outmaneuvered Millikan by using his friendship with Einstein to manipulate him into accepting the position at the Institute. Based on what we know about Einstein, he was probably not aware of the power play. Certainly had he known about it, he might have ended up in California instead of New Jersey.

With Einstein's future basically set, Albert and Elsa planned to leave Germany in December 1932. He had no idea at the time that this would be a permanent move. Albert planned to split his time between Princeton and Berlin, spending six months of the year at each location. But in a prophetic comment, the day he and Elsa left their newly built home, Albert turned to Elsa and said, "Before you leave our villa this time, take a good look at it." When she asked why, he said, "You will never see it again." He was right. Elsa died in Princeton in 1936, and Albert never set foot in Germany again.

Within a year of Einstein's decision, Hitler assumed total power in Germany. Now almost all of Germany, including the academic community, was against Einstein. Except for a few loyal friends who still defended him, including Max Planck who stood up to Hitler in a private interview, Einstein was not welcome in Germany, on the peril of his life. In the spring of 1933, an angry mob raided his house looking for him, but fortunately he was spending the winter in Pasadena.

Although most people in the United States liked Einstein and were interested in his work, a few groups like the National Patriot Council and the American Women's League felt that his ideas were worthless and that his political stance was nothing less than that of a communist. They requested that the State Department refuse him entry into the United States. Of course, Einstein didn't consider his views to be subversive or dangerous. In a typically humorous, and in this case sarcastic, response, he publicly stated:

> *But are they not right, these watchful citizens? Why should one open one's doors to a person who devours hard-boiled capitalists with as much appreciation and gusto as the Cretan Minotaur in the days gone by devoured luscious Greek maidens, and on top of that is low down enough to reject every sort of war, except the unavoidable war with one's own wife? Therefore give heed to your clever and patriotic womenfolk and remember that the capitol of mighty Rome was once saved by the cackling of its faithful geese.*

Albert Says

Every famous person is liked by some people and disliked by others. That could be said of most normal people, too. But when it comes to public figures, how does the public ever get to know who the person really is? Most of what we know about public figures gets filtered through the perspectives of others. The more public figures express their ideas and opinions, the more they open themselves to misinterpretation or becoming a representative of some group's or individual's philosophy. It took Einstein a long time to realize that the best policy was to remain silent, because in the end, people see only what they want to see.

Einstein never felt entirely at home in the United States. He came here in his early fifties, a late time for anyone to adjust to a whole new country and lifestyle. He learned to speak English pretty well, but he conducted most of his correspondences in German.

Prior to coming to the United States, Einstein spent some time in Belgium and became good friends with the queen. He wrote quite a few letters to her, many times sharing feelings that he didn't share with anyone else. In 1933, shortly after arriving in the United States, he wrote the following comments about his new home in a letter to the queen of Belgium:

> *Since I left Belgium I have been the recipient of many kindnesses, both direct and indirect. Insofar as possible I have taken to heart the wise counseling of those who urged me to observe silence in political and public affairs, not from fear for myself, but because I saw no opportunity for doing any good. Princeton is a wonderful little spot, a quaint and ceremonious village of puny demigods on stilts. Yet by ignoring certain social conventions, I have been able to create for myself an atmosphere conducive to study and free from distractions.*

In the following year, he wrote:

> *Among my European friends, I am now called the Great Stone Face, a title I well deserve for having been so completely silent. The gloomy and evil events in Europe have para-lyzed me to such an extent that words of a personal nature do not seem able to flow any more from my pen. Thus I have locked myself into quite hopeless scientific problems—more so since, as an elderly man, I have remained estranged from the society here.*

At the time Einstein wrote these letters, he was going through a tough moral dilemma. How could he continue to maintain his pacifist position in the face of Nazism? He realized that the Nazis were a severe threat to all of free Europe. Yet how could he go against his beliefs that violence and military action were inexcusable under any circumstances? Four years earlier, he had publicly stated:

> *In the event of war I would unconditionally refuse all war service, direct or indirect, and I would ask to persuade my friends to adopt the same position, regardless of how I might feel about the causes of any particular view.*

He had to resolve his dilemma. He did so by weighing one consequence against another. What is worse: war or letting the evils of Nazism go unfettered? He chose what he considered to be the lesser of two evils. Pacifism would not work as a tool against the Nazis. The only way they could be defeated was by fighting. For the world to be free and a safe place for everyone again, pacifism would have to be replaced by fighting. Here is an excerpt from a letter he wrote to a young French pacifist when he was asked to speak on the behalf of two Belgian conscientious objectors:

> *What I shall tell you will greatly surprise you. Until quite recently, we in Europe could assume that personal war resistance constituted an effective attack on militarism. Today we face an altogether different situation. In the heart of Europe lies a power, Germany, that is obviously pushing towards war with all available means … Imagine Belgium occupied by present-day Germany! Things would be far worse than in 1914, and they were bad enough even then. Hence I must tell you candidly; were I a Belgian, I should not, in the present circumstances, refuse military service, rather I should enter such service cheerfully in the belief that I would thereby be helping to save European civilization …*
>
> *This does not mean that I am surrendering the principle for which I have stood heretofore. I have no greater hope than that the time may not be far off when refusal of military service will once again be an effective method of serving the cause of human progress.*

Of course this view didn't sit well with Einstein's pacifist friends, who felt betrayed by one of their main leaders. However, when Bertrand Russell, who had been a pacifist all his life, saw the logic in Einstein's change of heart, it wasn't long before others followed suit.

At Home in Princeton

Albert and Elsa bought a house at 112 Mercer Street in Princeton, New Jersey. The small-town atmosphere suited him well. The townspeople were originally very curious about Einstein and somewhat surprised at his lack of formality and stuffiness. But they became a little more concerned when droves of European scientists began showing up

Mind Expansions

One of Einstein's longtime friends was the great English philosopher Bertrand Russell. They shared a love of logic and science and held almost identical views on pacifism and politics. Russell had been imprisoned by his own government during World War I for publicly denouncing its role in the war and speaking for pacifism. Until his death at the age of 92 in 1970, Russell continued to speak out for disarmament.

Mind Expansions

Even today, you can go to Princeton and run into someone with a grandfather or uncle who has a colorful tale to tell about a personal experience with Einstein. One of the most well-known stories is the story of the young girl who was having problems with her math homework, so she sought out Einstein for help. Her mother was getting concerned, because the little girl was disappearing for hours on end. One day she followed her and to her embarrassment found out that her daughter had the most famous scientist in the world helping her do her math homework. Smart kid.

at his door. The town wasn't completely free of anti-Semitism, and some people worried that all of these foreigners would disrupt their little community.

For the most part, the people were thrilled to have the famous scientist living in their little town. They observed his right to privacy and avoided bothering him or asking him questions. Einstein loved that, and without any teaching commitments, he was free to think, walk, and pursue his research at the Institute.

Abraham Flexner had been responsible for luring Einstein to the Institute in the first place and knew what a catch Einstein was. Probably Flexner was not as interested in what Einstein could contribute to the study of physics as he was in how much Einstein's name would help the Institute grow and attract still other big names. He was therefore somewhat overprotective of Einstein and a little paranoid that someone or something would make him a better offer and lure him away from the Institute (like he lured Einstein away from Millikan at Caltech).

Flexner checked all of Einstein's mail and phone calls, and anything that he felt could jeopardize Einstein's position at the Institute disappeared, including President Roosevelt's invitation to the White House. One of Einstein's friends heard about the invitation and was surprised that Einstein didn't want to go meet the president. When Einstein realized what had happened, he wrote a letter to the Board of Trustees listing all of Flexner's improprieties and calling Princeton a prison camp. In the end, Flexner relaxed his hold on Einstein, but Einstein was no longer running the Institute.

In the next three years, Albert experienced a number of personal tragedies. The year of his arrival at Princeton, he heard of the mental breakdown of his youngest son Eduard in Europe. Eduard never evidenced any symptoms of mental illness, so everyone was surprised when he very suddenly began exhibiting severe schizophrenia and had to be confined to a psychiatric hospital in Switzerland. There is no written evidence of Albert's feelings, but even though his two sons lived with Mileva, he wrote to them a few times a year. He was no doubt shaken at the news.

In the summer of 1934, Albert's stepdaughter in France became seriously ill. She was diagnosed with cancer. Albert stayed in the United States while Elsa visited her daughter. A number of specialists were sent for, but to no avail. Shortly after Elsa's arrival, her daughter died.

Einstein at home in Princeton.

Elsa herself had not been well. In the summer of 1935, she had her eyes examined because of swelling and inflammation in one of the eyes. She learned that she had a serious kidney and heart disorder. Surgery was the only possible cure, but Elsa refused this treatment. She spent her remaining months with Albert and died in December 1936.

There are a lot of interpretations about how Albert dealt with Elsa's death. Some people claim that he was rather aloof and surprisingly cold about the whole thing. Others feel that it affected him more than he could express, privately or publicly. Either way, Helen, his personal secretary, now became the main woman in his life. She became his cook, housekeeper, social organizer, and correspondence writer. She always made sure that he left the house with a hat and change in his pocket. She remained his closest helper until the day he died. The next chapter looks at the development of the atom bomb, the birth of Israel, and the death of Einstein.

The Least You Need to Know

➤ The seeds of World War II were sown at the end of the World War I. The Treaty of Versailles placed an excessive economic burden on Germany, and it couldn't recover.

➤ Einstein's pacifist ideology was seen as an extreme view by many, and some people considered him to be a communist at heart. Nothing could have been farther from the truth.

➤ Einstein almost wound up at Caltech, but because of events that played out behind his back, he ended up in Princeton.

➤ The inability to stop the Nazis through pacifist means caused Einstein to reverse his stance, and he came to support military intervention.

➤ After 20 good years of married life, Einstein's second wife Elsa died in 1936.

Chapter 22

Who's Got the Bomb?

In This Chapter

➤ The search for a unified field theory

➤ The discovery of nuclear fission

➤ Einstein's letter to President Roosevelt

➤ The communist witch hunts

➤ Einstein's return to pacifism

➤ The death of Albert Einstein

In the years following World War I, physicists looked to Einstein as a leader in their field. Not so much for his theories on relativity, but more for his help in the emerging field of quantum mechanics. Einstein was instrumental in the founding of quantum theories, but as evidenced by his famous debates with Niels Bohr, he didn't think much of the quantum theories that were developed thereafter. For the most part, Einstein turned his back on quantum mechanics, and this action ultimately removed him from much of mainstream physics. But his theories continued to play a key role in the larger study of the universe—the macrocosm, rather than the microcosm. Part 6, "Worlds Beyond Einstein," examines his influence in this area of study.

This chapter covers the last part of Einstein's life. It takes a brief look at his search for a unified field theory, his effort to convince President Roosevelt to develop the atom bomb, and his role in establishing a permanent home for the Jews.

Toward a Unification of Forces

By the time Einstein arrived in Princeton, he no longer questioned the predictions or accuracy of quantum mechanics. These had been verified through a number of experiments. But Einstein still felt that quantum mechanics was an incomplete system, just an intermediate step to a more complete understanding of the universe. The general theory of relativity was very useful in explaining reality on a large scale, the macrocosm, and quantum mechanics was very useful at explaining reality on a very small scale, the microcosm. Einstein wanted to develop a classical theory that could do both.

As early as 1918, Einstein was already thinking about developing a theory to bring together gravitational and electromagnetic forces. In July 1925, he submitted a paper to the academy in Berlin entitled, *The Unified Field Theory of Gravity and Electricity*. As he had been so many times in the past, he was convinced that he was on the right track. He was a little concerned as to whether his field theory could explain atoms and light quanta, but with his usual confidence and optimism, he was certain that he was pursuing the correct course.

At the heart of his search was the idea that this theory couldn't be founded on probabilities. It was this belief that would divide the rest of his life in physics into two main themes:

➤ He never agreed that quantum mechanics was a complete system.

➤ He unceasingly persevered to unite gravitation and electromagnetism into a single unified field.

Relatively Speaking

The **unified field theory** was Einstein's concept that all of physics could be brought together under one theory or unifying idea. He spent his last 40 years searching for this elusive theory and never found it. Today it's also called the **TOE,** or the **theory of everything.** In essence, it's any theory that can combine gravity with some form of the **grand unified theory,** which has attempted to combine all the known forces into a single package. However, each form of the grand unified theory developed so far has been flawed, so the search continues.

For over two decades, Einstein firmly believed that the answer could be found. He lost himself in abstractions and calculations that yielded no positive results. He studied advanced mathematics and tried to develop complex combinations of formulas that did nothing but get him more entangled in irresolvable dead ends. Throughout the course of these failures and letdowns, on a lonely road that everyone else considered folly, Einstein pursued his destiny with a smile. He never despaired, regardless of his failures, and maintained his almost inexhaustible optimism until his last breath.

A Theory of Everything

So what exactly is a *unified field theory* anyway? Einstein's idea was to develop a fundamental theory for the entire study of physics. It would not only bring together gravity and electromagnetism, but it would also explain the properties of elementary particles, quantum interactions, and the velocity of light and establish clear mathematical formulas that everyone could use. This very ambitious program has not been resolved even today. Today the unified field theory concept is synonymous with the TOE, better known as the *theory of everything*.

Einstein's search began with the two most fundamental fields known at the time: gravitational fields and electromagnetic fields. Do you remember what a field is? Basically, it's a concept that defines a region of space that contains some force, such as gravity, electricity, magnetism, or the combination of those two as electromagnetism.

Many classical physicists have pursued the unification of forces. This is one of the reasons why we spent time looking at how separate forces such as magnetism and electricity have historically been understood and unified. Einstein's special theory of relativity combined Maxwell's equations of electromagnetism with Newtonian motion into a new system of mechanics, called *relativistic mechanics*, that described the movement of forces through space without requiring a concept like ether. Finally, Einstein's general theory of relativity explained how the field theory of gravity takes a tangible form that distorts the very fabric of space and time.

Mind Expansions

After Einstein developed his general theory of relativity, he came to the realization that mathematics was the only true expression of physical phenomena. Many researchers have come to understand the profound connection between mathematics and physics. Theoretical physicists such as Einstein would not have advanced very far without an innate appreciation of the harmony and beauty that mathematics can convey in describing the natural world. A number of times, a mathematical structure or formula has predicted the existence of its physical counterpart. Paul Dirac's formulas are an example. His equations predicted the existence of antimatter, and not long afterward, it was discovered.

Trying to Make Two Become One

Let's look at some of the properties that distinguish gravitational and electromagnetic fields. Both can reach equally far in space, but their strengths are very different. Electromagnetism is much, much stronger than gravity. The ratio between the strengths of these two fields is a number with 40 zeroes at the end of it.

Another major difference is that electrical (or magnetic) forces between charged particles are attractive or repulsive. Gravity, on the other hand, has no duality. Masses are universally attracted to each other. There is no repulsive force of gravity.

Einstein wanted to unify the fields of gravity and electromagnetism. But he also had to deal with solid matter. At the time, the elementary building blocks of the universe were thought to be the proton and the electron. Every substance was thought to be composed of these two basic particles. This theory appeared rather orderly and natural to Einstein, as well as other physicists, so in the 1920s, the pursuit of a unifying theory seemed to be the next logical step.

By the 1930s the picture had begun to change, and many physicists couldn't believe that Einstein continued to cling to his belief in a single theory. Quantum mechanics was proving to be a powerful new tool, and new particles, new fields, and new forces were being discovered. But Einstein saw these new developments as elements of a grand theory that would incorporate everything. He hoped to show that quantum mechanics might just be a mathematical corollary that helped to explain a larger, more comprehensive theory. It never happened, and everyone tried to warn him that it wouldn't work. But you know Einstein—if he thought he was right, it was useless to attempt to persuade him otherwise.

Is It Fission Yet?

While Einstein continued to pursue his dream, mainstream physics was making enormous progress. Who knows what new discoveries could have been made if Einstein had added his own intellectual efforts instead of stubbornly refusing to participate? Experiments used the principles of quantum mechanics, and soon other physicists began making names for themselves with their discoveries. One of the most important contributions came from a young English physicist by the name of Paul Dirac.

In 1927, Dirac developed a set of equations for quantum theory that also meshed with relativity. He essentially unified the two systems to give a complete description of the electron. He was able to develop two sets of solutions: one for negative electrons and one for positive electrons. At this point you may be asking, "Don't electrons carry a negative charge? How can they be positive?" Your question is insightful, as usual. Dirac's equations predicted the existence of *antimatter*. Specifically, what he predicted was the existence of an antiparticle, which has come to be known today as a *positron*. Positrons were the first antiparticles to be discovered; Carl Anderson found them in cosmic rays in 1932. Dirac also opened the door to what would become an essential addition to quantum mechanics: *quantum field theory*.

Relatively Speaking

Antimatter is a form of matter in which the properties of a particle are the opposite of those of its counterpart. The most common example of antimatter is the antiparticle or antielectron, better known as the **positron**. This particle has the same mass as an electron, but the opposite electrical charge. It's positive instead of negative. There are antimatter counterparts to every particle known today. When matter and antimatter meet, they annihilate each other, and their masses are converted to energy via Einstein's famous equation. Antimatter has been artificially created using huge particle accelerators, but antimatter particles are very difficult to study because they exist for only about forty billionths of a second. Talk about a short lifetime!

Quantum mechanics was useful in explaining how chemical bonds happened. In chemical reactions, atoms bond together to form molecules. This interaction of atomic particles could be predicted and calculated very accurately using quantum theory. Some physicists believed that it wouldn't be long until all of chemistry and physics could be explained. They were either overconfident or just plain naive. Chemistry and physics were far from being completely solved.

Nuclear reactions were another area that quantum theory helped describe. In the early years of nuclear physics, one interesting feature that was discovered about the nucleus was that its mass and electrical charge didn't match. As you know, the simplest atomic nucleus is that of a hydrogen atom. It contains only one proton, which comprises its mass. The nucleus of a helium atom, the next heaviest after hydrogen, has two protons. However, the mass of the helium atom is equal to that of almost four protons. Where does this extra mass come from?

It was theorized that the nucleus must contain electrically neutral particles that account for the extra mass. The *neutron*, which I've already discussed, was the culprit. One result of this theory was the idea that a neutron could make an excellent probe into an atomic nucleus. It had no electrical charge, so it wouldn't be repelled or attracted by the protons or electrons. Before long, groups of physicists in different countries began probing the nuclei of various substances.

One group in particular, working out of Italy, used neutrons to probe the nuclei of uranium atoms. The leader of this group was Enrico Fermi. The scientists bombarded a shielded piece of uranium with neutrons and induced a process called *fission*. Fission occurs when a nucleus is broken up by a particle, in this case a neutron, and two lighter elements, in this case boron and krypton, are formed along with a number of other neutrons. What they didn't realize, because the uranium was shielded, was that the fission produced a strong energy-producing reaction. Einstein's famous formula had been demonstrated. The mass of a chunk of matter had been converted into energy through the process of fission.

If they had realized that energy had been released, the Fermi group would have been the first to detect nuclear fission. Instead, this discovery took place in Germany in 1938. Otto Hahn and Fritz Strassman in Berlin performed the same experiment and correctly interpreted the results.

Relatively Speaking

Fission is a process in which the nucleus of an unstable atom such as uranium splits into two or more parts, releasing energy and additional neutrons. Only certain heavy atoms (heavy referring to the number of protons and neutrons in the nucleus) can produce fission. When fission occurs in an unstable nucleus, it triggers a spontaneous chain reaction. Think of this as something like a row of dominoes. After one falls, the rest follow very quickly, one right after another. If this chain reaction is not controlled, a nuclear explosion results, like that produced by an atom bomb. If it's controlled, the energy from this chain reaction can be put to good use, like what happens in a nuclear reactor. In a reactor, the fission process is controlled by slowly inserting a material into the reaction chamber that absorbs some of the neutrons. This way, the chain reaction can't happen spontaneously.

The Race Is On

Niels Bohr brought news of the achievement of fission to Princeton. The news spread like wildfire. What were the implications? One of the first questions to be asked was whether additional neutrons were released during fission. If so, a chain reaction could occur, in which the extra neutrons that were released could break up another nucleus, free still more extra neutrons, which in turn would break up more nuclei, and so on. In this chain reaction, more and more energy would be released each time until blammo! you would have one massive explosion. If extra neutrons weren't produced, then there would be no such explosion. Fission would be an interesting experiment to write home to the family about and nothing more.

Relatively Speaking

An **isotope** is an atom of an element with a normal number of protons in its nucleus, but a different number of neutrons. For example, uranium 235 and uranium 238 have the same number of protons. They would have to—otherwise, they'd be different elements. But they have a different number of neutrons. Both have 92 protons, but U238 has 146 neutrons, and U235 has 143 neutrons. The number of the isotope signifies the total number of protons and neutrons in the nucleus. Many elements have isotopes. Some occur naturally, and others are created artificially in laboratories.

To answer this question, experiments had to be performed to monitor the output of additional neutrons, and a suitable element would have to be found to conduct the experiment on. Uranium seemed to be the only substance that allowed fission to take place. But some uranium worked, and some didn't. The key to solving this problem was the discovery of *isotopes*. Isotopes are different forms of the same element. For example, uranium naturally exists as several different isotopes. The only difference between them is the number of neutrons contained in their nucleus. Uranium 238 is the most common isotope; it contains 92 protons and 146 neutrons. The total number is 238, hence the name of the isotope. Only a very small amount of the Uranium 235 isotope exists in the world. This isotope is the one that produces fission. It has three fewer neutrons in its nucleus.

To make a nuclear weapon, the fuel would have to be Uranium 235. The process of separating it from other isotopes of uranium is difficult and very expensive. A tremendous amount of resources would be required to be able to obtain sufficient quantities of Uranium 235 for this purpose. As it turns out, developing the atomic bomb was the most expensive scientific project that the United States had funded up to that time.

Two large deposits of Uranium 235 were known to exist in the entire world. One was in Czechoslovakia, and the other was in the Belgian Congo. Hitler had already annexed the former to Germany, so it was important that the second source wasn't also lost. Because Einstein was a good friend of the queen of Belgium, the State Department thought that he should write a letter to her warning her of Hitler's possible desire to control the region and conveying the need for good diplomatic relations with the United States. However, the State Department felt it would be better for Einstein to contact President Roosevelt first.

The person responsible for establishing the Allies' atomic weapons program was a Hungarian Jewish physicist by the name of Leo Szilard. (I've mentioned that name before. He and Einstein had taken out a patent together for a noiseless refrigerator.

If you remember that, you'd do well in any trivia contest.) Back in 1938, Szilard wanted to alert Roosevelt to the possibility of Germany developing a nuclear weapon. He also knew of the large deposit of uranium in the Belgian Congo. But nobody knew anything about Szilard or his theories of fission, so he realized that if he wanted to get through to Roosevelt, he'd need someone like Einstein on his side.

What blows a lot of people away is that Einstein didn't figure out the potential development of nuclear fission himself. This isn't surprising though, for the following reasons:

➤ He had been working alone on his unified theory and was out of touch with mainstream physics.

➤ Although he read most of the scientific journals, he wasn't interested in anything that didn't relate to the work he was pursuing. Most of what he read dealt with quantum mechanics.

➤ He never thought there was much potential for nuclear energy.

But all this was about to change. Szilard, with the help of another Hungarian physicist, Edward Teller, convinced Einstein of the need to contact Roosevelt and to pursue research into an atomic weapon. On August 2, 1939, Einstein wrote the following letter to Roosevelt:

Mind Expansions

Frederic Joliot-Curie was a French physicist who married Marie Curie's daughter, Irene. His experiments with neutrons showed that they could be used to produce fission. He published his findings in open scientific literature, making it available for all to see. Leo Szilard, who was almost obsessed with making sure that the Allies developed a chain reaction fission process before the Germans did, just about went crazy when he saw that Joliot had published his findings in the open.

> Sir:
>
> Some recent work by E. Fermi and L. Szilard, which has been communicated to me in manuscript, leads me to expect that the element uranium may be turned into a new and important source of energy in the immediate future. Certain aspects of the situation seem to call for watchfulness and, if necessary, quick action on the part of the administration. I believe, therefore, that it is my duty to bring to your attention the following facts and recommendations.
>
> In the course of the last four months it has been made probable—through the work of Joliot in France as well as Fermi and Szilard in America—that it may become possible to set up nuclear chain reactions in a large mass of uranium, by which vast amounts of power and large quantities of new radium-like elements would be generated. Now it appears almost certain that this could be achieved in the immediate future.

This new phenomenon would also lead to the construction of bombs, and it is conceivable—though much less certain—extremely powerful bombs of a new type may thus be constructed. A single bomb of this type carried by boat or exploded in a port might very well destroy the whole port altogether with some surrounding territory. However, such bombs might very well prove to be too heavy for transportation by air.

The United States has only very poor ores of uranium in moderate quantities. There is some good ore in Canada and the former Czechoslovakia, while the most important source is in the Belgian Congo.

In view of this situation you may think it desirable to have some permanent contact maintained between the administration and the group of physicists working on the chain reaction in America. One possible way of achieving this might be for you to entrust with this task a person who has your confidence and who could perhaps serve in an unofficial capacity. His task might comprise the following:

a) To approach government departments, keep them informed of further developments, and put forward recommendations for government action, giving particular attention to the problem of securing a supply of uranium ore for the United States.

b) To speed up the experimental work which is at present being carried on within the limits of the budgets of the university laboratories, by providing funds, if such funds be required, through his contacts with private persons who are willing to make contributions for this cause, and perhaps also by obtaining the cooperation of industrial laboratories which have the necessary equipment.

I understand that Germany has actually stopped the sale of uranium from Czechoslovakian mines which she has taken over. That she should have taken such early action might perhaps be understood on the ground that the son of the German Undersecretary of state, von Weizsacker, is attached to the Kaiser Wilhelm Institute in Berlin, where some of the American work on uranium is now being repeated.

Yours very truly,

A. Einstein

Why didn't Hitler complete the bomb first? The German government was always quick to see a potential military use of a weapon, so why didn't they beat America to the punch? Here are some possible reasons:

➤ Hitler's military scientists were more focused on the development of rockets to deliver mass destruction. They spent a considerable amount of time developing the V1 and V2 rockets that they launched at England.

➤ Germany was more concerned with using fission as a source of power for industry rather than in war.

➤ Hitler seemed uninterested in what nuclear energy had to offer.

➤ Many of Germany's scientists involved in fission experiments suppressed news of discoveries, knowing what would happen if this information fell into Hitler's hands.

➤ The Allies had no idea that they had the advantage all along until after the war was over. But this lack of knowledge was at least partially responsible for the establishment of the *Manhattan Project* and the tremendous amount of energy and resources that went into the development of the atom bomb.

Einstein's involvement with the development of the atomic bomb was minimal. Despite what people tend to think, he wasn't the father of the atomic bomb. He made only two contributions to its production:

➤ His famous equation, $E = mc^2$, defined the convertibility of mass into energy, which partially explained what happened during nuclear fission.

➤ He wrote the letter to Roosevelt prompting him to take action before Germany did.

He was never involved in the Manhattan Project and was never present at any nuclear test. While physicists were working on the bomb in the rooms down the hall from him at the Institute, he was working alone on his unified field theory, oblivious to what was going on around him.

Although Einstein realized that notifying the president was an important matter, it hadn't been an easy decision for him. Even though he had reversed his position on pacifism, advocating atomic weaponry was another matter. But he knew that if the Allies didn't develop the bomb first, the Nazis probably would. As he had done before, he chose a course of action that went against his basic principles in favor of what he hoped would be the greater good.

Relatively Speaking

The **Manhattan Project** was the code name for the immense project that was undertaken by the United States in Work War II to build the first atomic bomb. Among the physicists heading up the project were Richard Feynman, Enrico Fermi, Edward Teller, and the man who many consider to be the father of the atom bomb, Robert Oppenheimer. The project was officially set up December 6, 1941, one day before Japan attacked Pearl Harbor. Three days later, Germany declared war on the United States.

Atom bomb explosion.

Einstein's Final Years

On August 6, 1945, the first atomic bomb was dropped on Hiroshima. About 70,000 people died instantly, and around 100,000 died later from radiation poisoning and burns. When Einstein heard the news of its use over the radio, his secretary Helen Dukas said that he cried out, "Oh horrible."

By the end of World War II, the secret of the atom bomb was in the possession of the United States, Britain, and a short time later, France. But within a few years, both Russia and China had built atomic bombs, too. Many of the physicists who had been involved in the development of the bomb formed a group called the Emergency Committee of Atomic Scientists. Einstein was the president of the committee. In a speech delivered in New York in this capacity, he said:

> We helped create this new weapon in order to prevent the enemies of mankind from achieving it first; given the mentality of the Nazis, this could have brought about untold destruction as well as enslavement of the peoples of the world. This weapon was delivered into the hands of the American and British nations in their role as trustees of all mankind and as fighters for peace and liberty; but so far we have no guarantee of peace nor any of the freedoms promised by the Atlantic Charter … The war is won—but the peace is not.

The existence of the atom bomb brought on the Cold War, which lasted almost 50 years. The distrust that existed between the superpowers confirmed for Einstein the need for a world government. Although the United Nations charter was signed in June 1945 by 51 countries, Einstein didn't believe that this was sufficient to guarantee peace. What he wanted was a Federal Constitution of the World, with a worldwide legal system.

Communist Witch Hunts

Einstein returned to pacifism with a vengeance. With the relationship between the United States and the Soviet Union slowly growing colder, he called for bold action and a radical change of our mentality to save civilization. But few heeded his call. The atomic age had ushered in a new fear and paranoia. Instead of disarmament and limitations upon nuclear development, the superpowers (the United States, the Soviet Union, and China) went in the opposite direction. The Soviet Union, the former ally of the United States, now became its most hated enemy. Communism was seen as the most serious threat that the free world had ever known, and with the Soviet Union's ability to destroy civilization, anti-Communist hysteria broke loose across the United States.

Unfortunately, Einstein's pacifism was no longer just seen as anti-war. Many viewed him as anti-American, even as a communist. Einstein had come under the scrutiny of the FBI as far back as 1939. By 1940, J. Edgar Hoover had a 1,500-page file complied on him. He was seen as an extreme radical, and it was due to Hoover's recommendation to President Roosevelt that Einstein was not involved in the Manhattan Project or anything else related to the development of the atomic bomb.

The idea that Einstein was a communist sympathizer borders on absurdity. However, with the rise of Senator Joseph McCarthy and his desire to purge America of any and all communists, Einstein was perceived as being in the enemy camp. Einstein denounced McCarthy's accusations and bullying manner, and this made him appear even more suspicious. Einstein was afraid that the United States was itself becoming a fascist nation. It was restricting civil liberties, mistreating black Americans, and beginning to support vicious dictators in other countries.

Things became even worse when Walter Winchell, a famous radio gossip columnist, sent Hoover a listener's letter that identified 17 different

Nuclear Meltdown

Figuratively speaking, McCarthyism was the nuclear meltdown of American government and society in the 1950s. McCarthy's brazen disregard for civil rights was equal to his political ambitions and instinct for putting himself in the spotlight. With the backing of a congressional committee, he proceeded to ruin the lives of many innocent people. This era of American history reflects a nation so fearful of Communism and nuclear war that even fanatical bullies such as McCarthy gained wide support. McCarthy retained his power for a number of years, and his legacy of fear and hatred influenced many lives for years after he was dethroned.

communist groups of which Einstein was allegedly a member. But in the end, there was no proof that Einstein supported any communist doctrines or groups. Hoover was afraid to confront the popular scientist, and all that ever came of it was that Einstein was kept under close watch.

Israel Becomes a State

With the end of the war came the horrible news of the Holocaust. The survivors of the Nazi concentration camps told the story of the killing of millions of innocent men, women, and children, most of them Jewish. The ghastly news shocked the world. From that point on, Einstein had a firm hatred of anything German. When he was asked to join a number of German academies again, he denounced their leaders as cowards. He said, "The crime of the Germans is truly the most abominable in the history of the so-called civilized nations."

This was a very troubled time for Einstein. He was battling accusations of being a communist and at the same time standing firm on his pacifism and support of nuclear disarmament. The news of the use of the atom bomb troubled him to no end, and the Holocaust made him bitter, sad, and angry at his old homeland. In 1948, his first wife Mileva died. His sister Maja passed away at his home a few years earlier. He was close to 70 years old and a sick man himself. He had had another operation to repair a weak area of his aorta, one of the main arteries of the heart. He was warned to take it easy, but he felt that there was still too much to be done. He said, "Death comes to everyone; it is an old debt that will soon be paid."

In November 1952, Einstein was given an honor that moved him greatly. In May 1948, Israel had become a state, and its first president was Einstein's Zionist friend, Chaim Weizmann. In November 1952, Weizmann died with no apparent successor. Einstein received an invitation to become the second president of Israel. The country was actually run by the prime minister, so the title was really just honorary with little political power. Einstein replied in a letter to the ambassador of Israel in Washington:

> I am deeply moved by the offer from our state of Israel, and at once saddened and ashamed that I cannot accept it. All my life I have dealt with objective matters, hence I lack both the natural aptitude and the experience to deal properly with people and to exercise official functions. For these reasons alone I would be unsuited to fulfill the duties of that high office, even if advancing age was not making increasing demands on my strength.
>
> I am the more distressed over these circumstances because my relationship to the Jewish people has become my strongest human bond, ever since I became fully aware of our precarious situation among the nations of the world.

All Great Ones Die, Too

By this time Einstein's health was quickly deteriorating. His weakened aorta had developed an aneurysm that was growing in size, and surgery would be required again to remove it. This time Einstein decided to let nature take its course. In April 1955, Einstein began to suffer severe stomach pains. The last few years of life were probably very painful, but he never complained, always responding when asked, "I'm well."

On April 12, 1955, Einstein collapsed in his home. Helen Dukas called the family doctor, who administered morphine injections. Two days later, Einstein's condition turned for the worse. It was thought that it would be best for him to be hospitalized, for better care. The doctors that took care of him during his last few days remember him being clear-headed and very calm. He knew that he was on the verge of death and was completely unafraid. Around 1:00 A.M. on April 18, his night nurse came in to check on him and found that his breathing was labored. She raised his head on a pillow and heard him mumble something in German. She didn't know the language, so she had no idea what he said. Shortly thereafter he died.

Einstein had been very clear about his funeral arrangements. He wanted no funeral, no grave, and no monument. He was to be cremated, and nothing of his was to become a shrine. After his death, his wishes were all carried out. His ashes were hidden, and it is thought that they were later thrown secretly into the Delaware River. The life of Albert Einstein came to an end, but his death is not the end of the story. The last part of this book shows how Einstein's theories influenced physics after his death. It goes beyond Einstein into the world of cosmology, where what we know about the universe changes everyday. Quantum theories are nothing compared to what's ahead.

Mind Expansions

Before Einstein was cremated, an autopsy was performed to learn the exact cause of death. With his family's approval, his brain was removed and carefully examined under a microscope to see if anything could be discovered that might give a physical clue to his genius. Nothing was found. It looked like the brain of any elderly man. Dr. Thomas Harvey, the Princeton pathologist who did the preliminary work, has kept Einstein's brain with him, at times in a jar behind a beer cooler in his office. In 1996 however, research done in Canada on some of Einstein's brain tissue showed that a key region was significantly larger than the same area in people with normal intelligence. The inferior parietal lobe, which processes mathematical thought, three dimensional visualization, spatial relationships and other processes was 15 percent wider. This could possibly account for his brilliance.

Newspaper headline announcing the death of Einstein.

The Least You Need to Know

➤ Einstein spent the remaining years of his life outside of mainstream physics, searching for a theory that would unify all the known forces in physics.

➤ Paul Dirac developed the first mathematical equations that unified elements in relativity with quantum mechanics.

➤ The use of neutrons to probe the nucleus of atoms led to the first experiments in fission.

➤ Einstein was involved in the development of the atom bomb in only two ways. First, his famous equation described the equivalence between matter and energy. Second, he wrote a letter to President Roosevelt encouraging him to pursue research into the development of atomic weapons.

➤ Einstein was offered the second presidency of Israel, and he humbly declined.

➤ Albert Einstein died on April 18, 1955.

Part 6
Worlds Beyond Einstein

We've almost come full circle in our understanding of the universe and how Einstein fit into that picture. We began our journey with a look at how ancient civilizations viewed the world in which they lived. The questions that they had about how the universe began were explained by creation myths. Some stories told of a fiery beginning; others explained creation from the waters of eternity. Some even had a divine being that brought forth the universe with his or her breath. Many of these ideas were incorporated into the stories of creation in both Eastern and Western religious traditions.

Here we are again, about to look at the same questions, mostly from the perspective of science. In this look at the latest theories and discoveries, you'll see how Einstein's theories helped to enhance our understanding of the galactic universe. I'll also try to present a larger perspective by including other views as well. Science and religion have their own views on how the universe began, and a dialogue between the two could prove interesting. Regardless of which approach you take, both are incorporated in the study of cosmology. Who knows, maybe someday science and religion may find a common ground that can explain creation using both systems of thought.

We'll close with a brief examination of some of the more interesting and fascinating ideas that have come out of the study of cosmology and human consciousness. Included are some Eastern philosophical interpretations that discuss how intimately connected we are to our world. You're on the final leg of the journey, the journey into the realms of understanding.

Cosmology and Consciousness

The desire to know how the universe began is as old as humanity itself. The obvious follow-up question is: how will it end? The study of cosmology tries to answer both of these questions, as well as others that deal with the role that humanity plays in the universe. Science seeks to provide answers based on observation and experimentation, mostly devoid of any participation by humanity. Religion approaches these questions from the other side, trying to provide an understanding of the universe as being purposeful and directly related to the existence of humankind.

This chapter takes a look at both of these perspectives, to see whether one can shed light upon the other and whether it's even feasible to answer these huge questions. My main goal is to provide you with an understanding that goes beyond the thinking exclusive to either science or religion.

What Price Glory?

What is it about human nature that causes us to identify so strongly with what we believe to be the truth? Is it that we need to feel that what we know somehow makes us better than others? Are we simply feeding our egos along the way, and thereby increasing our sense of self-worth? Or it could be more impersonal. Maybe it's the idea that what we believe to be true provides us with insights into the real nature of things. There are a lot of reasons why we believe what we believe so strongly. This chapter explores some of them.

Albert Says

Some of the discussion in this chapter seeks to explain various aspects of human consciousness. These ideas are taken from many different sources and are combined in a way that tries to present an overview of how consciousness is understood today. You may disagree with some of it, agree with other points, or find that your opinions fall somewhere in between. Whatever you think, be open to any insights that this chapter has to offer, regardless of your own personal paradigm.

This book has provided a number of examples of how beliefs act as filters of perceptions about reality. Because science and religion are both systems of thought that have been created by humankind's understanding of the world, each of these systems is worthy of a closer look.

Someone once said that you are what you eat. You are what you think is probably a little closer to the truth, although if you can't pass by a fast-food restaurant without going into withdrawal, you may in fact be closer to the first statement than the second. The ability to think about things gives people the opportunity to question the beliefs they hold so dear. But how often do they do that? Most people assume that because experience has taught them that certain things are true, that's that. This assumption pertains to beliefs about money, relationships, people, career, health, religion, and science, just to name a few.

But where are *we* in all of this? What is it that provides us with a sense of self-identity? Let's look at the different aspects of ourselves that go into defining who we are. These aspects fit into four main categories.

I Have a Body

The physical body is a good place to start. For many people, their bodies are their prime focus of awareness. The needs of the human body, on just a biological level, require you to feed it, keep it healthy, and provide it with rest and sleep. All of these things keep it running smoothly.

The quest to understand the human body on a biological level lies in the realm of medicine. The evolution of medicine is the story of science's progress in unfolding the underlying structure of how the human body functions. In the West, this structure is explained through a strictly scientific, biological study. In the East, the human body is understood as a system of energy. Both paradigms have worked well for their corresponding cultures, and an interesting dialogue is beginning to take place between the two.

We all have bodies, but aren't we more than that? Is your physical body the only thing you are aware of that comprises *you*? Of course not. We are more than just our bodies.

I Have Emotions

Emotions are one of the more difficult aspects of people to identify and define. Emotions provide us with our greatest pleasures as well as our most intense pains. Are they just a product of the physiological workings of the brain? They're surely a manifestation of our consciousness, and they don't belong solely to humans. We see other forms of life exhibiting emotions also, especially animals.

Where do emotions come from within yourself? Is there a point of origin that you can identify as the source of a particular emotion? Do you feel emotions in various places in your body, or is it an overall quality? Have you ever tried to describe in words what an emotion is? Define joy, or sadness, or anger. You can't. All you can say is what it's like or give an example. Or you can describe how it makes you feel, using other emotions in that description.

Psychology seeks to understand how the mind and emotions are linked. This concept was discussed briefly in Chapter 19, "How Conscious Are We?". We spend a lot of time with our emotions, and yet we understand very little about them. Some we can control, and we are at the mercy of others. But we are more than just our emotions.

Mind Expansions

Most of Western science seeks to define consciousness in terms of brain functions, so science's answer to the question of where emotions come from is the brain. However, many people feel emotions in various parts of their bodies. Of course, science would say that this is a result of the stimulation of nerves in that part of the body. But think about words in our culture such as heartbreak or getting cold feet. Are they just metaphors for how we feel, or do they reflect a deeper interconnectedness between mind and body that science is just starting to understand?

I Have a Mind

The third part of ourselves that helps provide us with a sense of identity is our mind. Science says this elusive, nonmaterial quality is merely a product of brain functions. That's been proven true experimentally to the extent that it's been shown that a definite relationship exists between the mind and the brain. Science claims that if the brain dies, the mind does, too. Whether that's ultimately true is still open to debate. Everybody has a unique perspective on that issue.

The mind is considered by many to be the seat of consciousness. As Sigmund Freud discovered, our minds contain more than one layer of consciousness. The mind and the emotions are also intimately linked. Some psychologists consider them to be one and the same. Contained within this system of mental and emotional interactions are our ideas, beliefs, information, and memories. These things further help to create our sense of self-identity. Are we just a swirling mass of physical sensations, emotional responses, and mental thoughts and ideas? Maybe there's more.

An Experience of the Self

The Self is what many consider to be the essence of the matter. The three other aspects of consciousness—body, emotions, and mind—make up our usual experience of ourselves. Have you ever closed your eyes and tried to find something inside of you that doesn't fall into one of these three categories? There always seems to be some physical sensation, emotional feeling, or thought popping into our awareness. Yet underlying all of these things there sometimes seems to be an awareness of something much deeper than the other three aspects. It's as though our bodies and emotions and thoughts are swirling around like a hurricane, but at the center, where it's peaceful and still, you find the eye of the storm.

Relatively Speaking

Of the four aspects of consciousness discussed, the concept of the Self is probably the most difficult to talk about. It means so many things in different paradigms that it's hard to define one way or another. Some other words that have been associated with it are higher self, true self, authentic self, inner self, soul, spirit, cosmic consciousness, oversoul, and atma (Sanskrit for "the divinity that dwells within"). There are many more words that mean the same thing. But most everyone is in agreement that it is something greater than what we normally experience as ourselves.

This center of being has been a topic in philosophy, psychology, and religion for centuries. Western religions often equate it with the soul. Eastern religions also acknowledge its existence, but they actively pursue it, with the idea that it can be known and experienced. Various psychological theories also consider it a real entity and try to understand its overall relationship with the other aspects of our identity. Science is the only area that still has problems admitting its existence, but that's to be expected. This aspect of ourselves defies scientific inquiry, at least for now. Let's call this part of consciousness our spiritual aspect.

The Spectrum of Consciousness

So what does all of this have to do with how we experience the world and the types of questions we ask? How does all of this relate to our beliefs about the world and self-identity? We are very complex beings. In the same way that no snowflake or fingerprint is the same as another, each of us is unique. This statement may seem quite obvious and simplistic, but it addresses a very important point. Our consciousness is made up of physical, emotional, mental, and spiritual aspects. Each of these four aspects manifests as part of consciousness all the time. You're only conscious of one aspect or the other depending on where your awareness is placed.

Normally you don't think of or aren't aware of the fact that you have the option to move your consciousness between these four aspects. You usually just sort of bounce back and forth among them without making any guiding effort. One moment your nose itches, the next you become angry because somebody cuts you off while you're driving to work, and the next moment you're thinking about what it would be like to be on vacation. You never stay with one aspect for very long. You just constantly shift back and forth.

When people do try to consciously choose a particular experience, they most often seek pleasure and avoid pain. They avoid experiences that make them feel uncomfortable and delight in those that give them pleasure. That pertains to all four aspects of identity: physical, emotional, mental, and spiritual. That is one of the reasons why the development of self-awareness can be such a useful tool. You can choose to change the focus of your awareness only if you know where your attention is right now. This is one thing that influences how you perceive the world.

Mind Expansions

One of the tests that you can do to find out how you perceive the world is called the Myers-Briggs Personality Profile. The profile can help you to evaluate which personality type you have. For example, it can give you a basic idea of whether you're an introvert or extrovert and can define you as a thinker, feeler, intuitive, or sensate person, or some combination of these types. If you haven't taken this profile, try it. You may be surprised at what you find out about yourself.

Each of us has a preference as to how we perceive and experience the world. Some of us like to consider the world strictly from a mental perspective, using the mental aspect of ourselves to think, analyze, and apply reason and logic to our perceptions. Others prefer to feel the world, experiencing it with emotions and intuition and savoring all the emotional tones and qualities that they evoke from the outside world. Still others prefer using their bodies to experience the world, welcoming and perceiving any and all bodily sensations that connect them to the physical universe. From the spiritual perspective, the world becomes an illusion, and the only thing that's real is the experience of the experience itself. Too cosmic? Let's bring it back to reality.

The sense of personal identity that gives you the perception of I-ness, as in when you say, "I feel tired," or, "I think that you're right," is not experienced as any permanent place within yourself. The way in which you experience the world is a complex combination of where your awareness is at any given moment, as well as your propensity to perceive the world in the way that feels most comfortable or familiar to you—in one of the four ways mentioned previously or using a combination of them. All of this gets combined with your upbringing, social environment, cultural tendencies, and inherited traits to make you different from anyone else. Humanity can be seen as an entire spectrum of consciousness, very much like a giant rainbow or like the electromagnetic spectrum, in which each individual falls into the color or vibration that is uniquely his or her own.

Social Collections

Although each of us is unique, we are at heart very social beings. For reasons as complex as human consciousness, we are drawn to groups of people who share similar traits, interests, and belief systems. These groups can fulfill some of our need to belong to something, or they can provide us with enjoyment simply because we are participating in social activities with other people. Although some people prefer less social interaction than others, and many people require time alone, everyone seeks to spend at least some important part of their time with at least one person.

The word *society* is used in a very general sense to refer to large collective groups of people living together. When we refer to the accepted behavior within a society, the morals of society, or the beliefs of society, we are referring to society very generally, not specifically. The same can be said of the smaller groups or social collections that make up the larger society. Scientific communities and religious communities are just two examples of an almost limitless number of communities and subcommunities that comprise the whole of society. For example, the scientific community can be thought of as consisting of all of the groups that are representative of science, including astronomers, biologists, physicists, geologists, and doctors. The religious community could include Christians, Jews, Buddhists, Hindus, and Moslems.

We can continue to break these groups down into even smaller units, because each is composed of smaller collections. Within the scientific community is the physicists group, in which there are hundreds of different areas of study that individuals can pursue or participate in. The same is true of religion. Christianity alone has over 156 main groupings, made up of 21,000 different denominations.

When we finally get to the smallest constituent, what do we find? The individual. A number of individuals within society don't belong to any particular group, but seek to define who they are through their own unique experiences, rather than within a group or collective experience. Many synthesize aspects of different groups into their world view, choosing some points of view and rejecting others.

Nuclear Meltdown

One of the most unfortunate aspects of a group paradigm can be a belief in its superiority to other paradigms. For whatever reason, this sense of superiority feeds the individual paradigms of its members, and everyone outside of the group is suspect or inferior. This happened in Nazi Germany and lies at the source of the cultural genocide happening in Eastern Europe. It's one of the most difficult aspects of any paradigm to change, because it involves detaching individuals from the group identity of which they are a part. Only a paradigm shift on a personal level can begin to break up the group identity.

Brother, Can You Paradigm?

One of the things I've talked about several times in this book is the concept of paradigms and paradigm shifts. Each of the larger groups in society has a paradigm. Usually, members of these groups adopt or accept some or all of the group paradigm as part of their own individual paradigms. Even people who prefer not to belong to any specific group have still formulated paradigms based on aspects of other individuals and groups within the society in which they live.

This personal paradigm, which is a collection of each person's knowledge, experiences, attitudes, beliefs, values, and convictions, is developed throughout a lifetime and is structured according to all the qualities and aspects of consciousness previously discussed. Each paradigm is uniquely possessed by the individual, and the degree to which you know its content reflects the level of self-awareness that you have.

So this sort of begs the question: If each of us has our own unique paradigm, and we are the individual members that comprise the larger collections within our society, is there really such a thing as a group paradigm? If you put 10 physicists in a room and ask them to define the quantum paradigm, you'll get agreement in certain areas and disagreement in others. Everyone brings a different interpretation to the table, and nobody can completely agree on what the quantum paradigm is.

The same is true of religions. Put 10 Christians or 10 Buddhists in a room, and ask them to define the paradigm of their religion. You'll get the same result you got with the physicists. The people will agree on some things and disagree on others. So is there such a thing as a quantum, or Christian, or Buddhist paradigm? When someone says, "I'm a Christian," does he mean that he is, but others aren't? What would a Catholic say about the paradigm of a Southern Baptist or vice versa? Which one reflects the real Christian paradigm? If a quantum experimentalist says that human consciousness plays a role in observing the behavior of a particle, and another says that human consciousness is immaterial, who represents the true quantum paradigm?

Paradigms, consciousness, and self-awareness are all intimately connected. A change in one spontaneously changes each of the others. An individual paradigm shift occurs when one no longer adheres to or participates in a collective paradigm. Remember what happened in ancient Greece, when some individuals began trying to understand the world through the application of reason rather than mythology? They experienced individual paradigm shifts that not only expanded their consciousness, but also increased their level of self-awareness. They saw themselves in a totally different relationship with the world around them and perceived things differently than before. They became aware of their own abilities to comprehend the world in a new way.

This kind of paradigm shift occurred again with the Copernican Revolution. Paradigm shifts also transpired individually and then on a mass scale very quickly with the birth of the Renaissance. As time has passed, both individual and collective

Mind Expansions

The rate at which new knowledge is being acquired seems to be growing almost exponentially. People now have access to tremendous amounts of information. Of course, what we do with this information is another matter. In one form, it's nothing but raw data, which can become useful knowledge when applied in the right way and can even lead to deep insights. Cosmology itself, because of the many recent innovations in technology, is changing daily. That's one of the reasons why it's not in the best interests of anyone to state that we're sure of how the universe operates. Tomorrow things could change significantly. Don't get attached to absolutes—they never stay the same for long. As a wise man once said, "Changes aren't permanent, but change is."

paradigm shifts have begun to occur at an increasing rate. There have been more changes in individuals and the understanding of the nature of reality in the last 200 years than in the previous 2,000. Our understanding of the world continues to change more and more rapidly.

Change Is Inevitable

One of the most significant views that changed in the understanding of the universe was the realization that it is not a fixed, static thing. Every single thing in the entire universe is constantly changing. From the smallest subatomic particle to the largest galaxies, the universe is dynamic, always in a state of constant flux. Yet we humans often wish to not go along with the rest of the universe. Many of us avoid change; others live for it. But change is not constant everywhere all of the time. It's more of a dynamic shift between stillness and movement, the duality of up-down, wave-particle, breathe in-breathe out, creative-receptive, Yin-Yang, the Tao, the balance.

But many of us look for that which doesn't change. Isn't there something in this world we can hang on to that can give us a sense of security? This is where our paradigms come in. Contained within them are beliefs that can give us a sense of security and provide us with a feeling that there is something that is constant, whether it's a belief in the exactness of science, the constancy of love, or the existence of a divine being. We all have something that offers us some sense of sameness. Some of us seek to find it within; others look outward to material possessions, money, other people, institutions, or collective social paradigms.

What happens when we find this sense of constancy? We identify with it completely and try to make it part of ourselves. It provides meaning and purpose to our existence and at the same time alleviates the underlying feeling of separateness that seems to be innate within human beings. A wise person once said that all of our desires, problems, issues, and searching for answers stems from one thing: a feeling of separateness. Whether it comes from our parents, family, society, ourselves, the world, or God, this feeling lies at the core of our being.

So we do anything that fills that void and end up making it a part of ourselves, our consciousness, our paradigm. It may only be a belief, but that belief contains within it what we are looking for. And this is when we run into problems. The belief works so well for us that we want it to work for everyone else, too. So collective paradigms are formed that seem to confirm for individuals as well as for others the truth of their beliefs. This process doesn't really account for the relative nature of belief systems. What is true for an individual cannot fill the void for others, no matter how strongly the individual insists that his or her paradigm is the one and only correct one for everybody.

This problem is what lies at the core of religious persecution and intolerance, as well as the tyranny of science. When a particular paradigm becomes static and unchanging, it goes against the very nature of the universe and the world in which we live. The belief that became part of our self-identity to fill the feeling of separateness now serves to do just the opposite of what it was intended to do in the first place. The belief has become a new source of separation, because no matter how hard we want to hang on to it, the universe and other people continue to change around us. We become strangers in a strange land.

Change is a constant in the universe. Our lives continue to change from day to day, and our inner worlds, which are made up of our beliefs, values, and paradigms, also need constant examination. If these inner parts of ourselves can't adapt to the changing demands of the outer world, we only become more separate. We can't turn back the hands of time, and nothing will ever be as it was, no matter how hard we wish it would be. Rigid paradigms can't grow, adapt, change, or expand in ways that would allow us to participate more fully in our daily lives. Instead of looking at life with blinders on and seeing only what we want to see, we could embrace it with a more inclusive, rather than exclusive, way of being in the world. Adaptibility, flexibility, and understanding are the tools that allow us to become comfortable with change and the unknown. In a world that offers less security than we'd like, the only real sense of security comes from within us. As long as we identify ourselves with limiting beliefs and paradigms self-awareness won't increase. If we want to argue for our limited perspectives, sure enough they're ours.

Albert Says

Some of the discussion in this chapter raises questions about beliefs that many of us hold dear. The purpose in discussing this material is not to show that what some people believe is right and what other people believe is wrong. The purpose is to show the dynamic process within which consciousness operates and that all of us have the option to look at ourselves and our beliefs in different ways. This self-examination can only be of benefit to all of us, because many personal and social problems are a direct result of people not knowing the contents of their own minds.

Cosmology 101

At various times in history, certain individuals' discoveries about the world in which they lived shifted the paradigm of their cultures, slowly in some cases and quickly in others. At the end of the nineteenth century, most physicists believed in the Newtonian paradigm, which stated that the workings of the world were all predetermined and operated like a big clockwork mechanism. Many were convinced that physics had discovered all there was to know about the universe. Or that if it hadn't, then it was about to.

When Albert Einstein proposed his relativity theory, it represented a significant paradigm shift from the accepted world view. As you know, most physicists scoffed at his theories. Why was there such an unwillingness to look at the world in a new way? In part, because the physics they had worked very well. It explained most of the physical phenomena that they could encounter. So the general feeling was: Why do we need something new? Now that you know how paradigms operate and the sense of personal investment and identification that goes along with them, you can probably understand why physicists felt this way.

Einstein himself couldn't move out of the paradigm he had created for himself. No matter how hard he wanted to extricate himself, he was locked in. Because of his inability to accept the new paradigm of quantum mechanics, the most interesting new problems in physics eventually passed him by, and he spent the rest of his life trying to resolve certain specific beliefs within his apparently outdated paradigm.

Many of the individuals in science today working in quantum mechanics and cosmology find themselves convinced of the truth of the paradigm to which they subscribe. Some are totally convinced that most of the answers about the universe are answered or at least are almost answered. Their personal convictions and beliefs about the power of the scientific method in its present form to solve the riddles of the universe leave no room to include or even to consider aspects of any other paradigm.

Of course, because humanity includes such a wide spectrum of consciousness represented by so many different people, there are always some who will be willing to question the assumed truths of a generally accepted paradigm. One of the key insights learned from the study of paradigms is that they all change over time. Physicists a hundred years from now will look back to these times, and some will marvel at how closed-minded some of us were about the nature of the universe. Ironically, some of them will be just as convinced that their current paradigm is the true understanding of how the universe operates.

Relatively Speaking

The **big bang** is the most popular current view in cosmology regarding the origin of the universe. It basically states the universe began approximately 15 billion years ago, out of a superhot, superdense fireball that has since spread out to form our galaxy, as well as all the other galaxies and materials in the known universe. There's some good confirmation that this is indeed what happened, but not everyone in the scientific community accepts it.

Everything that we have discussed about science can also be said about religion. When it comes to paradigms about the creation of the universe, there are many different beliefs that people accept as the truth. Some have identified themselves so strongly with their beliefs that anything that questions their paradigm is seen as a personal attack. Instead of showing a willingness to examine their paradigm, they lash out in defense of it.

In the West, the predominant religious paradigm is the Judeo-Christian-Islamic paradigm. Contained within that paradigm is teaching that the world was created by God in seven days. Some people believe this teaching to be the literal truth. Others see it as a metaphor, symbolic of how the universe was created. Still others view it as a creation myth, left over from ancient cultures that explained creation in terms of mythology. For some, the belief plays a significant part in their paradigm, and for others, it's less important. A lot depends on the degree to which we personally identify with our unique interpretation of it.

But this paradigm, just like the paradigms previously described in physics, has changed over the centuries. Although the basic teaching of the creation story is still the same, aspects within this paradigm have changed. That's why there are so many different sects, groups, branches, and denominations in all three of the Western religions. Religion, like science, continues to change over time.

Chapter 24 is devoted to cosmology from the point of view of physics, so you can understand where the latest theories and beliefs have come from and where they could be going. The predominant theory right now is the *big bang* theory. This theory states that there was a specific moment of creation and that it all started from a point in space. There was a gigantic explosion, and everything that exists in the universe is a direct result of it. For some people trying to synthesize certain aspects of science and religion, the big bang theory can go a long way towards satisfying both paradigms.

The Whole Truth

A basic perception contained within both the scientific and western religious paradigms is the role of humanity in relation to the world in which we live. In religion, humanity stands outside of God's creation and is given the responsibility to watch over and take care of it. Science approaches humanity's place in the grand scheme in the same way. Nature is meant to be studied, observed, and categorized. Both of these perceptions have ultimately led to the exploitation of nature, bringing us to the brink of ecological disaster.

The aspects of the paradigm that reflect this mindset are an inherent part of Western culture. Our desire to analyze, categorize, pick apart, and label everything we see is a direct result of the paradigm developed by the ancient Greek philosophers. The absence of any interconnectedness to our world has turned it into a fragmented system of parts and pieces. It's only been in the last 20 years or so that systems theory has shown that our lives on this planet are intimately connected in ways that we never understood before.

That's not to say that there haven't been other paradigms that have known of the interconnectedness all along. Some of the eastern religions have had that concept as part of their paradigm from the start, and so have indigenous tribal cultures in many parts of the world.

But this realization has directly impacted some of the new paradigms emerging in cosmology. Some of the new ideas and concepts being developed by physicists and cosmologists see this interconnectedness as a necessary principle that makes us an integral part of any process of measurement. Rather than just being an observer, separate from that which we are measuring, we are part of the whole system and therefore influence the measurement. Certain aspects of this paradigm are already affecting other areas of study, including medicine, psychology, and education.

The next chapter specifically focuses on some of the theories of cosmology in science. The last few chapters will bring back some of the concepts developed here as we explore alternate realities, probable worlds, and parallel universes.

The Least You Need to Know

➤ Human consciousness expresses itself through the physical, emotional, mental, and spiritual aspects of ourselves.

➤ Self-awareness is directly related to the type of paradigm you have, as well as the way in which your consciousness manifests through that paradigm.

➤ The whole of humanity can be seen as a large spectrum of consciousness, with each individual occupying a specific place within it.

➤ Fixed paradigms don't allow room to grow and change. Change is a natural process of the universe, and inflexible paradigms have difficulty adapting to change.

➤ The study of cosmology and the study of human consciousness are interconnected in ways that we're just beginning to understand.

Big Bang or Bust

Chapter 23, "Cosmology and Consciousness," covered some ideas related to the study of human consciousness. These ideas stated that there are a number of complex, dynamic processes at work every time you interact with the world around you. Also, the degree to which you identify with your core beliefs either permits you or prevents you from accepting new paradigms. The chapter concluded with a description of different paradigms about the creation of the universe, with the understanding that our concept of it continues to change and evolve.

This chapter will examine the history of paradigms and ideas that led up to the popular theory of the big bang. It takes a look at Albert Einstein's role in the development of this theory, as well as the philosophical and scientific evidence that has led many cosmologists to accept it. Although the subject is still open to debate, many significant experiments have provided convincing evidence that the big bang explains how we all got here. Are you ready for a bang-up time?

Are We Talking About the Same Thing?

The 1993 annual meeting of the American Association for the Advancement of Science involved a panel discussion called *The Theological Significance of Big Bang Cosmology.*

Scientists and theologians met to discuss the possible connections between the theory of creation put forth by science, known as the big bang, and the Christian story of creation found in the Bible.

Mind Expansions

One of the most well-known conflicts between evolution theory and creationism was the famous *Scopes Monkey Trial* of 1925. John Scopes, a teacher in Tennessee, was put on trial for teaching evolution theory in his science class. The trial was made into a Broadway play and then into the movie *Inherit the Wind*. Although based upon the trial, the play and the movie took liberties in portraying the conflict. They were somewhat accurate in their account of the trial, but represented both lawyers as being more extreme in their positions than they were. However, both the play and movie did accurately represent the traditional paradigms of evolution theory and religious fundamentalism.

Years earlier, in 1951, Pope Pius XII made one of the first official statements of the Catholic Church regarding the big bang theory. He stated that, "Scientists are beginning to find the fingers of God in the creation of the universe." Does this mean that science and religion were finally beginning to find some agreement about the origin of the universe? There was at least a willingness to explore both sides of the issue to some degree. Science and religion don't necessarily have to be in opposite corners when it comes to paradigms about creation.

The evolution of man is a much bigger debate. Evolution theory versus creationism is still a source of contention within our society. In many ways, these are two of the most conflicting paradigms that exist between science and religion. Creationists believe God created humanity instantaneously, but evolutionists believe mankind gradually developed to be more and more advanced and intelligent over the course of millions of years.

The religious view, for the most part, is based on faith and belief, whereas the scientific concept is based on empirical evidence. Even though there is good evidence to support evolution, scientists are the first to point out that it's still just a theory and that it's far from being a complete explanation of how and why living things have developed the way they have so far. The previous discussion of paradigms can possibly add insight into both views.

Some of us identify with the scientific explanation, and some of us identify more with the religious concept. Each paradigm provides us with answers that satisfy and fulfill what we're looking for. They provide a structure through which other aspects of our lives and the world in which we live can be understood. Because we identify so strongly with the beliefs that make up the larger paradigm, we sometimes assume that because something is true for us, it should be true for others as well. The problem here is that the two paradigms are mutually exclusive. Each states that it is the only true way to interpret the world, so they can't be reconciled with each other.

If each paradigm considers itself to be exclusive of all others, it separates itself, and in the very act of separation, it destroys what it sought to do in the first place, which was to provide an inclusive sense of meaning and connectedness to the world in which we live. The paradigm becomes less and less dynamic, more unchanging. There is no room for growth or expansion. Instead the paradigm feeds upon itself, absorbing only those qualities that make it appealing to those who already believe that the paradigm is true.

We end up seeing ourselves as representatives of the one true way and are not open to seeing other points of view. We become defensive and occasionally self-righteous. Our continued separation can fuel more anger and frustration because we can't understand why others don't see the correctness of our position. We continue to defend, argue, and criticize things outside of the paradigm; inside the paradigm, we continue to create the illusion of truth, self-importance, and goodness.

This is an extreme example of how a paradigm works, but unfortunately this example is not uncommon in the world today. It can be found in science, religion, politics, medicine, and government and can manifest through sexuality, race, and culture. But this is only one way in which a paradigm can exist. The fact that humanity is made up of an entire spectrum of consciousness prevents any one paradigm, no matter how big, to influence others for very long. Some have been around for a very long time, but because of the dynamic process of the universe, they eventually end up dissolving away if they don't change with the times.

Although divergent paradigms concerning the evolution of humanity still exist, views about creation itself are less acrimonious. The dialogues taking place between science and religion in this area demonstrate more open-mindedness. Many people are seeking a larger perspective, a more inclusive rather than exclusive paradigm, to provide insight and a greater understanding into the parallels that exist between science and religion.

Einstein Revisited

After Einstein put forth his general theory of relativity, he turned his attention to cosmology and the development of a unified field theory. His general theory of relativity had a number of consequences for cosmology. It predicted:

➤ **The existence of gravitational lenses.** This phenomenon relates to how gravity can bend light waves. Although Einstein's theory was proven correct based on evidence obtained by observing solar eclipses, *gravitational lenses* are huge astronomical objects, such as other galaxies and *quasars*, which weren't discovered until years later.

➤ **The existence of black holes.** This prediction was one of the more famous ones that came out of relativity theory. Black holes are stars that have collapsed into themselves. Their gravitational fields are so powerful that they warp space very deeply, to the point where nothing can escape from them, including light.

➤ **The dynamic nature of the universe.** The universe is not static, but it is either contracting or expanding.

This last implication of relativity theory, that the universe was dynamic and either expanding or contracting, directly conflicted with the traditional view that the universe was static, finite, and unchanging. Einstein himself didn't believe that the universe was dynamic. Although he had introduced the world to many revolutionary theories, he went along with the traditional view that the universe was static. It wasn't until the 1930s that evidence first began accumulating to support the big bang theory.

One of the main reasons that Einstein wouldn't accept the fact that the universe was expanding was that this fact implied that at some point long ago, the universe must have started from a single point. This meant that all of space and time had to have been bound up in what is called a *singularity*, an infinitely dense and infinitely small point. The resulting inability to calculate anything before this singularity presented the ultimate barrier to human knowledge.

Einstein considered this barrier nonsensical, so he fudged his equations by adding a force of unknown nature called a *cosmological constant*. This constant represented a force that could counteract the gravitational forces in the universe. In essence, these gravitational forces and the cosmological constant cancelled each other out, giving us a static universe that is neither expanding nor contracting. The paradigm of how Einstein understood the universe would not allow the idea that it might be dynamic. Years later, Einstein acknowledged his introduction of the cosmological constant as, "the biggest blunder of my life." But it took some convincing evidence for him to admit that.

The Birth of the Big Bang Theory

The earliest version of the big bang was first proposed by a Belgian Catholic priest, Georges-Henri Lemaître (1894–1966), who was also an astrophysicist. Interestingly, this priest/scientist represented a synthesis of the paradigms of science and religion into a larger perspective. He saw no conflict between science and religion.

In 1927, Lemaître formulated a theory called the hypothesis of the primordial atom, which suggested that the universe began from a single, primeval atom of

Relatively Speaking

Quasar is short for quasi-stellar source. These bodies look like stars, but they emit far more radiation than the largest stars. They have been shown to be bright cores of forming galaxies.

Gravitational lenses are huge astronomical objects, like galaxies, which are so massive (because they contain billions of stars) that they bend light from stars very dramatically, much more than is observed during a solar eclipse. They create celestial versions of atmospheric phenomena like mirages and multiple images. One of the best-known examples of this phenomenon is called Einstein's Cross. A gravitational lens has caused light from a single quasar to split into five separate images, forming a cross pattern.

A **singularity** is the theoretical single point from which the big bang occurred. It is so small that it has no size. It is infinitely small, and at the same time it is infinitely dense. It's the ultimate barrier to human knowledge, because there are no laws of physics that can explain exactly what it is. Space and time do not exist there, so consequently our understanding of physics breaks down.

energy. This atom began by dividing repeatedly, in much the same way that cells divide and multiply in the body. From this one atom came all of the matter in the universe, with space and time unfolding as matter spread out from the original atom.

At the fifth Solvay conference, where Einstein and Bohr first began their famous debates, Lemaître approached Einstein in an attempt to interest him in his primordial atom. He did not succeed. Einstein replied in an unusually brusque manner, saying, "Your calculations are correct, but your physical insight is abominable." Sounds like he didn't think too highly of Lemaître's idea!

The Infinite Universe

Along with Einstein and Lemaître, Edwin Hubble (1889–1953) was a member of the triumvirate responsible for laying the groundwork for the big bang theory. The Hubble telescope currently orbiting Earth and providing astronomers with new information about the universe on a daily basis was named after him. His contributions and dedication to his work made him one of the most significant astronomers of all time.

Hubble's first discovery forever changed the concept of the size of the universe. One thing that has frustrated astronomers for years has been the measurement of astronomical distances. For example, how do scientists know that Earth is 93 million miles from the Sun? For that matter, how does anyone know the distances between any planets or stars? The system of measurement used to determine these distances is called *parallax*.

Hold a finger up close to your eyes, and then close one eye. Open that eye and close the other eye. You'll notice that your finger seems to move back and forth in relation to the background. The closer you hold the finger to your eyes, the bigger the shift. This shift is called parallax. The farther away you hold your finger, the smaller the parallax. This parallax effect is caused by seeing the position of your finger from two different perspectives: one from your left eye and the other from your right eye. You can use simple geometry to calculate the distance from your eyes to your finger using the triangle that is formed between the three points of your finger and your two eyes.

Astronomers have used this method to figure out the distance of heavenly objects. When you observe an object from two different observatories on Earth, say one in California and one in Japan, you can determine the parallax of that object against the background of the other stars.

Mind Expansions

Edwin Hubble observed the heavens from Mt. Wilson in southern California, which at the time was equipped with the largest reflecting telescope in the world. The 100-inch telescope provided him with images no one had ever seen before. Hubble was so passionate about his work that he spent hundreds of bone-chilling hours in the observatory while controlling any shivers that might send the slightest vibration to the photographic plates he was exposing throughout the night. His exactness and dedication transformed astronomy as Galileo and Copernicus did.

Relatively Speaking

The **parallax effect** is a system of measurement used to calculate astronomical distances within our solar system. It's related to the apparent shift in position of a heavenly body when it is observed from two different positions. It works pretty well for heavenly bodies within our solar system, but it breaks down when measurement is needed for very distant objects.

The word **nebula** originally only pertained to galaxies, but now it only refers to clouds of gas in space. Some nebulae are dark. Others are bright because nearby stars or stars embedded in them make them glow.

Relatively Speaking

The **Doppler effect** or **Doppler shift** occurs when a moving object emits a sound. As the object approaches the observer, the sound it emits is of a high pitch. After the object passes the observer and is moving away, the sound becomes lower in pitch. You hear the Doppler effect every time an ambulance passes by. It also applies to light waves, in which case the color changes instead of the pitch.

A triangle is formed between the three points of the heavenly object and the two observatories. Again, using simple geometry, the distance between the object and Earth can be calculated accurately.

This method of measurement works out fine for planets and other objects in our solar system, but it breaks down for stars much farther away. You can't form a big enough triangle to permit accurate calculation using any two observatories, because the distance between them is too small relative to an object very far away.

In 1924, Hubble developed another technique to measure the distance to the star that he was observing in the Andromeda *galaxy*. He was able to calculate the distance by showing that the intensity of light diminishes with the square of the distance. (Remember Galileo's calculation that the distance an object travels down an inclined plane is proportional to the square of the amount of time elapsed? This is the same relationship, only it's using the intensity or brightness of a star instead of the distance on the inclined plane.) In other words, Hubble could determine how far away an object was according to how bright its light was. (The calculation also factors in the star's color, absolute brightness and apparent brightness, and special types of stars called *variable stars* because their light intensity changes cyclically, but that's beyond the scope of our discussion.)

Using this new system of measurement, Hubble concluded that the star in the Andromeda galaxy was 800,000 light years away, which is more than 10 times the distance from Earth to the farthest stars in our galaxy. (Today this figure has been updated to 2 million light years away.) So what did this mean? It showed that the Andromeda Nebula was far beyond the Milky Way galaxy. In fact, it was a distant galaxy all its own. This discovery blew everyone away. It was always thought that our galaxy was the extent of the universe, and that nothing existed outside or beyond it. Hubble's discovery gave cosmologists a whole new paradigm to work with. The universe was no longer seen as finite. Instead, it appeared to be infinite. We know today that there are billions of galaxies besides our own, and more are being discovered all the time.

And It's Getting Bigger, Too

Hubble's second discovery overthrew the long-held belief that the universe was static. Einstein and many other physicists held this view. This section explains how Hubble did away with that view.

The *Doppler effect* occurs when a train or car passes by with its horn blowing. As the vehicle approaches you, the sound of its horn has a higher pitch, and as it passes you by and goes off into the distance, the pitch becomes lower. The sound waves emitted from a stationary vehicle pass you at a constant rate. But when the vehicle is moving toward you, each new sound wave that is created starts off closer to you than the one before, and they end up packed together due to the motion of the vehicle toward you, thereby increasing the frequency of the wave and increasing the pitch. As the vehicle moves away from you, the reverse happens. Each sound wave is emitted from a point farther away, and takes a longer time to reach our ears, which results in a decreased frequency and a lower pitch.

This effect also applies to light emitted by moving objects. The main difference is that instead of the pitch changing, the color changes. Remember our old friend the electromagnetic spectrum? Visible light is only a small part of it, but it's the part we're concerned with here. In this part of the spectrum, the color of light corresponds to its frequency. Red has a lower frequency, like the lower pitch of the horn, and blue has a higher frequency, like the higher pitch of the horn. Using this information, light from a star moving toward you would appear bluer, but light from a star moving away from you would appear redder. In scientific terms, an object is called either *blueshifted* or *redshifted*, depending on whether it's approaching you or moving away from you.

When Hubble combined his methods for determining the distance of stars and accounting for the Doppler effect, he found that almost every galaxy was redshifted. In other words, the galaxies are moving away from each other! The universe is expanding in all directions and is not static after all, as everyone once thought.

Mind Expansions

The redshift or blueshift of a heavenly body can be measured due to the discoveries of two men who lived in the nineteenth century. A German physicist by the name of Gustav Kirchoff (1824–1887) discovered the law that all substances absorb the same light frequencies that they can emit. Another German physicist, Joseph von Fraunhofer (1787–1826), repeated Newton's experiments on the solar spectrum, using prisms to break the light down into the colors of the rainbow. With prisms of much better quality, he was surprised to see that the rainbow was intersected by a large number of very thin black lines. These lines, called **Fraunhofer lines,** are caused by absorption of very narrow wavelength bands by atoms in the path of the light. Kirchoff's law and Fraunhofer's lines are invaluable for spectroanalysis because they not only show the composition of stars and other objects in the sky, but also allow astronomers to see the degree to which the wavelengths shift to either red or blue, thereby telling them if the objects are moving toward or away from them.

Taking this discovery one step farther, Hubble developed what is considered to be the first cosmological law, known as Hubble's law. It states that the further a galaxy (or star or whatever) is away from us, the larger its redshift is. This law is important in that it tells how fast the universe has expanded, giving astronomers the tools to determine how old the universe is. The numerical value has been difficult to estimate. The universe could be as old as 20 billion years or less than 10 billion years old. The closest estimate right now is between 12 and 15 billion years old.

In 1931, while Albert Einstein was staying in California, he had an opportunity to spend some time with Hubble at the Mt. Wilson observatory. Georges Lemaître heard about Einstein's visit and thought that he might try one more time to discuss his theory of the primordial atom with him and in this case with Hubble, too. It would've been great to have been there. Einstein announced that the cosmological constant had been his greatest blunder, and this trio of legendary minds came to an understanding that laid the foundation of cosmology for years to come. It is said that after hours of discussion, Einstein rose from the table with tears in his eyes and stated that this was the most pleasant, beautiful, and satisfying interpretation of astronomical science he had ever run across.

So Where's the Bang?

By the late 1930s, many physicists began accepting the big bang as a workable model for how the universe was created. However, not everyone agreed with it. Another popular theory that some physicists adhered to was called the steady state theory. For over two decades, arguments between champions of these two theories waged on. Even today, while nearly all physicists accept the big bang theory, in some circles, the steady state theory is gaining new followers.

The steady state theory basically says that the universe is in a steady state—in other words, that it's essentially the same as it was 15 billion years ago and will continue to be the same in the future. The universe has no beginning and no end. It is eternal and in a process of continuous creation.

A famous physicist by the name of Fred Hoyle was the main proponent of the steady state theory. He totally disagreed with the big bang theory for two main reasons. Each of his objections was related to the implications that the big bang theory seemed to suggest:

Nuclear Meltdown

There are two common misconceptions about the big bang theory. First of all, the classical bang didn't happen at a specific point in space within an infinite void. It happened everywhere, because it was everything. There was nothing outside of it, not even empty space. Secondly, the big bang was not an explosion that spewed matter into a preexisting space. There was nothing to spew into. When you blow up a balloon, the space inside expands as you continue to fill it. That's what's happening with the expansion of the universe. The expansion is creating space as it expands—it's not moving into space that's already there. That's also how the big bang worked. Space and time unfolded as the universe kept growing larger and larger in size. If you can't quite wrap your mind around this concept, don't worry; it's a big conceptual stretch. It's one of the reasons why cosmology can be so mind-expanding.

➤ The second law of thermodynamics basically states that all closed systems eventually decline into entropy, which as you know means disorder, and this appeared to be the fate of our universe. If no new matter was being introduced since the big bang, eventually the whole universe, like a big mechanism, would run down and become a cold, lifeless, nothing. This concept was known as the heat death of the universe. All the stars and suns would burn out, leaving no light or warmth anywhere.

➤ Considering the amount of mass in the universe, the gravitational forces will eventually have to stop expanding, and the whole process will begin to reverse, ultimately ending in the big crunch. The entire universe will end up contracting back in on itself.

Neither of these prospects was pleasing to the steady state theorists. Their theory was based on a more philosophically pleasing concept of the origin and fate of the universe. Ironically, Hoyle was the person who coined the term big bang. He meant it as a derogatory expression for the expanding universe theory, but it ended up having the opposite effect. It gave physicists a term that fueled the imagination and led to increased pursuit of scientific proof supporting the big bang theory.

Albert Says

Much of the material presented in this book is designed to give you a basic understanding of only some of the concepts found in physics, philosophy, and psychology. This is especially true of cosmology. If you wish to pursue a deeper understanding of some of the ideas presented here, read some of the books listed in Appendix B.

But the big bang was still far from achieving wide acceptance during the 40s and 50s. Even into the 60s, the steady state theory was considered just as plausible as the big bang. The steady state had a number of things going for it. Because there was no real solid evidence for the big bang, it was much more compelling to accept a theory in which the universe didn't have a beginning. Why? Because the concept of the universe having a temporal beginning begs the question: What existed before the big bang? Science has a hard time dealing with questions about the physical universe that it doesn't have answers to.

The big bang has as its beginning a singularity of zero size and infinite density. All known laws of physics break down in trying to explain how this point came into being, what was going on inside of it, and why it exploded. On the other hand, the steady state theory posits a universe in which the laws of physics are constant. The expansion of the universe that Hubble discovered could be explained by the creation of new matter, which caused the universe to continuously expand. As galaxies receded farther away, new matter continued to pour into the universe from some other source, such as through points of singularity at the center of galaxies.

The proponents of these two theories continued to argue over which theory could supply the best solutions to some of the following questions:

➤ How old is the universe?

➤ Where did all the elements come from?

➤ How did matter get distributed across space and time?

➤ What is the average temperature of outer space?

In the end, the big bang theory supplied the best answers to these questions. It's taken almost 60 years to provide some of these answers, and many are still just theories, but the acquisition of some evidence for the big bang finally shifted the tide of disagreement.

Now We Know, or Do We?

One of the predictions of the big bang theory was that there should be residual radiation left over, permeating all of space. This radiation is called the cosmic microwave background radiation. For many years, astronomers had been recording unusual background radiation from the sky, but no one could guess its source. Most thought that it was some form of stray radio noise or instrument error. In 1946, a sensitive new instrument was invented that could precisely measure the cosmic microwave background radiation. Over the next 20 years, this instrument was further refined. In a parallel development, a pair of researchers at Bell Labs in New Jersey, Penzias and Wilson, were working on a similar instrument to make better radar systems. They puzzled over excess noise in their radar system for some time. At first, they thought the noise was due to a large amount of bird droppings in the radar antenna. However, after they cleaned their antenna, the noise remained. At this point, they talked with the astrophycists who were working to detect the cosmic microwave background radiation. Sure enough, Penzias and Wilson had measured the exact amount of radiation predicted by the big bang theory. Penzias and Wilson had scooped several researchers, even though, initially, they weren't even looking for the cosmic microwave background radiation! The discovery of the cosmic microwave background radiation more or less put to rest the steady state theory.

This evidence of background radiation is probably the strongest reason why the big bang theory is the most popular one around. Steady state theorists argue that one of the reasons why the big bang is so popular is because it resembles the creation talked of in the Judeo/Christian/Islamic tradition. After all, one of the first advocates of the big bang theory, Georges Lemaître , was also a priest. They accuse the critics of the steady state theory of being swayed by nonscientific motives. (Actually, Lemaître was very upset that Pope Pius XII cited the big bang as evidence of divine creation.) Be that as it may, there is more scientific evidence to support the big bang theory than there is to support the steady state theory.

But this isn't the end of the story. A number of new questions have arisen, and another branch of cosmology, *particle physics*, can help answer those questions. Particle physics is the study of the most elemental forms of matter. Their interactions can be induced and studied in the large particle accelerators found at many universities and research centers around the world. Much of the next chapter is devoted to exploring this branch of cosmology. But for now, you may be asking yourself, "What does particle physics have to do with the big bang theory?"

For the last 20 years or so, the study of cosmology has been approached from two areas of inquiry: the microcosm and the macrocosm. The study of the subatomic world, or the microcosm, looks at how matter is created and transformed in the initial fireball, just after the big bang. It incorporates quantum mechanics as the means to study and comprehend what goes on at this level of reality. Astrophysics, which is the area of study that combines astronomy and physics, deals with the macrocosm. It looks at how galaxies are formed, as well as the physical properties and phenomena of the stars, planets, and other heavenly bodies.

Each of these areas of study has enriched the other, helping to explain or answer questions that the other can't answer on its own. But in this process, other questions are raised as well. The big bang theory is one area in which particle physics has supplied answers to questions that the theory itself couldn't explain. The next chapter describes particle physics and how it affects astrophysics.

Only a few chapters are left in your excursion through time, space, and the mind, but some of the best stuff has been saved for last. Although Einstein's theories continue to influence and help explain phenomena both in the microcosm and the macrocosm, there are some areas where even his theories have a hard time explaining anything. You're definitely going to go where few mortals have gone before. Next stop: the first three minutes of creation, the standard model, and then in the last chapter, dark matter, black holes, and probable universes.

The Least You Need to Know

➤ The paradigms of science and religion tend to be exclusive of one another, but finding common ground for discussion is possible if people are willing to seek a deeper mutual understanding between the two.

➤ Georges Lemaître was a priest and astrophysicist who put forth the first theory of a big bang, which he called the primordial atom theory.

➤ Edwin Hubble discovered two important things: that the universe was larger than our Milky Way galaxy and that the universe was expanding.

➤ Lemaître, Einstein, and Hubble are the trio of individuals who laid the foundation for the development of the big bang theory.

➤ The two most prominent theories for the creation of the universe are the big bang theory and the steady state theory.

➤ The big bang theory is the most widely accepted scientific theory of creation today.

Chapter 25

Worlds Within Worlds

In This Chapter

➤ The first three minutes after the big bang

➤ The standard model of particle physics

➤ The four forces found in nature

➤ The grand unified theories

When astronomers point their powerful telescopes at the night sky, they are looking back in time at all of the objects they see. The light they see from a galaxy 21,000 light years away left that galaxy 21,000 years ago. The more distant an object is, the farther back in time you see when you look at it. The Hubble telescope has allowed astronomers to look so far back in time, because of the tremendous distances it can see, that they are close to knowing exactly how long ago the universe began. But once they know when it began, they still need to try to figure out what went on at the moment of creation, if that is even possible.

To figure out the beginnings of creation, you need to leave behind astronomy and astrophysics, and look at particle physics. The last chapter explained that cosmology became a marriage of astrophysics and particle physics. Both fields deal with the study of the nature and origin of the universe. Whereas astrophysics deals with the macrocosm, particle physics explores the microcosm. Amazingly, the key to understanding what went on a fraction of a second after the big bang, as well as what happened for the first billion years after that, can be found by studying the interactions between subatomic particles. The world of the infinitely small awaits us.

The First Three Minutes

In the seventeenth century, an Irish bishop by the name of James Ussher wrote a chronology of the Old Testament in which he added up all the generations of men and women in the Bible since Adam and Eve and proclaimed that the world was created in 4004 B.C.E., at 2:30 P.M. on Sunday, October 23. When one of his parishioners asked him what God was doing before he created the universe, the Bishop impatiently replied, "Creating hell for those who ask such questions!"

Physics has done basically what the bishop did to figure out what happened in the beginning. Only instead of using the Old Testament, physicists can mathematically calculate the expansion of the universe in reverse. Working backwards from where the universe is right now, all of the galaxies and matter in the universe recede simultaneously and meet at a point: the big bang. When this process is complete, we can move forward again and get a glimpse of what the first few moments after the big bang were like. Steven Weinberg, the 1979 Nobel prize winner for his work in this field, describes what it was like:

> In the beginning, there was an explosion ... At about one hundredth of a second later ... the temperature of the universe was about one hundred trillion degrees centigrade. This is much hotter than in the center of even the hottest star, so hot, in fact, that none of the components of ordinary matter, molecules, or atoms, or even the nuclei of atoms, could have held together. Instead the matter rushing apart in this explosion consisted of various types of the ... elementary particles.

> As the explosion continued, the temperature dropped ... reaching one trillion degrees at the end of the first three minutes. It was then cool enough for the protons and neutrons to begin to form into complex nuclei, starting with the nucleus of heavy hydrogen, which consists of one proton and one neutron. The density was still high enough ... so that these light nuclei were able to rapidly assemble themselves into the most stable light nucleus, that of helium, consisting of two protons and two neutrons ... At the end of the first three minutes, the contents of the universe were mostly in the form of light, neutrinos, and anti-neutrinos ... This matter continues to rush apart, becoming steadily cooler and less dense. Much later, after a few hundred thousand years, it would become cool enough for electrons to join with nuclei to form atoms of hydrogen and helium.

The big bang occurred roughly 15 billion years ago. It took several billion years after that for the galaxies to form in their present configuration. Within the Milky Way, a cloud of hydrogen and helium began to condense about 4.5 billion years ago, and its gravitational pull slowly began to attract more and more material. This object formed into the most compact geometric shape possible, the sphere, and became so massive that its interior was put under tremendous pressure. This pressure caused the heat inside to reach 20 million degrees centigrade, which in turn caused the continuous fusion of hydrogen into helium, giving birth to the Sun. This process occurred in countless places, giving birth to an unlimited number of stars in billions of galaxies throughout the universe.

Hydrogen and helium comprise 99 percent of all of the matter in the universe. Over hundreds of millions of years, heavier elements were formed as offshoots of the early formation of stars. In the time right after the big bang, there was relatively little change in the elemental makeup of the universe. Fusion of heavier elements with each other in the cores of extremely hot stars and during supernova (exploding stars) are responsible for creating the remaining 90 natural elements which make up only 1 percent of the universe.

Around the same time that the Sun was forming, dust and gas that included heavier elements were attracted by the gravitational pull of the forming Sun. As the Sun condensed, so did these other masses of matter circling around it, eventually forming more solid bodies. Thus, Earth and the other planets in the solar system were born.

So in a page and a half you have the creation of the universe. Not bad, considering we've covered billions of years in the process. But what does all of this have to do with particle physics? In Steven Weinberg's description of the first three minutes of existence, he talked about the formation of the earliest forms of matter, and this is where the study of particle physics comes in. How did these particles interact to form more complex types of matter? What forces allowed this to happen? These interactions, along with the myriad particles involved, lie at the core of the study of particle physics. To know how this process operates is to understand how the universe was formed.

It's Raining Particles

As you know, Einstein's famous equation showed that matter and energy are one and the same. Mass is a form of energy, and a very small amount of mass can produce an enormous amount of energy. Because mass and energy are equivalent, it's possible to describe particles in terms of their energy. Therefore, physicists measure particles by their energy content.

Mind Expansions

Ninety-nine percent of the matter in the universe is hydrogen or helium. Much smaller quantities of elements such as lithium, beryllium, and boron are left over from the big bang. However, the heavier elements necessary for life, carbon, oxygen, and nitrogen, were created at the cores of enormous stars. So every molecule in your body is composed of star stuff.

Relatively Speaking

The **electron volt** (eV) is the unit of measurement that physicists use to calculate the energy content of subatomic particles. KeV is the abbreviation for a kiloelectron volt, representing one thousand eV. Next comes MeV, or a million electron volts, the *GeV* or gigaelectron volt, which is a billion electron volts, and the TeV or teraelectron volt, which is a trillion electron volts.

Decay in physics refers to the breakdown of a particle into smaller ones. This breakdown occurs naturally in radioactive elements and as a result of collisions of cosmic rays in the atmosphere, and occurs artificially in particle accelerators.

The basic unit of energy that's used for such measurement is the *electron volt* (eV). This is a very small amount of energy, smaller than the amount expended by a bug flapping its wings. Subatomic particles have an energy content much greater than one electron volt. For example, an electron's mass is equal to 511,000 eV. Most other particles, because their masses are much greater than that of the electron, have much higher energies. For example, the proton mass equals an energy of about 938,000,000 eV.

Previous chapters discussed the discovery of the main particles found in the atom: the proton, the neutron, and the electron. I also discussed the discovery of antimatter, specifically the antimatter counterpart of the electron: the anti-electron or positron. The positron was not the only new particle discovered during studies of cosmic rays using cloud and bubble chambers. In 1936, two particles were discovered called *mu-mesons*. These particles were found to have equal masses, but opposite charges, one negative and one positive. They were called mesons, because with a mass 210 times that of the electron, they were about midway between the lightest and heaviest known particles at the time. Meson means intermediate one. The name mu-meson was later shortened to *muon*. *Mu* is a letter in the Greek alphabet and is written μ. Therefore, μ was adopted as the symbol for the muon.

In 1947, a slightly more massive meson was discovered. It was called the *pi-meson*, shortened later to *pion*. There are three kinds of pions: negative, positive, and neutral. Pi is also a Greek letter and is used to represent pion particles.

Neither muons or pions exist for very long. They are created and then decay in a tiny fraction of a second. When physicists talk about *decay*, they refer to the breaking down of a particle into smaller particles. For example, the muon decays into an electron and two *neutrinos*, which are particles with no mass. The speed at which muons and pions decay is so rapid that they can be detected only by the tracks they leave when passing through a detection device such as a bubble chamber or cloud chamber.

The discovery of these particles was only the beginning. In 1952, an uncharged particle called a *lambda*, named after another Greek letter, was discovered in a cloud chamber. Because the lambda has no electrical charge, it leaves no trail of its own. But as it passed through the cloud chamber, it decayed into a proton and a negative pion. These two particles could be identified by their trails. By working backward from the point of decay,

Mind Expansions

The first accelerators were called linear accelerators because they accelerated particles down long, straight tunnels. One of the largest is the Stanford Linear Accelerator Center (SLAC) at Stanford University. Built in the 1960s, it is two miles long and can produce a beam of electrons equivalent to 30 GeV. The drawback to linear accelerators is that the energy levels are dependent on the length of the system. This is why Ernest Lawrence built a circular accelerator called a **cyclotron.** It was only a foot in diameter, but the particles could be whirled around in a circle many times, picking up more and more energy with each revolution. The largest circular accelerator is a four-mile ring at Fermilab outside Chicago. It can produce energies up to 1 TeV. The SSC, or Superconducting Super Collider, would have been able to produce energies around 2 TeV, but Congress canceled funding for it a few years ago.

physicists could determine the particle's characteristics. In other words, because the particle broke up into a proton and a pion, the scientists were able to conclude that it must have been more massive than a proton.

The lambda particle also had a much longer lifetime than any other particles similar to it. (A longer lifetime consists of fractions of a second, which in the subatomic world is a very long time.) This fact was determined by figuring out how long it was able to travel through the chamber before it decayed. This particle was grouped together with other particles that were discovered to have relatively long lifetimes. This property was called *strangeness*. Strange particles took longer to decay than those that didn't have this property. (During this discussion of particles, you'll notice that physicists came up with some terms to define specific characteristics of particles that were strange. Humor even exists in the subatomic world.)

In order to study these subatomic particles in greater depth, physicists realized that they couldn't sit around waiting for them to show up in their detectors. They needed some sort of device that could artificially produce these particles. If enough energy could be produced with these devices, even more massive particles could be created. For example, to produce an *antiproton*, which has the same mass as a proton but the opposite electrical charge, 1,836 times more energy was needed than that required to produce an electron or positron. One way to achieve such high energy levels was by accelerating particles to high velocities and smashing them into some-thing, just like how cosmic rays smash into Earth's atmosphere. Enter the atom smashers, better known as particle accelerators.

Albert Says

The study of particle physics can be very confusing and even a little intimidating. Don't try to remember all of the names or what the particles do. The tables included in this chapter break down the information as simply as possible. The main thing to remember is that matter is composed of quarks and leptons, and there are four forces that explain their interactions.

The Development of the Standard Model

The construction of the first accelerators began in 1928. These machines were capable of producing energies between 400,000 and 750,000 eV. If the SSC, or Superconducting Super Collider, had been completed (funding for it was canceled by Congress), it would have been capable of creating two trillion eV. Currently, the accelerator at Fermilab in Illinois is capable of producing one trillion eV. That's quite a jump in magnitude from the accelerators built over 70 years ago.

Over the years, hundreds of particles have been discovered. It's hard enough for physicists to keep track of all their names, so I won't list all of the particles, what they are, what they do, or how they are created. The following table lists some of the particles that have been discovered and describes how particle physics was structured prior to the 1960s. After that, particles were categorized differently due to the discovery of quarks.

Fermions		Bosons	
Baryons (Hadrons)	Leptons	Mesons (Hadrons)	Gauge Bosons
Proton	Electron	Pion	Photon
Neutron	Muon	Kaon	Gluon
Sigma	Tau	Eta	W+, W-, Z-zero
Xi	Electron neutrino	(many more)	Graviton
Lambda	Muon neutrino		X-particles
(many more)	Tau neutrino		

Relatively Speaking

A **neutrino**, which means "little neutral one," is a particle that has neither mass nor charge and is therefore very unusual. They're very hard to detect, because they seldom interact with anything. They constantly pass through our bodies without having any effect on us. Even if you had a lead plate a trillion miles thick, a neutrino might pass right through it.

Hadrons, whose name comes from the Greek word meaning strong, are the particles that respond to strong force. Hundreds of hadrons have been discovered.

Leptons, whose name comes from the Greek word for small, are the particles that are involved in electromagnetic and weak interactions. The electron is the most well-known lepton and has the smallest mass of any charged particle.

In 1964, a physicist by the name of Murray Gell-Mann proposed the existence of a new kind of particle. He drew a parallel between hadrons and molecules. As you know, all matter is made up of molecules, and each molecule is made up of two or more atoms. There is an unlimited number of possible combinations of atoms that make up different molecules, but there are only a little over 100 different kinds of atoms. Gell-Mann theorized that in the same way, there must be an unlimited number of *hadrons*, which are groups of particles that respond to the strong nuclear force, but that they are all made up of a very small number of more fundamental particles. He called these particles *quarks*, from a line in James Joyce's novel *Finnegan's Wake*: "Three quarks for Muster Mark." Quark is a German name for a kind of cheese. This term is another example of physics whimsy.

When Gell-Mann first named these tiny particles, he thought that, like the three quarks mentioned in Joyce's book, they came in three *flavors*. Flavor refers to a very special property that all quarks have. Physicists were going to use names of ice cream for the different flavors, but they never did. (Who said we left Alice back in Wonderland?!)

Just like atoms and molecules, there are many different ways in which quarks can arrange themselves within hadrons. This explains to physicists why the possible number of hadrons is very large, if not unlimited. The quark theory revealed an entirely new level of matter. The particles inside an atom were now found to be made up of even smaller particles. For example, a proton in the nucleus of an atom is composed of three quarks, plus some particles called gluons, which like glue, carry the forces that keep the whole thing stuck together. None of these particles would have been discovered without the use of huge particle accelerators. It seems ironic that the study of the smallest particles in the universe requires the largest machines ever made by humanity.

The discovery of quarks led to the formation of the standard model, which is outlined in the following chart. It seems a little weird to show the essence of the universe in a chart, but at least you'll get an idea of what is known at the present time. Included in this chart are the forces that I'll be talking about shortly. The only one not included is gravity, because nobody knows how that force is carried yet.

The chart is basically broken into three parts. The first two are what matter is made up of: quarks and leptons. These are divided into generations based on their mass. The letters that represent the different flavors of quarks are abbreviations for the following:

➤ *u* stands for up

➤ *d* stands for down

➤ *c* stands for charm

➤ *s* stands for strange

➤ *t* stands for top

➤ *b* stands for bottom

Next come the messenger particles, or *gauge bosons*, that carry the three forces. Quarks also come in three colors, so if you factor that in, you get 18 quarks, 6 leptons, and 12 gauge boson force carriers.

Matter

First generation	Second generation	Third generation
	Quarks	
u	c	t
d	s	b
	Leptons	
v(e)	v(μ)	v(t)
e	μ	T
	Forces	
	Gauge bosons	
Electromagnetism		Photon
Weak force		W-, W+, Z-zero
Strong force		Eight gluons

There's also the antitable, which has all the matter particles listed as antiparticles. That's 60 particles in all—a very confusing business. All you really need to know is that all matter is made up of quarks and leptons. These are the smallest known particles, but someday we may discover even smaller ones. In the same way that the universe is infinitely large in the macrocosm, the subatomic world could be infinitely small in the microcosm.

The Four Forces

You're already familiar with two of the four forces this section covers: electromagnetism and gravity. Einstein sought to unite these two forces in the early years of his search for a unified field theory. With the discovery of the other two forces, unifying all of the forces became even more difficult, and as you know, he never accomplished it.

Nuclear Meltdown

Cosmology can get confusing when physicists refer to the standard model, because they're referring to two different things. On the macrocosmic scale, the standard model means the big bang, the most popular accepted theory of the origin of the universe. On the microcosmic scale, the standard model refers to the classification of forces and particles that are currently known in the subatomic world. As you know, not all cosmologists agree with the big bang theory, even though it's the most widely accepted at this time. Although the world of particle physics has put together its own standard model, many physicists think that there are way too many parameters and parts to it; it's just not simple enough. This yearning for simplicity reflects an underlying bias in particle physics that the correct theory must be simple and elegant. What if it just plain isn't? Maybe the subatomic world is far more complex than we are willing to accept.

Electromagnetism explains how the atom's negative electrons are held in place, because they are attracted to the positively charged protons of the nucleus. But what held the nucleus together? The protons in the nucleus, which are positively charged, are packed much closer to each other than they are to any electron. According to the fundamental laws of electricity and magnetism, they should repel each other, making the formation of the nucleus impossible. Because this doesn't happen, there must be a force stronger than electromagnetism that binds protons together. This force is called the *strong nuclear force* and was found to be 130 times more powerful than electromagnetism.

Physicists found that, unlike gravity and electromagnetism, this force is exerted only over very short distances. For example, the protons in an average nucleus are only 10^{-13} centimeters apart (that's 0.0000000000001 centimeters!). At that distance, the strong nuclear force works very well in keeping the nucleus together and is able to overcome electromagnetic force. But if the protons are separated any farther apart than their own diameter, the force has no effect.

A fourth force, discovered by Enrico Fermi, is called the *weak force*. Instead of holding systems together, this force causes the slow decay of particles into smaller, more stable pieces of matter. It is essentially responsible for radioactive decay, the decay of many particles created in accelerators, and decay caused by cosmic rays. The weak force is like the strong force, in that its effect is extremely short-range. Without this force, the inability of atoms to get rid of all the particles that rain down upon us in the form of cosmic rays would make the world a very different place. The following table compares the four forces.

The Four Forces

Name	Strength Compared to Strong Force	Effective Range In centimeters	What It Does
Gravitation	6 x 10 to the minus 39	Infinite	Holds planets, stars, galaxies together
Weak	1 x 10 to the minus 5	10 to the minus 15	Causes particles to decay
Electromagnetism	7 x 10 to the minus 3	Infinite	Holds atoms together
Strong	1 x 10 to the 0	10 to the minus 13	Holds atomic nucleus together

Remember our friend Werner Heisenberg? He developed a theory that turned out to be fairly accurate in explaining how the strong force operates. He thought that the protons in the nucleus were held together by constantly exchanging a special kind of particle with the atom's neutrons. These particles came to be known as gauge bosons. They were named after an Indian physicist, Satyendra Bose, who did a lot of work in this field. The exchange causes the protons to change into neutrons, and vice versa, in quick succession. This happens so quickly that before the proton is repelled by a neighboring proton, it receives an exchange particle from a neutron, which reverses their identities. Before the newly created proton can be repelled, it switches back to being a neutron by the same process. So at any given instant, the atom has the required number of protons and neutrons in its nucleus, but they keep changing back and forth. The time it takes to make this exchange is unimaginably small, about a trillionth of a trillionth of a second.

These four forces came to be known more as interactions, and that is how they are thought of today. This brief overview of particles and their interactions just about brings us back to the first moments after the big bang. You need to understand just one more concept, and that concept is something called symmetry.

Back to the Beginning

Symmetry refers to similarity of form or arrangement on either side of a dividing line, where there is a correspondence between the opposite sides in size, shape, and position. For example, our bodies are for the most part symmetrical. Our left side is the same as our right side. But the symmetry is not perfect. You might have a mole or freckle on one side of your face, but not on the other. One foot may be slightly bigger

Relatively Speaking

Gauge bosons are the particles that mediate particle interactions. In other words, they are the carriers of the force. The photon is the particle that carries the force in electromagnetism, the **W** and **Z particles** do it for the weak interaction, the **gluons** for the strong interaction, and the **gravitons** for gravity. The graviton is still an unproved theory so far, but it fits the model with the other particles.

Symmetry refers to the geometric correspondences of opposite sides of an object or system. Any pattern that looks the same on one half as on the other half is symmetrical. Symmetry is also built into the laws of nature in a deep way. What this means is that the laws of nature are the same at every place in the universe and are the same at all times.

than the other, and most people favor the use of one hand over the other. These differences are examples of *broken symmetry*.

Broken symmetry is all around us. The petals on most flowers are not arranged in perfect symmetry. Doors have knobs on one side and hinges on the other. Yet many examples of almost perfect symmetry exist, too. Snowflakes are very symmetrical, as are wallpaper patterns. Both of these types of symmetry are also found in the subatomic world. Many physicists think that the reason so many particles exhibit varying forces is because of broken symmetry.

By working backwards to the beginning of the big bang, physicists have theorized that all four forces, which are now distinct from each other, were unified into one single dominant force. This was possible because of the immensely high concentration of energy in those very early moments. This single force is considered to have exhibited perfect symmetry.

To get an idea of what this may have been like, imagine swimming underwater in an ocean, lake, or pool. Imagine that you're at a depth where you can't see the bottom or the surface. As a matter of fact, no matter which way you look, everything looks the same. You have no sense of direction, time, or space. Everything appears to be equal. That's sort of what the universe was like at the moment of the big bang. No time, no space, no direction, everything the same: perfect symmetry.

When physicists talk about the moments right after the big bang, they use a term called *phase transition*. A simple example of a phase transition is when hot steam cools and condenses into water. When the water cools enough, it becomes ice. Each of these changes is a phase transition. When the big bang occurred, the temperature was incredibly hot, but as time went on, the temperature went through a series of phase transitions. Compared to what it was back then, the universe today is frozen, even though it's warm enough for us to live in.

The perfect symmetry of the universe at the time of the big bang was broken by these phase transitions. As it cooled off, the forces, or interactions, began to break away from the symmetry, one after another. Gravitation was the first to break away and exist as a separate interaction. This occurred sometime before the first 10^{-43} of a second. That is as close to the moment of the big bang as physicists can currently estimate. This moment of time is called Planck time, named after you-know-who.

This moment is so close to the beginning that it's important to realize how much went on in that first second. So much happened that it's necessary to divide that first second into segments and examine each separately. At Planck time, the universe had a diameter of about 10^{-28} centimeters and a temperature of 10^{32} degrees Kelvin. The diameter of an atom is only 10^{-8} centimeters, which is much bigger than the universe at Planck time. Particle accelerators are trying to recreate these first moments, but the energy required to power even the most powerful ones, like Fermilab, fall short by 10^{14} zeros of energy.

As the universe cooled and expanded a bit further, more phase transitions took place. Before the universe was a second old, the strong, weak, and electromagnetic interactions had broken free from the original symmetry. Quarks and leptons no longer interchanged, but it would be quite some time before the universe cooled off enough for the first atomic particles to form.

Physicists are trying to understand the conditions right after the big bang in two ways. One is by trying to recreate the conditions using massive particle accelerators that can produce some of the high energies needed to make the forces recombine, so the first particles can be examined. Because of the high energies needed, it may be a while before this approach is successful.

The second way is to do it mathematically. Mathematical descriptions such as these are called *Grand Unified Theories*, or *GUT*. The ultimate goal is to combine all four forces into one simple theory that will show that each force is the same phenomenon appearing differently because of the freezing of the universe. Because each force can be described as a field, the mathematical technique used to explain these fields is called *gauge symmetry* and the theories are therefore called *unified field gauge theories*.

Relatively Speaking

A **phase transition** is the change that matter goes through as it cools from a very hot temperature to a cold one. In physics, it describes the various states and the breaking away of forces during the first microseconds after the big bang.

Planck time is the closest that physicists can get to the moment of the Big Bang. It is a moment 10^{-43} of a second after it. It was named after the famous father of quantum mechanics, Max Planck. It is the smallest measurement of time that has any meaning, even though a smaller division of time is possible. It is based on the relationships between the speed of light, Planck's constant, and the universal constant of gravity. A similar quantity is Planck length, which in the quantum world is the smallest measurement of length that has any meaning.

So far, physicists have been able to combine the electromagnetic and weak interactions. These two interactions were the last to break away from the original perfect symmetry. In 1979, the Nobel prize in physics was awarded to Steven Weinberg, Abdus Salam, and Sheldon Glashow for their work in unifying these two forces. This discovery reduces the number of interactions to three, the electroweak, which is the combination of electromagnetism and the weak force, gravitation, and the strong force.

Ultimately, theorists want to combine the strong and gravitational forces into this unification, but they must first combine the strong with the electroweak. Doing so will yield a grand unifying theory, or GUT. When gravity, which is the most difficult and least understood of all the forces, can also be included, we will have a Super GUT. Some physicists believe that this will be some form of a quantum gravitational theory. Many theories out there are vying for predominance. Here are just a few:

➤ **Supersymmetry.** In the 1970s, physicists tried to develop a geometrical description of everything by using what was called *gauge symmetry*. Later, supersymmetry was developed by adding an additional four dimensions to the four we currently have in ordinary space and time. The resulting eight-dimensional geometry is known as *superspace*. Because space and time were nonexistent at the time of the big bang, it is unknown how many dimensions of space and time there were. These additional dimensions allow for the interactions between particles to take place in a way that could provide a key in understanding their unification.

➤ **String theory and superstring theory.** These theories were developed in the 1980s and pertain to any theories that describe elementary particles and their interactions in terms of tiny one-dimensional entities, or strings. These strings form loops, which are much smaller than particles such as protons. Particles are usually thought of as being just points that have no extension in any direction. In string theory, these point-like entities are replaced by the idea of particles as objects that have extension in one dimension. This can profoundly affect mathematical equations and helps to explain many of the features observed in the particle world. The exciting part about this theory is that it includes gravity automatically within the same framework as the other forces.

➤ **Theory of everything (TOE).** You've already been introduced to this theory, but just to refresh your memory, it applies to any theory that attempts to combine gravity and some form of a GUT. Right now, the most likely candidate will be some form of superstring theory or a possible combination with supersymmetry.

That brings this chapter to a close. The final chapter looks at some of the strangest things known about the universe, including probable worlds, dark matter, some interpretations of Eastern philosophy, and also how Einstein's theories are still being applied today.

The Least You Need to Know

➤ Ninety-nine percent of all the known matter in the universe is composed of hydrogen and helium.

➤ Physicists measure the energy of subatomic particles in electron volts, or eV.

➤ Matter is composed of two fundamental groups of particles: quarks and leptons.

➤ The four forces found in the universe are gravity, electromagnetism, the strong force, and the weak force.

➤ The only forces that have been combined so far are the electromagnetic and the weak forces, now called the electroweak force.

➤ The best candidate to provide a theory of everything is superstring theory.

Is the Truth Out There or In Here?

It's been quite a sojourn through the world of ideas, including the discoveries, the paradigms, and the individuals whose contributions both before and after Einstein left their imprint on the study of science, philosophy, and human consciousness. This chapter adds a few more fascinating ideas, discoveries, and mind-expanding theories to expand your big picture of the universe, including some of the more unusual findings that cosmologists have come across and other concepts that at one time belonged only to the realm of science fiction. A brief overview of Eastern philosophy will shed more light on the structure of the universe, and finally you'll see how physicists are continuing to apply Einstein's theories in their quest for a deeper understanding into the nature of the world in which we live.

It's What We Can't See That Counts

Remember our friend Isaac Newton? His law of gravity helps to predict the speed at which planets orbit the Sun and can even be applied to the velocity at which stars orbit around the center of their galaxies. Basically, Newton's law of gravity predicts that planets on the edge of our solar system, like Pluto, orbit at speeds slower than those planets that are close to the Sun, like Mercury. This law was shown to be correct, because Mercury orbits the Sun at a velocity about 10 times faster than Pluto does. According to the theory, the same should be true for spiral galaxies. The more distant a star is from the center of the galaxy, the slower its orbital velocity should be.

As far back as 1950, an astronomer by the name of Vera Rubin tried to persuade fellow astronomers that the stars on the periphery of many galaxies displayed unusual motion. She discovered that the predicted relationship between the speed at which a star traveled and its distance from the center of the galaxy didn't work out. Her results showed that stars at the edges of their galaxies orbited at the same speed as those closer to the center. This result either means that Newton's law of gravity fails at a galactic level or that galaxies have a lot more matter than we can see. Newton's law of gravity and later Einstein's general theory of relativity worked so well at describing observations in our galaxy that no one was willing to believe these laws were different in far away galaxies. So the search for the mysterious missing matter was on.

If you were to look at a galaxy through a telescope, you would see that most of the mass of the galaxy is concentrated toward the center and that the galaxy becomes less dense as you look out to the edges. But according to Rubin's discovery, which has since been verified by many astronomers, the stars at the periphery move as if they were embedded in a much greater mass—so much mass that it must extend way beyond the edge of the galaxy. If this information is correct, and most astronomers agree with her findings, then the galaxies are not at all what they appear to be. The stars we see must be swamped in an immense quantity of some kind of invisible mass.

Data was collected on hundreds of galaxies, and they all behaved the same way. This evidence led to the conclusion that the major component of galaxies is *dark matter*. This matter is dark, because it can't be seen. This revelation became an obsession among cosmologists. Some of the questions that they asked were:

➤ What is dark matter made of?

➤ How much of it is there?

➤ How will it impact the fate of the universe?

➤ What role does it play in the formation of galaxies?

➤ How does it affect our understanding of the origin of the universe?

For cosmologists, these are some pretty profound questions to ask, and very few of them have been answered. The one thing we know fairly accurately is that what we can see—in other words, all of the mass from the billions of galaxies that have been observed—accounts for only 1 percent of all the mass in the universe. As much as 99 percent of the mass in the universe is dark matter. You can't see it, but it lies at the very heart of how the universe is structured.

The issue of what dark matter is made of gets a little more difficult. Scientists can account only for 10 percent of the total amount of what dark matter is. The other 90 percent, and scientists hate to admit this, is some kind of exotic material, and there are only a couple of clues as to what it is. The best clues may come from particle physics. Some theories predict the existence of a particle that has the characteristics that physicists have defined as being required of dark matter. But more powerful accelerators are needed to artificially create this particle, because its energy content is very high, and our existing accelerators aren't powerful enough. Also, another type of particle, called the axion, may account for a lot of the dark matter. However, axions are even more elusive than the neutrino. Researchers have been looking for just one axion for several years, and have not seen even one!

Relatively Speaking

Dark matter is an invisible material that constitutes a large percentage of the mass of the universe. It's revealed by its gravitational influence on visible matter such as stars and galaxies.

Holes in Space

For now, the riddle of dark matter remains unanswered. Another unusual discovery, and a topic of science fiction for years, is the existence of *black holes*. These holes are some of the most difficult objects in space to detect and measure. Their existence was speculated about as far back as the eighteenth century, when scientists imagined the possibility of worlds so massive that nothing could escape their gravitational grip, including light. In 1939, Robert Oppenheimer, the father of the atomic bomb, used Einstein's general theory of relativity to explain how a black hole might form. He showed that a black hole warps space so entirely that not even light can escape.

These holes may inhabit the centers of galaxies, including the Milky Way. The center of our galaxy emits intense gamma radiation, which could be the result of stars falling into a black hole. But for several decades, black holes were merely thought to be a mathematical curiosity, because no one thought that it was possible that physical objects could collapse to the state of extreme density that would be required to make a black hole.

Today it's known that neutron stars are produced when a massive star explodes as a supernova. Black holes are thought to be produced by the same process. If this type of explosion were to take place close to another star, for example when two stars orbit each other, it would strip matter from the other star to form a disc of hot material that would funnel into the black hole. The material would be so hot that it would radiate x-rays,

Relatively Speaking

A **black hole** is a concentration of matter so dense that its gravitational field is strong enough to curve the space-time continuum completely around itself to the point where nothing, not even light, can escape.

Neutron stars are stars made almost entirely of neutrons. Their mass is very dense because they burned themselves out as regular stars. They often become black holes.

thereby making it detectable. Since the early 1970s, many objects like this have been found. Based on the number that have been detected so far, it is estimated that there are roughly 100 million black holes in our galaxy, which contains about 100 billion visible stars.

Worms in Space

Other speculation has suggested the existence of *wormholes*, which are tiny black holes that may form tunnels through the space-time continuum. You can think of a wormhole as a shortcut through the space-time continuum, a sort of cosmic subway that connects two black holes. The other end could be anywhere.

Many scientists have scoffed at the concept of wormholes, but in the 1980s physicists at Caltech showed that they could exist. Using Einstein's general theory of relativity, they found solutions to his equations that theoretically allow for the existence of wormholes. Einstein himself, along with Nathan Rosen at Princeton, discovered certain equations based on his general theory that represented what could be a black hole connecting two regions of flat space and time.

The difference between a black hole and a wormhole is simple to understand. A black hole is a one-way ticket. You can get in, but you can't get out. On the other hand, a wormhole allows for two-way traffic. It's essentially two black holes connected together. The possible existence of naturally occurring wormholes suggests some interesting possibilities to physicists:

➤ If wormholes exist on the scale of the Planck length, which is the smallest measurement of length that has any meaning, they could provide a sort of foam-like structure of space and time, weaving the very fabric of space and time out of wormhole strands.

➤ These ultra-subatomic, very, very, *very* small wormholes could link distant parts of space together and leak the laws of physics to all parts of the universe, thereby assuring that the principles of physics work everywhere.

➤ The small wormholes could also be seen as equivalent to the tiny strings theorized in superstring theory, which I mentioned in the last chapter. They could help explain the structure of matter on the smallest scale, possibly providing the missing link to the theory of everything.

Mind Expansions

The fate of our universe will be determined by its contents. If it contains enough mass, then its gravitational forces will be strong enough to halt and then reverse the expansion of the big bang. If there isn't enough mass, then the expansion will go on forever, with the universe eventually cooling off and ending in the big chill. What happens will also be related to the shape of the universe. There are three possible shapes. The universe could be flat, open, or closed. A flat universe is flat, with ripples in it caused by the local concentrations of mass by planets, stars, and other celestial bodies. An open universe is shaped like a saddle, curved such that it never closes in on itself. A closed universe looks like a sphere. Right now we are pretty sure that the universe is either flat or saddle-shaped.

Time Travel

One of the other intriguing implications of wormholes is the idea of time travel. For the most part, this topic has been dealt with only in the realm of science fiction. Until recently, that's where it seemed like it was going to stay. But research carried out in the 1980s revealed that time travel isn't forbidden by the laws of physics. This discovery doesn't mean that we can go out and build a time machine tomorrow, but eventually, we may be able to utilize wormholes as the means to travel either forward or backward in time.

Not only can wormholes connect different areas of space together, they can also connect different times. Space and time are both involved in Einstein's equations. His general theory of relativity deals with the space-time continuum. His equations show that a wormhole that takes a shortcut through space and time could just as well link two different times as two different places. So a naturally occurring wormhole could provide the means to travel to a different time. This isn't just science fiction. Some insightful physicists have pushed Einstein's equations to extremes and found that an intimate connection exists between time travel and the quantum mechanical description of the universe.

Albert Says

Time travel has been the topic of many science fiction stories, but it was never considered possible. In order to travel back in time, it would be necessary to travel faster than the speed of light, and as far as we know, that's impossible. Einstein's theory of relativity states that as an object approaches the speed of light, its mass increases. The speed of light is never attainable, because at that speed the mass would become infinitely large, therefore requiring infinite energy to make it accelerate. But now, with the theories of wormholes, time travel may become a possibility, using the wormhole as a gateway through time. It's still all just theory, but at least there's nothing in the laws of physics that would prevent it from happening. Anyway, don't count on it happening any time soon.

Probable and Parallel Worlds

Dark matter, black holes, and wormholes show that the macrocosm and microcosm are indeed very closely connected. Although these phenomena are predicted to exist throughout the universe on the macrocosmic level, their fundamental structure is being explored on the microcosmic level. For this reason, cosmology has become a combination of astrophysics and particle physics.

Any thorough study of cosmology includes three areas of inquiry: physics, philosophy, and metaphysics. The last two chapters have explored how physics interprets cosmology, so let's return to philosophy and metaphysics for the remainder of our discussion.

A number of fascinating ideas have come out of the weird things that occur at the quantum level of reality. These theories are intimately connected to how we as humans perceive the world. Niels Bohr's Copenhagen interpretation, which was explained in Chapter 18, "Uncertainty Is Certain," combines uncertainty, complementarity, and probability into one relationship. Very simply stated, it says that there is no meaning to the objective existence of an electron at a specific point in space, for example at one of the two holes in the double-slit experiment, independent of observation. The electron seems to spring into existence as a real object only when we observe it. This means that reality is in part created by the observer.

One of the popular ideas that came out of this interpretation is called the *many worlds interpretation*. This idea states that whenever the quantum world is faced with a choice, for example when the electron has a choice of which hole to go through in the double-slit experiment, the universe divides into two, or as many parts as there are choices, so that all possible options are followed. In the case of the double-slit experiment, the electron goes through slit A in one world and goes through slit B in another world.

This interpretation says that an infinite number of alternate realities, or probable or parallel realities, exist alongside our own. Every possible choice is realized in one or another of these alternate realities. We enter the picture when we observe what the outcome is, thereby influencing which world becomes manifest into our reality. Our participation in the experiment makes one particular world become the reality that we are physically in and the others fade into parallel worlds that don't include our physical selves.

If this is true, why is it that everyone observes the same thing? A group of physicists watching to see which slit the electron goes through all see it go through the same slit. Why don't some see it go through slit A, and others see it go through slit B? If this were the case, we would all end up experiencing different realities from each other, and we'd never be able to agree upon what was really occurring. (To a certain extent, this is already true. Have you ever tried to get a room full of people to agree upon what they saw?) Somehow there seems to be a collective, unconscious agreement, by which all of us end up seeing the same thing.

When we look at an object, such as a tree or a car, we all agree upon what we see, at least in a general way. There are definitely slight differences in our perception of what we see, based on our position in relation to the object. Because no one can occupy the exact same space as someone else, their frame of reference with which they perceive the object is unique. But this observation deals with everyday objects. The connection between consciousness and quantum events seems to be much more intimate than simply agreeing upon what we all see.

Relatively Speaking

The **many worlds interpretation** states that whenever the quantum world is faced with a choice, the universe divides into as many parts as there are choices, such that all possible paths are followed. This idea is one of a number of interpretations that have been conceived in an effort to understand what happens at the quantum level. Of all the quantum interpretations, it is the easiest to understand. Some of the names of other interpretations are the sum over histories approach, the transactional interpretation, consistent histories interpretation, decoherence, and the ensemble interpretation. Most are too involved to cover here, but go check them out in any good introductory book on quantum theory.

Some philosophers think that the very nature of human consciousness influences quantum events. If this is true, then part of our consciousness must reside outside of us so that it can interact with the objective world. In the same way that the electron appears to make a choice as to which hole it goes through, each of us makes choices when we take specific actions at every moment of the day. Some of them are unconscious choices, and others are more conscious.

If the quantum world breaks into as many parts as there are choices for particles to make, that can also mean that each time we make a choice to pursue a particular path, a part of our consciousness pursues the choice that we didn't make, carrying out and following through with the other option. This would mean that we have a large number of *probable selves* existing in parallel worlds, living lives based upon the choices we didn't follow in our reality. Now of course this is all very speculative, but it is one of the logical conclusions of this particular interpretation. As strange as it sounds, some interesting work in consciousness research has shown that human consciousness has the capacity to experience more than one level of reality at the same time. Who knows what more we'll discover in the future?

Along with all of this come questions about the nature of reality. Is reality in the eye of the beholder? Earlier, I noted that basically we all agree upon the general nature of objective reality. A tree is a tree. But as I've mentioned many times, our beliefs act as filters on our perceptions. As we move toward more complex perceptions, like people and events rather than trees and cars, our experience of reality can become very different than that of the person next to us who is witnessing the same thing.

The point of this discussion of probable selves, parallel universes, and alternate realities is that there is so much more to who we are than meets the eye. Whether these theories are true is not what's important. What is important is to understand that not all the answers have been found, and that when they are, they don't have to conform to our present understanding. The quantum world, the structure of the cosmos, and the role of human consciousness in participating in all of this are very far from being understood.

Albert Says

One of the reasons there has been a lot of interest in drawing analogies between quantum interpretations and consciousness is because the two are so fundamentally linked at the quantum level. Because we don't have a good understanding of the exact nature of human consciousness, nor of the events that unfold at the quantum level, an investigation into the similarities between them might help shed light on both at the same time. Because observation lies at the heart of science, and we're the ones doing the observing, it's only natural to try to see how the two may be connected.

Eastern Cosmology

Our interpretations of cosmology here in the West are biased by the cultural paradigms that influence our perspectives. Science and all it has to offer, along with the Western religious traditions, are the predominant views through which many of us understand the world in which we live. But the Eastern perspectives, which are thousands of years older than our paradigms, are steeped in interpretations of the universe that are based on intuitive insight and are filtered through their belief systems. Because science as we know it was first introduced to the East only a few decades ago, the East's original foundation for comprehending the world was based on an inner knowledge that came from centuries of exploring the inner world rather than the outer world.

The three philosophical traditions that I'm going to briefly cover are Hinduism, Buddhism, and Taoism. An important point to mention is that each of these systems of thought and belief are much more than mere philosophical traditions. They encompass aspects that are religious, psychological, cultural, and somewhat mystical. I'm just going to focus on some of their views on the nature of the universe.

The greater Eastern paradigm, of which the specific paradigms of the three traditions are a part, has at its core the concept that there is a basic unity and interdependence of all things and events in the universe. When we experience this awareness, we recognize the world as a manifestation of a fundamental oneness. Also contained within these systems of thought are teachings that help us attain and experience the awareness that everything in the world is a manifestation of the same ultimate reality.

Nuclear Meltdown

Some physicists have a meltdown when their physics is compared to Eastern traditions. They get incensed at the concept that science can be compared or thought of as similar to a subjective experience. They have a point. Science doesn't need religion or mysticism, any more than religions or systems of inner development need science. But as human beings, we need both. So it's important not to get caught up in defending a paradigm that doesn't need defending. In the end, it's only a lack of self-awareness that prevents us from reaching a deeper level of understanding of the dynamic interplay between the two.

What the Eastern traditions seek to do on an inner level by transforming our consciousness, physics seeks to do on the objective level by finding the theory of everything. The goal of each is to supply us with an understanding of the fundamental unity of the universe. The difference between them is that one is *in here* (subjective), and the other is *out there* (objective). As we've seen, the difference between the two is just a matter of which one we believe offers the best explanation to fit our system of thought, based on our own experience. One is an *experience* of unity, and the other is an *explanation* of unity. But whichever one we experience as being the truth, this doesn't mean that the other is not the truth. To quote an Eastern saying, "That which is One is one in which that which is not One is also One."

In other words, unity doesn't exclude anything from it. Regardless of whether it's an experience of unity or an explanation of it, both are part of the greater whole. It's all included in the big picture.

Hinduism and the Dance of Shiva

Hindu cosmology sees the universe as a manifestation of *Brahman*, the indescribable, impersonal, and absolute source of all there is. One of the many forms that Brahman can take is that of *Shiva*, the cosmic dancer, which symbolically creates and destroys the universe through its rhythmic dance. All of creation is seen as a manifestation of cycles, excluding nothing. We cycle through our days and nights, which occur within our yearly cycle. We even cycle through our lives, which is why Hindus believe in reincarnation. Why should everything in the universe operate on a cyclic basis except for us?

Nature and the universe cycle through ever larger periods of time. Our solar system cycles through the galaxy, and our galaxy turns on its spiral. Even the creation of the entire universe isn't seen as a one-time linear event. The universe expands as Brahman exhales and contracts as Brahman inhales. Our big bang is the exhalation, and if there is a big crunch in 50 billion years, that will be the inhalation. But this cycle has occurred countless times. This could be the 10,000th time or more that the big bang has occurred.

Hindu cosmologists have even broken down these inhalations and exhalations into specific lengths of time called *yugas*. Each yuga lasts for a specific period of time. The total time is equal to 80,640,000,000 years, roughly the amount of time equal to Western estimates if the universe doesn't keep on expanding and ends up contracting back into the big crunch.

Relatively Speaking

Yuga is a Sanskrit word that defines a given length of time in the cycle of the universe. There are four yugas for each **kalpa**, which is equivalent in time to 4,320,000 years. There are 2,000 kalpas in one full expansion and contraction of the universe. The entire cycle is called a day and night of Brahman or one cosmic day. There have been many cosmic days in Hindu cosmology. The one we're in now could be one of many that have already occurred, with many more to come.

Buddhism

One of the most significant aspects of Buddhism is the idea that everything is impermanent. The Buddha taught that all of the suffering in the world comes from our desire to cling to objects, people, or ideas, instead of accepting the universe as it flows, moves, and changes. Adaptability, along with detachment from material desires, reveals the dynamic quality of the universe. One of the goals of Buddhism is to move and change with life, instead of resisting change and clinging to things that we think give us a sense of sameness and permanence in a world whose essential quality is change.

These concepts within Buddhism are not only part of how we can psychologically adapt to the flow of life, but they also lie at the heart of the quantum world. Space and time are just seen as reflections of

Albert Says

An important point in our discussion of the state of consciousness in which the universe is experienced as a unity is that only a handful of individuals have ever achieved this profound state of awareness. The experience of it takes a tremendous amount of effort, and only those individuals who devote their life to its attainment ever get even close to it.

341

states of consciousness. Subatomic particles move, interact, transform, and are continually changing and impermanent. In the end, Buddhists, like physicists, see objects in the world not as things, but as dynamic processes, participating in a universal movement that is constantly in a state of transition.

Taoism

In Chinese philosophy, the dynamic movement of the universe is called the *Tao* (pronounced *Dow*, like in Dow Jones). It lies at the core of everything, because it puts into motion the active balancing of complementary opposites, called *yin* and *yang*. The universe is seen as a system of energy, constantly changing and flowing. This idea of flowing energy, called *chi*, is seen as the life force and the principle that underlies all of the Chinese systems of medicine, cooking, architecture, art, and martial arts.

The Tao is the one unifying force in all of creation, sort of like the Force in *Star Wars*. Einstein's famous equation showed that matter and energy are equivalent, so it's understandable to view the universe as a system of energy rather than as a system of material forms.

The *I Ching* (pronounced *yee ching*), or *Book of Changes*, is one of the oldest books in the world, dating back to around 1500 B.C.E. or earlier. It explains the dynamic patterns that underlie all of nature, human interactions, and natural processes. Some scholars and scientists have linked its explanation of patterns to the genetic code, Jungian psychology, the binary basis of our computer systems, and the interaction of forces in the macrocosm.

As you can see, many parallels exist between physics and Eastern thought. Even though they are two very different approaches to understanding the universe, they can supply us with a paradigm that is inclusive of both.

Einstein's Legacy

With all that's been covered in this book, it's appropriate to finish up with one last look at Einstein, the man who forever changed the face of physics and whose contributions continue to influence many areas of science, now and into the future. His fame rests on his theories of relativity, even though he wrote prolifically on numerous scientific subjects. Altogether, his scientific publications numbered about 350.

Einstein's two theories of relativity play fundamental roles in the study of both the microcosm and the

Mind Expansions

Carl Jung, whom you met in Chapter 19, "How Conscious Are We?," devoted a great deal of time to the study of the *I Ching*. He gained so much insight from his investigations that some say it was the main source for the development of the ideas that led to his theory of synchronicity, the archetypes, and the collective unconscious. The *I Ching* is a symbolic system that helps to reveal to the conscious mind, the unconscious patterns that play out in our lives. In China, it's been used as a system of divination, so it's often been considered as nothing more than a fortune-telling game. But Confucian and Taoist scholars have studied it more for its revealing philosophical worth, and here in the West, it has become associated with psychology and inner development.

macrocosm. From particle physics to astrophysics, physicists continue to apply or expound on his formulas, coming up with deeper insights into the workings of the universe.

Most recently, at Stanford University in Palo Alto, California, physicists have been using his equations to check for something called the *frame-dragging effect*. Einstein's general theory of relativity predicts that a large spinning object like Earth does not simply warp space time, as do all heavenly bodies, but it actually causes a spiraling effect. If this effect is indeed taking place, a telescope in orbit around Earth pointing at a star would be affected by these spirals. To test for this effect, gyroscopes mounted within the frame of the telescope are used to point the telescope. After some time, if the frame-dragging effect is taking place, the telescope will no longer be pointing at the star. This is because the spiraling space-time causes the gyroscopes to point differently, relative to the star. The Stanford group will launch a satellite telescope with very precise gyroscopes early in the year 2000 to test for the frame-dragging effect.

This is just one example of the continued influence of Einstein's theories and equations. Although other physicists have contributed many important ideas and theories to physics, Einstein, like Newton, brought humanity's understanding of the nature of the universe to a whole new level.

With that, this exploration into the life, times, and mind of Einstein comes to a close. You may have gotten more than you bargained for in this effort to understand all that makes up an individual such as Einstein. This excursion into the exciting world of ideas has linked many areas of human endeavor. The bridges of understanding that we've made in connecting it all together will hopefully continue to inspire us to search for insights into ourselves and the world around us.

The Least You Need to Know

➤ Ninety-nine percent of the universe is made up of dark matter, and the remaining 1 percent comprises everything that is visible.

➤ Black holes are produced by the collapse of burnt-out stars, whose gravitational force has become so strong that nothing can escape it, not even light.

➤ Theories of the existence of alternate realities are a direct result of the unusual phenomena that occur in the quantum world.

➤ The goal of the Eastern traditions is parallel to the goal of theoretical physics. Both seek to achieve an understanding of the unity of universe. The main difference between them is that one is a subjective experience, and the other is an objective explanation.

➤ Einstein's theories and equations continue to influence physics as they are applied to new areas and explored in more depth.

Relatively Speaking

a priori   A Latin term that means "prior to having an experience."

absolute brightness   In astronomy, the true brightness of a star, as it would appear to a nearby space traveler.

abstraction   Refers to a type of thinking, abstract thinking, that deals with pattern recognition or seeing the similarity between different kinds of patterns. For example, the game of chess develops abstract thinking because it incorporates movements that have specific patterns.

alchemy   From the Greek word "to pour," it was an early form of chemistry practiced in the Middle Ages that attempted to change baser metals into gold.

algebra   A mathematical system that uses symbols to represent numbers.

allegory   A story in which people, events, and things have hidden or symbolic meanings.

amber   A form of petrified fossilized tree resin.

amp   The strength of an electrical current.

analytical geometry   A branch of mathematics that deals with the combination of algebra and geometry.

angular momentum   In classical physics, it defines the movement of any object on a curved path (such as the Moon as it orbits around Earth). In quantum physics, it is related to movement within the atomic world, such as electrons in orbits.

anthropomorphism   Conceiving or representing a god with human attributes or assigning human qualities to nonhuman things.

antimatter   A form of matter in which a particle has the opposite properties of its counterpart. For example, the anti-electron (or positron) has the same mass as an electron, but it has the opposite electrical charge.

apeiron   A Greek word for describing something that is boundless and limitless, almost infinite in size.

apparent brightness   How bright a star appears to us on Earth, from however far away it is.

archetype   The term used by Carl Jung to describe the symbols and mythic figures that reveal inherent psychological processes of a universal nature.

astrophysics   The combination of physics and astronomy that studies the physical properties of galaxies, stars, planets, and all other heavenly bodies.

axiom   A universally accepted principle assumed to be true without being tested.

Bar Mitzvah   A Judaic ceremony for 13-year-old boys that celebrates their coming of age and responsibility. It literally translates as "son of the commandment."

big bang theory   The popular view in cosmology that states that the universe began 15 billion years ago when a fireball spread out to form all of the galaxies, including our own.

binomial theorem   In algebra, a general formula for writing any power of a binomial, meaning two terms, without multiplying it out. It was discovered by Omar Khayyam, and then later generalized by Isaac Newton.

blackbody problem   The problem is that if radiation is explained in terms of waves, using the same theory which describes sound waves, only the brightness of the light should change with temperature. The color should remain the same! Classical theory could not explain why the color of a blackbody changed as it got hotter.

Black Death   A name for the bubonic plague, a deadly disease that devastated Europe and Asia in the fourteenth century.

black hole   A hole in space and time. It is a concentration of matter so dense that the gravitational field created by it cannot be escaped, even by light.

blueshift   The effect seen in a spectrometer created by the wavelength of light, which shows that the object being measured is moving toward you.

Boyle's law   Named after Robert Boyle, it states that the volume of a given amount of any gas at a set temperature is inversely proportional to the pressure applied.

Brahman   Also Brahma. In Hinduism, the supreme, absolute, and eternal essence or spirit of the universe.

Brownian motion   Named after Robert Brown, it is the motion of particles when they collide with other molecules suspended in a liquid substance.

bubble chamber   Invented by Donald Glasser in the early 1950s, this device tracks high-energy particles that pass through a fluid and make a trail of bubbles along their line of flight.

Cabala   An esoteric religious philosophy that developed out of Judaism, that provides a mystical interpretation to the scriptures. It can also be spelled Qabalah, or Kabalah; each spelling pertaining to a specific way of applying the mystical teachings.

calculus   A mathematical system of calculation using a special system of symbolic notation. It combines both differential and integral calculations.

calorie   The amount of heat needed to raise 1 gram of water 1 degree centigrade.

capacitor   A very small container that can hold an electric charge; it's used in electronics.

Cartesian coordinate   Named after the French philosopher René Descartes, a graphic mathematical system for plotting the coordinates of any linear equation onto a plane. It is used in surveying, navigation, and engineering.

cartography   The art of making maps or charts.

cathode ray tube   A device for producing a visual display using electronic circuitry, without which we would not have television, radar displays, or oscilloscopes. Otherwise known as a CRT.

Celsius or centigrade scale   A temperature scale developed by Anders Celsius based on a division of 100 equal degrees, with the freezing point of water at 0 degrees and the boiling point of water at 100.

Charles's Law   Named for Jacques Charles, it states that the pressure of any gas increases by 1/273 of its original value for every 1 degree centigrade the temperature is raised.

chi   In Taoism, it is life force or vital energy, the cosmic spirit that pervades and enlivens all things.

collective unconscious   A Jungian term, a level of the human unconscious made up of universal concepts, including archetypal symbols and images. It's accessible only indirectly, for example, through dreams.

compound   A combination of two or more elements in which the individual properties of the elements are lost, but the combination has new properties. For example, salt is a compound of sodium and potassium.

Copenhagen interpretation   The standard explanation for understanding quantum mechanics that says there is no meaning to the objective existence of a quantum particle unless it is observed. It led to Einstein's question, "Does the Moon disappear once we stop looking at it?"

corpuscle   A very small particle. The name originally given to particles of light by Newton.

cosmological constant   Einstein's mathematical equation that he added to his general theory of relativity to counteract the gravitation forces in the universe so the universe wouldn't appear to be expanding or contracting.

cosmology   A branch of philosophy and science that studies the universe as a whole, its form, nature, and other characteristics.

counter-intuitive   A characteristic of an idea that is contrary to what you would normally expect and makes you stretch your concepts to grasp its meaning.

cyclotron   A circular particle accelerator first invented by Ernest Lawrence that can produce high-energy interactions. As the particles move around the circle, they pick up more energy on each revolution.

dark matter   An invisible or dark form of matter in the universe, revealed to exist because of its influence on bright matter, the stuff that can be seen.

decay   In physics, it refers to the breakdown of a particle into smaller particles.

deductive logic   Reasoning derived from basic principles or theories that leads to new conclusions. For example, if your spouse likes to read science fiction, you might deduce that he or she would like to see a science fiction movie.

diffraction   A phenomenon that occurs when rays of light interfere with one another, creating light and dark bands.

distillation   The process of heating a mixture to separate its more volatile parts from its less volatile parts, cooling it, and then condensing the vapor to produce a more refined substance.

dogma   A personal or collective system of beliefs that inherently defines itself as true and exclusive of other systems, which it considers inherently false.

Doppler effect   The change in frequency of light or sound waves, which varies with the position of the observer in relation to the moving source.

earthshine   The faint illumination of the dark areas of the Moon by the sunlight reflected from Earth.

electrodynamics   A branch of physics that deals with electrical currents and magnetic forces.

electrolysis   A term used in chemistry that refers to passing an electric current through a substance so that compounds can be broken down into their individual elements.

electron   A negatively charged particle that orbits the nucleus in an atom.

electron volt   The unit of measurement that physicists use to calculate the energy content of subatomic particles. It's abbreviated as eV.

element   The simplest form in which matter can exist. For example, carbon is an element found in all human beings.

Elysian Fields   From Greek mythology and found in Homer's *Odyssey*, these fields were found at the end of the world. Those favored by the gods would go there after death.

empiricism   The search for knowledge through observation and experimentation.

entropy   A measure of the degree of disorder in a substance or system; entropy always increases and available energy decreases in a closed system.

entropy thermoscope   A device that measures the amount of energy available to do work in a system.

epicycle   The convoluted system of planetary orbits developed by Ptolemy in which a body rotates around a point that moves around the circumference of a larger circle.

epistemology   One of the main branches of philosophy, it seeks to discover the origin, nature, methods, and limits of human knowledge.

equipartition   The even distribution of energy among vibrating bodies.

ether   Scientists once believed that this substance filled the space between the Moon and other planets in the solar system.

Fahrenheit scale   This temperature scale developed by Daniel Fahrenheit uses the melting point of water and the temperature of the human body as reference points.

field   In physics, a region of space that contains a force such as gravity, electricity, or magnetism, or a combination of these forces, such as electromagnetism.

fission   The process by which a nucleus of an unstable atom (such as uranium) splits into two or more parts, releasing energy.

flavor   A property that distinguishes one quark from another. Quarks come in six flavors: up, down, top, bottom, charm, and strange.

fluxion   A term used by Newton to refer to his mathematical system of calculus.

frequency   A term used in electricity and electronics that refers to the number of cycles per second that electricity oscillates in reference to the current delivered. It also refers to electromagnetic waves such as radio, television, x-rays, and light.

frequency threshold   Refers to the way a plate material reacts to light by emitting a certain number of electrons. If only a few electrons are emitted, the frequency threshold of the substance is very low.

galvanization   Named after Luigi Galvani, it's the process used to plate metal with zinc. A galvanometer measures and detects small electrical currents.

gamma ray   A form of electromagnetic radiation that has tremendous penetrating power and is emitted by the nucleus of a radioactive substance.

gauge boson   Any particle that carries one of the four forces in nature and thereby mediates particle interactions.

geocentric theory of the universe   A system of thought that says Earth is the center of the solar system.

gravitational lensing   The phenomenon that occurs when light from a distant astronomical object is bent around an intervening galaxy so that we see two or more images of that object. Einstein theorized that a strong gravitational field could bend light waves around it.

gravity   A term adopted by Newton from the Latin word *gravitas*, meaning heaviness or weight.

Gregorian calendar   The calendar in use today, developed during the time of Pope Gregory XIII. Each year is equal to 365 and 1/4 days, which is why we have a leap year every four years consisting of 366 days.

GUT   Stands for grand unified theory. Any theory that can mathematically combine the electromagnetic force, the weak force, and the strong force into one unified theory.

hadron   A particle that responds to the strong force. It comes from the Greek word meaning strong.

heliocentric   A term derived from the Greek word for sun, *Helios,* that refers to the concept of our solar system as sun-centered.

heuristic   A term used in computer programming to describe how computer systems learn. It comes from German and means to discover or to learn.

I Ching   *The Book of Changes.* A Chinese book of wisdom and oracles, essentially Confucian but containing Taoist elements.

inductive logic   A type of reasoning based upon probability. For example, if you observe a certain event over and over, you might infer that it will turn out the same way the next time it occurs.

inertia   The resistance of a body to acceleration.

isotope   An atom of an element with a normal number of protons in its nucleus, but a different number of neutrons. For example, uranium 235 and uranium 238 have the same number of protons (92), but U235 has 143 neutrons, and U238 has 146 neutrons. Their identifying numbers represent the sum total of their protons and neutrons.

joule   A physics term for a unit of energy, named after James Prescott Joule for his discovery of the conversion of work into heat.

kalpa   From Sanskrit, meaning world cycle or world age. It's made up of four yugas and covers 4,320,000 years.

kinetic energy   The measurement of energy from a body in motion that is equal to half the product of its mass and the square of its velocity.

Knights Templar   Established during the Crusades in the twelfth century, they were a military and religious order with branches that still exist today.

lambda   One of the first particles discovered to have a longer lifetime than the other subatomic particles known at the time. It is electrically neutral and decays into a proton and pion.

lepton   Any particle involved with electromagnetic and weak interactions. The most well-known lepton is the electron.

linear accelerator   The first type of particle accelerator built, in which particles are accelerated down long, straight tunnels.

luminiferous   From the Latin term *lumen*, which means light, this term refers to anything related to light or some aspect or quality of light.

macrocosm   A term meaning big universe that is used to describe the relationship between large and small systems. For example, the planetary orbit system of our galaxy is a macrocosm of the atomic world.

Manhattan Project   The code name of the immense project that was undertaken in the United States during World War II to build the first atomic bomb.

many worlds interpretation   One of the many interpretations of the quantum world, having its roots in the Copenhagen interpretation. It states that whenever the quantum world is faced with a choice or choices, the universe divides into as many parts as there are choices so that all possible options are followed.

mass   The amount of substance or stuff that there is in an object.

mathematical function   A quantity whose value depends on another quantity or quantities. For example, temperature may depend upon time as when you are heating water on the stove.

matrix mechanics   A mathematical system developed by Werner Heisenberg to describe states of quantum entities by using a matrix of numbers similar to a chessboard. It was used to locate probable positions and other properties in the subatomic world.

metaphysics   A branch of philosophy that asks questions about the nature of reality, especially those that are considered unanswerable, such as: Is there a God? What is the soul? Is there life after death?

mica   The name given to a group of minerals that crystallize in thin layers and are resistant to heat and electricity.

microcosm   A term meaning little universe that is used to describe the relationship between large and small systems. For example, the atomic world is a microcosm of the universe.

microwave radiation   The part of the electromagnetic spectrum lying between the far infrared and lower frequencies. It is also the wavelength of the background radiation left over from the big bang.

model   Used in science to help visualize an idea or concept, it is not necessarily a reflection of reality, but it helps explain how things work.

muon   A particle identical to the electron, except that its mass is 206 times greater. One of the first subatomic particles discovered.

Natural philosophy   A branch of ancient philosophy devoted to the study of the natural world, today known as science and more especially, physics.

nebula   This term originally pertained only to galaxies. Today, nebulae are vast, cloudlike patches consisting of groups of stars too far away to be seen singularly.

neutrino   A particle that has zero charge and possibly zero mass. It interacts with other particles only through the weak interaction.

neutron   One of the three elementary particles found in the atom, it resides in the nucleus along with the proton. It has a neutral electrical charge and has almost the same mass as the proton.

neutron star   A collapsed star made almost entirely of neutrons, formed after it exhausted its supply of hydrogen and helium.

nucleus   The central part of the atom containing two fundamental particles, the proton and neutron. It carries a positive electrical charge equal to the total negative electrical charge of the orbiting electrons.

ohm   A unit that measures the impedance of the flow of electrical current.

optics   During the time of Aristotle, a branch of natural philosophy. It later became an area of study in physics that deals with the nature and properties of light and vision.

oscilloscope   An instrument that displays an electrical wave.

Ottoman Empire   From the thirteenth century until 1918, a Turkish empire that covered southeast Europe and parts of Asia and Africa.

pantheon   All the gods and goddesses of a culture or people.

parallax   The apparent change in the position of an object when viewed from two different locations. Used in astronomy to measure the distance from Earth to heavenly bodies within our solar system.

perfidious   Having the characteristic of breaking of faith or trust.

Periodic Table of Elements   The table of all the known elements, which now number 113.

personal consciousness   A Jungian term, the level of consciousness where nothing is permanently retained, including our sense of self and ego. It's a sliding frame of reference occasionally lit by awareness.

personal unconsciousness   A Jungian term, the level of the unconscious that belongs to each individual where images are found that are unique to that individual.

phase transition   The change in a physical system from one state or phase to another without any change in its chemical composition, such as water freezing into ice or heating into steam.

phlogiston From the Greek world meaning flammable, it was a theory developed to explain why substances burned before the element oxygen was discovered.

photoelectric effect A function that occurs when a beam of light that acts as part of an electronic circuit is broken. For example, when you walk into a store, you break the beam of light, and a doorbell rings.

photon A particle of light.

physics A term that comes from the Greek word *physika,* meaning natural things. It is the study of the properties, changes, and interactions of matter and energy.

pion One of the particles that helps bind the nucleus together by carrying the strong force between protons and neutrons.

Planck length A quantum of length, the shortest measurement of length that has any meaning.

Planck time The time it would take light to travel one Planck length, the shortest amount of time that can be measured that has any meaning.

Planck's constant A fundamental constant, denoted by h, found by the German physicist Max Planck. It relates electromagnetic radiation to a frequency, as described in the formula $E = hf$.

Plato's Academy A famous school founded by Plato in ancient Greece where young men could study philosophy, mathematics, and music.

Primordial atom The name given by Georges Lemaître to the single, primeval atom that was the source of the big bang.

principle of complementarity Developed by Niels Bohr, it explains the dual nature of some quantum phenomena. It states that waves and particles represent complementary aspects of the same physical phenomenon. Without one, you could not have the other.

principle of equivalence One of the two foundational ideas that make up Einstein's general relativity theory. It states that acceleration is equivalent to gravity, because without a frame of reference, you cannot distinguish between the two.

privatdozent Meaning *private teacher*, it's a term that dates back to the Middle Ages. It is a form of education used in Europe in which the teacher is paid by the students, not by the university.

probability function A type of mathematical function in which the likelihood that a certain event will occur is determined. Max Born used space as a function to predict the probable location of a single electron.

propagation A word in physics that means the transmission of something through space, particularly light waves.

proton A particle found in the nucleus of an atom that has a positive electrical charge.

Pythagorean solids   Named after the Greek mathematician Pythagoras, who proved that there were only five perfect solid shapes besides the sphere. The Pythagorean solids are the tetrahedron (4 sides/faces), the cube (6), the octahedron (8), the dodecahedron (12), and the iscosahedron (20).

quantum leap   A discontinuous transition between quantum states that occurs when an electron leaps from one energy level into another.

quantum mechanics   From the Latin word *quanta*, which means how much. It deals with the laws of mechanics in relation to the subatomic world and its structure. Also called quantum physics.

quark   The subatomic particle that forms the building blocks of all hadrons. It's a level of matter below protons and neutrons.

quasar   Short for quasi-stellar source. Refers to heavenly bodies whose emission of radiation is too high to be given off by just one star. They're most likely cores of galaxies.

redshift   The effect seen in a spectrometer, created by a wavelength of light, that shows that the object being measured is moving away from you.

refraction   An optical illusion that occurs when light rays are bent as they pass through water or any medium of a different density.

Reich   A term that refers to a period of rule in Germany. The First Reich was during the Holy Roman Empire, the Second Reich was from 1871 to 1919, and the Third Reich was from 1933 to 1945.

relativistic mechanics   A system combining Maxwell's equation of electromagnetism with Newtonion motion.

Renaissance ideal   A development from the realization that knowledge of the world is based on perception and that perceptions needed to be questioned and tested.

resonance   The ability of one vibrating body to set in motion or amplify another body.

Scholastic philosophy   A branch of philosophy practiced during the Middle Ages that used reason to enhance faith.

Self   A level of consciousness that connects the personal consciousness to the soul. Can also refer to the deepest unifying aspect within a person.

Semitic   Anyone descended from Shem, the eldest of Noah's three sons, including the Hebrews, Arabs, Assyrians, and Phoenicians. Today this term mostly refers to followers of the Jewish faith.

Shiva   The third member of the Hindu trinity, who functions as the destroyer of ignorance.

singularity   Refers to the single point, infinitely small and infinitely dense, that was the beginning point of the big bang.

Socratic method   A method of teaching developed by Socrates that employed a dialectical approach of asking questions that led the answerer to a logical conclusion.

SSC   Short for superconducting supercollider. A state-of-the-art particle accelerator that was never built because of funding cutbacks by Congress.

standard model   In particle physics, the current description of the known subatomic particles and their interactions. In cosmology, the most widely accepted theory for the creation of the universe: the big bang.

states of matter   The most common forms of physical existence, including gas, liquid, and solid.

steady state theory   A theory of the origin of the universe that states that the universe is in a process of continuous creation with no beginning and no end.

strangeness   One of the six flavors that a quark can have. A property defined by the lifetime of the particle.

string theory   Any type of theory that describes subatomic particles and their interactions in terms of very small, one-dimensional things called strings.

symmetry   The correspondence between opposite sides of a geometric shape in size, shape, and position. In physics, this term pertains to geometrical descriptions of the relationships between forces and particles in the quantum world.

synapse   The place in the body where nerve impulses are transmitted through neurons.

synchronicity   The relationship between events that are not based on cause and effect. For example, when you are singing a song in the shower, and afterwards turn on the radio and hear the song you were singing. It transcends the notion of coincidence.

Talmud   The sixth century C.E. collection of writings that constitutes Jewish civil and religious law. It is divided into two parts, the Misnah, or text, and the Gemara, or commentaries.

terminal velocity   A term in physics that describes the velocity reached by a falling body when the resistance of the enveloping medium is equal to the force of gravity.

thermodynamic equilibrium   When a body absorbs as much energy in the form of heat as it radiates out in the form of electromagnetic waves.

thermodynamics   The study of heat. The term is derived from the Greek word for heat, *thermos*.

thought experiment   A tool used by Einstein that helped him formulate his theories by turning them into visual images.

time dilation   A term for what happens to time as an object approaches the speed of light.

TOE   Also called the theory of everything. Refers to any theory that attempts to combine gravity and some form of a GUT.

torr   The unit of measure for atmospheric pressure. It is the amount of pressure needed to support a column of mercury at a height of 1 millimeter.

trigonometry   The branch of mathematics that deals with the ratios between the sides of a right triangle with reference to either acute angle, the relations between these ratios, and the application of this knowledge to finding the unknown sides or angles of any triangle (as in surveying, navigation and engineering).

triumvirate   A group consisting of three members.

Teutonic   A group of northern Europeans, including Germans, Scandinavians, Dutch, and English. Commonly used in reference to Germans.

Tychonic theory   Theory developed by Tycho Brahe that put Earth at the center of the solar system, with the Sun and Moon orbiting Earth and the other five planets orbiting the Sun.

ultraviolet catastrophe   Classical wave theory predicted that most of the radiation emitted from a blackbody was high frequency in the ultraviolet, x-ray, and gamma ray region of the spectrum. In cases where classical theory predicted excess ultraviolet rays emitted from a blackbody, there were none. This discovery was a catastrophe for classical physics theory.

unified field theory   A concept that all of physics can be brought together under one theory or unifying idea. Although Einstein spent his life in pursuit of this theory, he never found it.

variable stars   Stars that vary or change their brightness periodically.

volt   A unit of electromotive force. It is the energy required to move from a point of lower potential to a point of higher potential.

wavelength   The physical length of a wave measured in meters from crest to crest. For example, a low-frequency signal may have a wavelength of 6 miles, but a microwave is about 4 inches long.

wormhole   Formed when two black holes are connected together, wormholes can connect various parts of space and time.

yuga   In Hinduism, any one of four stages or time periods in one kalpa.

Zionism   A movement in the twentieth century that helped re-establish the state of Israel and continues to support it.

Zeno's paradox   A term named after Zeno, a Greek philosopher who lived during the time of Aristotle. It is the concept that if you cut a distance in half, and continue this process, you will never get to the end, because there will always be some amount left to cut in half again, even if it's extremely small.

List of Resources, Web Sites, and Videos

History and Philosophy of Science

Cohen, I. Bernard. *The Birth of a New Physics*. New York: W.W. Norton & Co., 1985.

Gamow, George. *The Great Physicists from Galileo to Einstein*. Toronto: General Publishing Company, 1961.

Kuhn, Thomas S. *The Structure of Scientific Revolutions, International Encyclopedia of Unified Science, Vol 2, No. 2*. Chicago: University of Chicago Press, 1970.

Lindberg, David C. *The Beginnings of Western Science: The European Scientific Tradition in Philosophical, Religious, and Institutional Context, 600 B.C. to A.D. 1450*. Chicago: University of Chicago Press, 1992.

Park, David. *The Fire Within the Eye: The Historical Essay on the Nature and Meaning of Light*. Princeton, New Jersey: Princeton University Press, 1997.

Rankin, William. *Introducing Newton*. Cambridge, England: Icon Books, Ltd., 1994.

Further Reading on Einstein

Bernstein, Jeremy. *Einstein*. New York: Viking, 1973.

Brian, Denis. *Einstein, a Life*. New York: Wiley, 1996.

Calder, Nigel. *Einstein's Universe*. New York: Random House, 1982.

Clark, Ronald. *Einstein: The Life and Times*. New York: World Publishing, 1971.

Dank, Milton. *Albert Einstein*. New York: Watts, 1983.

Dukas, Helen and Banesh Hoffmann. *Albert Einstein, the Human Side*. Princeton, New Jersey: Princeton University Press, 1979.

Einstein, Albert. *Ideas and Opinions*. New York: Random House, 1954.

Einstein, Albert, et al. *The Principle of Relativity*. New York: Dover Books, 1952.

Einstein, Albert. *The World As I See It*. New York: Carol Publishing Group, 1984.

Folsing, Albrecht. *Albert Einstein*. New York: Penguin, 1997.

Frank, Philipp. *Einstein: His Life and Times*. New York: Alfred Knopf, 1947.

Gardner, Martin. *Relativity Simply Explained*. Toronto: General Publishing Company, 1997.

Goldsmith, Dr. Donald. *The Ultimate Einstein*. New York: Byron Press, 1997. (It comes with an excellent CD.)

Hilton, Gerald. *Einstein, History, and Other Passions*. New York: Addison-Wesley, 1997.

Hoffman, Banesh. *Albert Einstein: Creator and Rebel*. New York: Viking, 1972.

Mayer, Jerry and John P. Holms. *Bite-Size Einstein: Quotations on Just About Everything from the Greatest Mind of the Twentieth Century*. New York: St. Martin's Press, 1996.

Mermin, N. David. *Space and Time in Special Relativity*. New York: McGraw-Hill, 1968.

Sayen, Jamie. *Einstein in America*. New York: Crown Publishers, 1985.

Schilpp, Paul, editor. *Albert Einstein: Philosopher-Scientist*. New York: Tudor Publishing, 1951.

Schwartz, Joseph and Michael McGuinness. *Einstein for Beginners*. New York: Random House, Ltd., 1979.

Talmey, Max. *The Relativity Theory Simplified and the Formative Period of Its Inventor*. New York: The Darwin Press, 1932.

Thorne, Kip. *Black Holes and Time Warps: Einstein's Outrageous Legacy*. New York: W.W. Norton & Co., 1994.

White, Michael and John Gribbin. *Einstein, a Life in Science*. New York: Penguin Books, 1993.

Cosmology, Quantum Mechanics, and Consciousness

Goswami, Amit, Ph.D. *The Self-Aware Universe*. New York: Putnam, 1993.

Gribbin, John. *Spacewarps*. New York: Delacorte Press, 1983.

Guillemin, Victor. *The Story of Quantum Mechanics*. New York: Charles Scribner's Sons, 1968.

Jespersen, James and Jane Fitz-Randolph. *From Quarks to Quasars, a Tour of the Universe*. New York: Atheneum, 1987.

Kaku, Michio. *Beyond Einstein*. New York: Doubleday, 1987.

McEvoy, J. P. and Oscar Zarate. *Introducing Quantum Theory*. New York: Totem Books, 1997.

Talbot, Michael. *The Holographic Universe*. New York: HarperCollins Publishers, 1991.

Weinberg, Steven. *The First Three Minutes*. New York: Basic Books, 1997.

Wilbur, Ken, editor. *Quantum Questions*. Boston: New Science Library, 1984.

Zohar, Danah. *The Quantum Self: Human Nature and Consciousness Defined by the New Physics*. New York: William Morrow & Company, 1990.

Videos

A. Einstein, How I See the World, PBS Home Video, 1995.

Einstein Revealed, NOVA, 1996.

The Quantum Universe, Smithsonian Video, 1996.

Stephen Hawking's Universe, PBS Home Video, 1997. (Three volumes)

Web Sites

There are thousands of Web sites and links to pages about Albert Einstein. Many of them have photos that you can download for private use, as well as collections of his original works, and quotes by him. You name it; you can find it. The following short list contains some of the best sites:

www.westegg.com/Einstein   Albert Einstein Online is probably the best resource online for anything and everything related to Einstein.

www.home.earthlink.net/~brimc/einlinks.htm   Albert Einstein, General Links has information on time travel and Einstein and the Einstein Papers Project.

www.aip.org/history   AIP's Einstein Links is the biggest set of links; it also includes NOVA public TV's Einstein.

www.humboldt.com/~gralsto/einstein/links/html   The Albert Einstein Links site has photos and collected quotes from Albert Einstein and is indexed.

www.kemp.net/~einstein/links.htm   Albert Einstein Online Links has biographical information about Einstein and the MacTutor History of Mathematics Archive.

www.pbs.org/wgbh/nova/einstein   NOVA Online has Einstein links, photos, and collected quotes.

Timeline

1879   March 14, Albert Einstein is born in Ulm, Germany to Hermann Einstein (1847–1902) and Pauline Koch (1858–1920).

1880   Einstein's family moves to Munich.

1881   Einstein's sister, Maja, is born (she dies in 1951).

1888   Einstein enters the Luitpold Gymnasium in Munich.

1894   The Einsteins move to Milan, Italy; Albert remains in Munich.

1895   Einstein joins family in Pavia and goes to school in Aarau, Switzerland.

1896   Einstein enters the ETH in Zurich, graduates in 1900, and renounces his German citizenship.

1901   Einstein becomes a Swiss citizen.

1902   Einstein begins work as a patent examiner in Bern, Switzerland; his father, Hermann, dies.

1903   Einstein marries Mileva Maric (1875–1948). They have two sons, Hans Albert (1904–1973) and Eduard (1910–1965), and a daughter, Liserl (born 1902), who was apparently put up for adoption. Her fate is unknown.

1905   Einstein publishes his four famous papers in the *Annalen der Physik*, the foremost physics journal in Germany.

1907   Einstein publishes his quantum theory for solids (specific heats).

1909   Einstein becomes an associate professor at the University of Zurich.

1911   Einstein becomes a full professor at Karl-Ferdinand University in Prague, Czechoslovakia.

1912   Einstein becomes a professor at the ETH in Zurich.

1914   Einstein becomes a professor at University of Berlin; he separates from Mileva and sons.

1916   Einstein publishes paper on general relativity and gravitation.

1919   Einstein and Mileva divorce; he marries his cousin, Elsa.

1920   Einstein and his relativity theory are publicly attacked by anti-Semites.

1921   Einstein visits the United States for the first time.

1922   Einstein wins the Nobel prize for his discovery of the laws of the photoelectric effect.

1925   Einstein submits paper entitled "The Unified Field Theory of Gravity and Electricity."

1927   Einstein and Niels Bohr began their famous debates about quantum mechanics at the fifth Solvay conference.

1930   Einstein takes a three-month teaching position at Caltech in Pasadena, California, meets with Edwin Hubble, returns to Germany in 1931.

1932   Einstein is appointed professor at Institute for Advanced Study in Princeton, New Jersey; he leaves Germany.

1933   The Nazis come to power in Germany; Einstein permanently moves to the United States.

1936   Elsa Einstein dies.

1939   World War II breaks out (August 2); Einstein writes his famous letter to President Roosevelt warning of the dangers of the atomic bomb.

1940   Einsteinbecomes a citizen of the United States (retains Swiss citizenship).

1945   August 6, first atomic bomb is dropped on Hiroshima.

1946   Einstein heads the Emergency Committee of Atomic Scientists.

1948   Einstein's first wife Mileva dies.

1952   Einstein is offered the presidency of Israel, but he declines.

1955   Albert Einstein dies on April 18 in a Princeton hospital.

Index

D

E